THE DUCHESS'S SHELLS

BETH FOWKES TOBIN

The DUCHESS'S *Shells*

NATURAL HISTORY COLLECTING IN THE
AGE OF COOK'S VOYAGES

PUBLISHED FOR THE PAUL MELLON CENTRE FOR STUDIES IN BRITISH ART BY
YALE UNIVERSITY PRESS ~ NEW HAVEN AND LONDON

Copyright © 2014 by Beth Fowkes Tobin

All rights reserved.
This book may not be reproduced, in whole or in part,
in any form (beyond that copying permitted by
Sections 107 and 108 of the U.S. Copyright Law
and except by reviewers for the public press),
without written permission from the publishers.

Designed by Emily Lees
Printed in China

Library of Congress Cataloging-in-Publication Data

Tobin, Beth Fowkes, author.
The Duchess's shells : natural history collecting in the age of Cook's voyages / Beth Fowkes Tobin.
pages cm
Includes bibliographical references and index.
ISBN 978–0–300–19223–0 (hardback)
1. Shells – Collectors and collecting – England – History – 18th century. 2. Shells – Private collections – England. 3. Portland, Margaret Cavendish Holles Harley Bentinck, Duchess of, 1715–1785 – Natural history collections. 4. Natural history – England – History – 18th century. I. Title.
QL406.2.T63 2014
594'.680941 – dc23
2013042768

A catalogue record for this book is available from the British Library

Frontispiece: Charles Grignion after E. F. Burney, frontispiece to *A Catalogue of the Portland Museum*, 1786 (detail of fig. 23).

CONTENTS

Acknowledgments   vi

Introduction: The Catalogue and the Collection   1

1. The Duchess, Natural History, and Cultures of Collecting   25
2. The Duchess and Shell Collecting   71
3. Patronage, Brokers, and Networks of Exchange   113
4. Pacific Shells, Cook's Voyages, and the Question of Value   145
5. Patronage, Publication, and the Illustrated Conchology Book   181
6. The Dispersal of the Collection   215
7. Lost Legacies: The Duchess's Shells in the Nineteenth Century   247

Notes   268

Index   304

ACKNOWLEDGMENTS

My interest in natural history collectors began at the Linnean Society in London, while reading the letters of the Society's founder and first president, Sir James Edward Smith, whose illustrated annual, *Exotic Botany*, was the subject of the final chapter of my book *Colonizing Nature* (2005). I found myself fascinated by the emotional tenor of the letters and enthralled by the world they conjured. I was startled by a sense of recognition, since in them Smith shared with his fellow naturalists his frustrations and concerns about his position as an amateur naturalist within the scientific community. I felt the ground beneath me shift as I saw my decades-long fascination with the representational strategies within British art and literature give way to the power of letters, especially those written by individuals speaking for themselves rather than in an official capacity. Of course, I had read letters before for my work on the representation of the tropics and in my study of colonial painting, but these were mostly official correspondence between colonial administrators and government bureaucrats discussing policy and the administration of subject territories. It was this encounter with Smith's letters that changed the direction of my work; I was captivated by their enticing handmade materiality and by the powerful utterance of a complex, historical subject.

While I was sitting at a desk in the Linnean Society's lovely library and reading Smith's correspondence, Gina Douglas, the head librarian at the time, plopped down in front of me an ordinary-looking folder containing letters – they were from Margaret Cavendish Bentinck (1715–1785), duchess of Portland, to the physician and botanist Richard

Pulteney — and said that she thought I might like to take a look at them. For this I will be forever grateful, since the short stack of letters became the object of my study for the next eight years and the center of this book.

The paucity of ready information on the subject of the duchess of Portland's natural history collections is surprising, considering that her collection of shells, let alone antiquities and decorative objects, was one of the largest and finest in Europe during the late eighteenth century. Until 2006, when "The Duchess of Curiosities" exhibition opened at the Harley Gallery, Welbeck, the duchess's collections had received little attention. This is again surprising, since the natural history collections of her contemporaries Sir Ashton Lever and Sir Joseph Banks have garnered much attention among scholars in anthropology, Pacific studies, British imperial and maritime history, and museum studies, although, to be fair, this has focused on artifacts from the Pacific rather than natural history specimens. Peter Dance's definitive book on shell collectors, *A History of Shell Collecting* (revised 1986), includes an overview of the duchess's importance to the field and provides crucial information on members of her shell network. In addition, he studied the links between Captain Cook's voyages and shell collecting in Europe, as well as conclusively establishing the authorship of *A Catalogue of the Portland Museum* (1786), the auction catalogue of the duchess's collection. My study does not seek to compete with or replace Dance's invaluable and informative history of European shell collecting. Instead, I emphasize the cultural practices of collectors, placing their activities within the social, economic, and scientific contexts of the Enlightenment. Apart from Dance's important work and Rebecca Stott's booklet accompanying the Harley Gallery exhibition, and articles by Alexandra Cook and Stacey Sloboda, the duchess has usually appeared in the scholarly literature in the context of the Portland Vase, or as the friend of the Bluestocking Elizabeth Montagu and of Mary Delany, whose artistic legacy has caught the attention of feminist scholars and art historians.[1] One suspects that Margaret Cavendish Bentinck would be better known today as an important figure in the history of Enlightenment science and the history of museums if she had been a man or had insisted on keeping her collection together as a monument to her endeavors — or at least had made provisions in her will either to donate it to the British Museum or to sell it at a very reduced rate (as she arranged for her father's books and manuscripts to be purchased by Parliament on behalf of the nation).

The key to recovering the duchess's shell collecting practices is Richard Pulteney's correspondence, much of which is held at the Linnean Society, for among the hundreds of letters he received from his many correspondents over his long life is a clutch from Margaret Cavendish Bentinck devoted to the topic of shells. These letters are invaluable since they are nearly unique in their focus on shells. Longleat, the ancestral home of Lady Elizabeth, the duchess's daughter who married the first marquess of Bath, and the University of Nottingham library both contain vast quantities of letters written by and to the duchess, and yet there is little in them about her shell collecting. What natural history material emerges in the correspondence held at Nottingham centers on her interest in botany and her activities at Bulstrode, which included a menagerie, a rose garden, a hothouse, a botanic garden, and an aviary. These activities are important and should not be overlooked by scholars, but they are outside the scope of this book, with its focus on her shell collection. The question of why these archives contain so little on shell collecting may find an answer in the fact that the duchess expressly asked her shell correspondents to address their letters to her London townhouse in Whitehall, presumably because this was where the shell collection was stored. This may explain how documentation about her shells (letters, bills of sale, lists of shells, notes, catalogues) became separated from the papers that were at Bulstrode when she died, and then made their way into the Longleat archives via her daughter and into the Nottingham collection via her eldest son and the ducal seat at nearby Welbeck.

Given this relative paucity of letters about the duchess of Portland's shell collecting, which was clearly a large part of her life, I have had to reconstruct her actions from various other textual sources, primarily letters, manuscripts, and books by naturalists who mention her. In this group of correspondents, Richard Pulteney's letters are the most helpful, followed by Emanuel Mendes da Costa's correspondence, which fills eleven massive volumes in the British Library and which provides essential background to shell collecting practices and shell book publication. Both da Costa's correspondence and that of Pulteney, divided between the Linnean Society and the Natural History Museum, are invaluable documents in reconstructing this world of shell collecting, as are the letters of the larger network of natural history collectors, including Henry Seymer, Thomas Pennant, John Lightfoot, John Timothy Swainson, Sir Joseph Banks, Daniel Solander, and George Humphrey. Also crucial are the letters of the duchess's best friend, Mary Delany, who would note in passing some of the natural history activities that she and the duchess were engaged in, such as Sir Joseph Banks's visit to Bulstrode in the fall of 1771, just after his return from Cook's first circumnavigation. In reconstructing the collecting and selling of shells gathered by the various participants in Cook's voyages, I have found that the *Resolution Journal* of Johann Reinhold Forster and his son Georg's *A Voyage Around the World* (1777) are far more informative than the official logs, journals, and correspondence generated by Cook and his officers, primarily because most natural history collecting, with the major exception of that of Banks and the Forsters, was done on the side and not officially sanctioned by the Admiralty. Even naturalists' descriptions of collecting shells in the Pacific are slight, often mentioned in passing without detailing much of the actual collecting processes. As with the duchess's collecting practices, I have had to turn to other kinds of texts to compensate for the gap in the archive; in this case, pamphlets both in manuscript and printed, addressed to ship captains and other enterprising travelers, coaching them how to collect and preserve natural history specimens. All in all, the numbers of letters written by those in the duchess of Portland's network that describe shell collecting are far fewer than I had hoped to find when I began this project, making the work of reconstructing the social world of the shell exchange – and specifically the duchess's activities as a skilled conchologist, field collector, and patron of shell books – an effort and an exercise in the creative use of available resources.[2]

To the Linnean Society, and Gina Douglas specifically, I owe an enormous debt of gratitude. Likewise, the librarians at the Natural History Museum, London, in particular Paul Cooper of the Zoology Library, have been ever gracious, patient, and encouraging, making me feel welcome in such an awe-inspiring place. The British Library, of course, always amazes me with its generosity, sharing its incredible treasures with the world, and there I examined Emanuel Mendes da Costa's correspondence and many of Sir Joseph Banks's natural history books, among many other items including auction catalogues. Lisa Gee, Director of the Harley Gallery at Welbeck, Kim Sloan, Curator of Prints and Drawings at the British Museum, and Maggie Reilly, Curator of Zoology at the Hunterian Museum in Glasgow, offered much encouragement and helpful advice early on in this project. I am grateful to Kathy Way, Collection Manager of the Division of Higher Invertebrates at the Natural History Museum, who took time to show me Cracherode's and Banks's shell collections and to discuss the possible present whereabouts of the duchess's shells, and to Richard Preece, Watson Curator of Malacology, who was kind enough to take me down into the basement storerooms of the University Museum of Zoology at Cambridge University to look for an oyster shell that Henry Seymer had owned. The staffs of other archives and libraries have proven equally gracious: the University of Nottingham's library and rare book room, the Captain Cook Memorial Museum in Whitby, the National Maritime Museum in Greenwich, the National Archives at Kew, the Warwickshire County Record Office, the John Crerar Science Library at the University of Chicago, Arizona State University Libraries, the Huntington Library in San Marino, the Honolulu Art Academy, the Bishop Museum in Honolulu, Special Collections of the University of Hawai'i at Manoa's Hamilton Library, and especially the research library of the American Museum of Natural History in New

York, where Tom Baione welcomed me as I spent many happy weeks poring over its magnificent collection of eighteenth-century conchology books and travel narratives. Without the support of the British Academy and the Huntington Library's Exchange Fellowship, I would have been unable to spend as much time in Britain as I needed for this project, and without the support of the National Science Foundation this book would not exist in its present form. The NSF's Scholars Award combined with a sabbatical leave and a senior research leave from Arizona State University enabled me to devote the time needed to complete this project. Finally, the Institute for Women's Studies at the University of Georgia provided support that was crucial in the final stages of preparing this book for publication. I owe thanks to the museums, galleries, and libraries that supplied the images for this book, several graciously waiving or reducing reproduction fees: The Harley Gallery and Portland Collection of Welbeck Abbey, Nottinghamshire; Arizona State University Libraries; Lewis Walpole Library, Farmington, CT; The Hunterian, University of Glasgow, and the University of Glasgow Library, Special Collections; the National Library of Australia, Canberra; and, in London the British Museum, the Linnean Society, the National Gallery, the National Portrait Gallery, the Natural History Museum, and the Victoria and Albert Museum.

Generous with advice, encouragement, helpful information, and feedback on drafts were S. Peter Dance, Kathie Way, Kim Sloan, Maggie Reilly, Geoff Hancock, Lynda Brooks, Ilya Tëmkin, Quentin Wheeler, Andrew Hamilton, Ann B. Shteir, Joan Landes, James Delbourgo, Daniela Bleichmar, Stacey Sloboda, Elizabeth Heckendorn Cook, Marissa Petrou, Joan Landes, Paula Lee, Elizabeth Eger, Tim Erwin, Ruth Dawson, Laura Lyons, Cindy Franklin, Ayanna Thompson, Roxanne Eberle, Maureen Daly Goggin, and an anonymous reader for Yale University Press, as well as audiences at the various conferences and colloquia where I presented talks on the duchess of Portland, Cook's voyages, and natural history collecting: The Early Modern Center at the University of California Santa Barbara; the Comparative Colonialisms Workshop at the University of Chicago; the Global History Colloquium at UCLA; "Finding Animals" conference at Penn State; the Glasscock Center for Interdisciplinary Humanities at Texas A&M; the History of Science Society meeting in Vancouver; the New York Shell Club, in particular Florence Fearrington; the annual conferences of the American Society for Eighteenth-century Studies in Montreal and Vancouver; and CIAH2012, the International Committee of the History of Art. Earlier versions of portions of Chapters One and Two have appeared in print: "The Duchess's Shells: Natural History, Gender, and Scientific Practice," in *Material Women: Consuming Desires and Collecting Practices, 1750–1950*, co-edited with Maureen Daly Goggin (Farnham, Surrey: Ashgate, 2009), 247–63; and "Revisiting the Virtuoso:

Natural History Collectors and their Passionate Engagement with Nature," in *Environmental Criticism for the 21st Century*, ed. Stephanie LeMenager, Teresa Shewry, and Ken Hiltner (New York and London: Routledge, 2011), 49–60. To historical taxonomists past and present, I owe a large debt; their work on descriptive bibliography is exemplary. I often felt as if I was traveling in good company as I followed in the tracks of T. Iredale, J. W. Jackson, E. A. Kay, Nora F. McMillan, P. J. P. Whitehead, and Guy L. Wilkins, who provided models for the kind of writing I aspired to do in this book, one that tracked specimens and their textual representations over time and through various hands. It has always been a dream of mine to work with Gillian Malpass of Yale University Press because she makes such beautiful books. I am grateful that she took my project on and, together with Emily Lees, brought this book into the world. I am also grateful to the Paul Mellon Centre for Studies in British Art for its generosity in supporting the publication of this book.

Provenance hangs over this project as a silent aspiration, for although I can trace the paths by which some of the shells entered and exited the duchess's collection, only three of her shells out of the thousands she owned can be traced with any certainty into the present. I was drawn to this project, in part, by the hope that I could trace the paths by which the shells entered and left her collection, tracking them up to the present. While I have learned that this is an impossible task, I have kept that movement into and out of the collection as my narrative arc, supplying as much as I know about who, what, where, and when in the lives of these shells as they made their way from the beaches of Weymouth and the far reaches of the globe into the duchess's cabinet and out again into the world and to who knows where.

INTRODUCTION

A

CATALOGUE

OF THE

PORTLAND MUSEUM,

LATELY THE PROPERTY OF

The Duchess Dowager *of* Portland,

𝔇𝔢𝔠𝔢𝔞𝔰𝔢𝔡:

Which will be SOLD by AUCTION,

BY

Mr. SKINNER *and* Co.

On MONDAY the 24th of APRIL, 1786,

AND THE

THIRTY-SEVEN FOLLOWING DAYS,

AT TWELVE O'CLOCK,

SUNDAYS, and the 5th of JUNE, (the Day his MAJESTY'S BIRTH-DAY is kept) excepted;

At her late DWELLING-HOUSE,

In *PRIVY-GARDEN, WHITEHALL*;

BY ORDER OF THE ACTING EXECUTRIX, *Lady Weymouth, eldest daughter of the Duchess.*

To be viewed Ten Days preceding the Sale.

CATALOGUES may now be had on the PREMISES, and of Mr. SKINNER and Co., ALDERSGATE-STREET, Price FIVE SHILLINGS, which will admit the Bearer during the Time of Exhibition and Sale.

N°. 753

2000 Catalogues were sold before the Sale began.

1   Title page, *A Catalogue of the Portland Museum* (London, 1786). Courtesy of The Lewis Walpole Library, Yale University

THE CATALOGUE AND THE COLLECTION

At noon on Monday, 24 April 1786 the long-awaited auction of the duchess of Portland's collection of fine and decorative arts and natural history specimens began. Covered in the press as a major event, the auction was advertised heavily in several newspapers, both in the preceding month and during its run of thirty-eight days. An advertisement printed repeatedly throughout the sale in the *Gazetteer and New Daily Advertiser* provides a glimpse of the range of objects that were to be sold:

> This CABINET is well known to contain the most copious and splendid COLLECTION of SHELLS, both native and exotic, in Europe, many of them uniques, and most of them named from the system of Linnaeus, or the descriptions of Dr. Solander . . . Also, large quantities of fine OLD CHINA, matchless pieces of fine OLD JAPAN; a number of elegant DRAWINGS of British and exotic plants, most of them by the celebrated Ehret; a few miniature Pictures of great value; a great number of valuable SNUFF-BOXES . . . Likewise, a most curious ANTIENT SCULPTURE of the head of JUPITER SERAPIS, in a kind of basalts brought from Italy, the workmanship superlatively fine.[1]

The advertisements seem to have been effective, because the Portland auction drew large crowds, some of whom were potential buyers of the duchess's china, prints, and snuffboxes. Others were merely spectators, eager to see her beautiful things and to gain entrance into her London townhouse, which was the site of the auction, while others, including naturalists, natural history collectors, and agents, came to purchase some of the thousands of her natural history specimens.

Complaints about the crowded conditions surfaced immediately in newspapers, and reports circulated about women fainting from the packed and over-heated rooms. Journalists offered advice on ways to alleviate the crowding: they urged that the limited schedule set by Mr. Skinner of noon to four should be extended, and suggested that bidding should take place in the courtyard, where carpets could be laid on the damp ground so that ladies would not get their feet wet. The crowded conditions however deterred neither observers nor buyers – nor pickpockets, who found the crush of the crowded rooms congenial to their trade.

The excitement that such an auction generated is captured in Frances Burney's novel *Cecilia* (1782), when the silly Miss Larolles, who rushes from one fashionable event to another, describes the joys of attending a sale: "All the world will be there . . . there'll be such a monstrous crowd as you never saw in your life. I dare say we shall be half squeezed to death." When asked whether she would buy anything at Lord Belgrave's auction, she answered: "Lord, no; but one likes to see the people's things."[2] As a form of polite entertainment and a social event not to be missed, the Portland auction attracted not only Londoners but also visitors from the country, as one stop on an itinerary that might include visits to the British Museum, the opera and ballet, and hot-air balloon demonstrations.

In the weeks leading up to the auction, Skinner, the auctioneer, sold more than 1,700 copies of the catalogue of what was known as the Portland Museum (pl. 1). So popular was this catalogue that numerous copies – some even annotated by participants and observers – have survived, now housed in a range of art and science museums, in the rare book rooms of academic and national libraries, and in the private collections of book collectors, malacologists, and shell collectors.[3] The catalogue is organized by day and describes in brief the contents of each lot. For instance, the "Twenty-Seventh Day's Sale" was a big day for drawings, prints, miniatures, Queen Elizabeth I's Prayer Book (containing prayers she had composed), and illuminated missals, one of which was purchased by Horace Walpole for the large sum of £169. This day's sale began with nearly fifty drawings of plants by James Bolton and 500 drawings of flowers by G. D. Ehret, and ended with "the heads of twelve Caesars, in alabaster." Lot 2948, a miniature by Holbein, sold for £10. Lot 2918, "The Works of Hollar, comprised in 13 folio volumes, of the most beautiful impressions, collected by her Grace in the most liberal manner, at an immense expense, with a variety of proofs and variations, in fine preservation," sold for £385. While the lots on this day were devoted primarily to fine and decorative art, the following day's sale was exclusively natural history objects with 125 lots to be sold, ranging from lot 2954 to lot 3079. Lots 3005 to 3018 were minerals, "Spars, Ores, Crystals," while the rest were a mixture of "Shells, Corals, Petrifactions," the latter meaning fossils. Most of these lots consisted of shells from China, the West Indies, the Pacific and Indian Oceans, and Britain. Lot 2971, for instance, was made up of "A great variety of British Bivalves and Multivalves of different genera, many of them labeled" (pl. 2), and lot 2969 is described as containing "Sixteen cards, including a variety of Haliotis, among which are Asinina, tuberculata of *China*, one figured in *Humph, Conch. pl. 9. fig. 5*. the scarlet-ridged, from the *Cape*, and other rare species" (pl. 3).[4] Thirty days of the auction were devoted entirely or largely to shells. Approximately 50 percent of the 4,263 lots consisted of shells, with each lot containing anything from one to dozens of shells.[5]

Clearly, Margaret Cavendish Holles Harley Bentinck, the dowager duchess of Portland, had amassed a huge shell collection (pl. 4). She had gathered some herself along the coast of southern England, while others had been received as gifts from sea captains. The haliotis shells from the Cape may have been a gift from Captain James Cook or his lieutenant Charles Clerke, and those from China may have come from Elizabeth Montagu's brother, who was a sea captain in the East India fleet. The duchess also benefited from the efforts of friends and acquaintances who had traveled abroad, picking up exotic shells in the fish markets of Italy and gathering snail shells while touring the Swiss countryside. She also purchased shells from natural history dealers who operated as brokers between collectors and sailors who had just returned from such places as the West Indies and the Cape of Good Hope. She also funded expeditions – for example, Henry Smeathman's to Sierra Leone – contributing to the purchase of equipment and the costs of the passage, all in exchange for specimens.

This book tells the story of the formation and dispersal of the duchess of Portland's shell collection: how she acquired the shells; what she did with them; and what happened to them when she died. It was one of the largest – if not *the* largest – shell collections of its time, rivaling those of Queen Louisa Ulrika of Sweden and Charles Alexandre de Calonne, the French minister of finance. According to Thomas Martyn, author of *The Universal Conchologist* (1784–89), the duchess had the largest and finest shell collection in Britain, if not all of Europe:

> The first praise is confessedly due to the superb collection of the Dutchess Dowager of Portland; so rich a display in the number as well as rarity and perfection of these subjects, together with species of marine productions, perhaps is not to be equalled. In this branch of science her Grace's superior knowledge is as well known as it is

2   J. Agnew, "Mya gigas," in *British Conchology or Original Drawings of Fresh & Salt Water Shells*, circa 1786–1818. © Natural History Museum, London

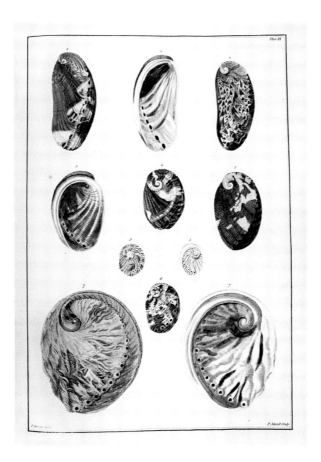

3 Emanuel Mendes da Costa, Plate 9 in *Conchology* (1770–71).
By permission of University of Glasgow Library, Special Collections

eminently demonstrated, in the critical arrangement of this immense cabinet, which altogether justifies the very great expense of time and money employed in the formation of it.[6]

Amassing the collection took decades, and cataloguing it took years. The duchess's plan was to publish a scientific catalogue of the shells, complete with the most current Linnaean names and descriptions, but she died before this project was completed, and with her death all the natural history specimens that she had worked so hard to acquire were sold. If she had lived to see the catalogue into print, it would have made an enormous contribution to the natural history of conchology (as the study of mollusks was called during the Enlightenment).

4   C. F. Zincke, *Margaret, Duchess of Portland*, enamel miniature. Private Collection

Why write a book about a shell collection, and the duchess of Portland's in particular? Shells are subject to multiple discursive and semiotic systems, and though they are natural objects, they are also material culture and as such are social and cultural texts rich in meaning. Shells are attractive; they are delicate, fragile, and possess intricate and patterned shapes, and the complex combination of delicacy and durability invites our admiration. Recent work such as Alfred Gell's writings on objects, in particular aesthetically pleasing ones, in which he posits that objects possess agency, may lead us to conclude that shells make us *do* things, such as picking them up on the beach, and that for this reason they are not merely empty receptacles or blank slates on which to impose systems of thought.[7] That shells can elicit a certain kind of behavior from humans is possible; but an object's physical attributes can never secure the way in which it will be used and what significances it will have. The physical attributes of shells, however pleasing, cannot explain fully the uses to which they were put and the meanings with which they were imbued in the eighteenth century.[8] A shell could not determine whether it would be made into a button, a fishhook, a bead on a wampum belt, a scientific specimen emblematic of its species, a piece of trash after the mollusk had been consumed as food, or a decorative ornament, polished and shining in a glass-faced cabinet. This very mutability makes shells ideal for studying those social processes by which natural objects are transformed into material culture and are exchanged and circulated within social networks as commodities, gifts, decorative pieces, and scientific specimens, or removed from exchange circuits and expelled as refuse. What is intriguing about this interaction of materiality with social significance is the way in which things – solid, tangible, and finite – can be so mutable, possessing an array of social meanings both simultaneously and consecutively.

Shells as material culture have been investigated by anthropologists studying Pacific Island cultures, beginning with Bronislaw Malinowski's study of the shell exchange among the Trobriand Islanders. Taking up Malinowski's observations, Marcel Mauss elaborated on the significance of this Pacific shell network (or the Kula exchange as it is called), analyzing the importance of the rituals surrounding the exchange of shells and the social implications of shells as gifts. This Kula exchange has been reinterpreted by contemporary economic anthropologists such as Marilyn Strathern, Annette Weiner, and Nicholas Thomas, who have examined the Trobriand shell network in terms of theories of reciprocity, social obligation, and inalienable possessions.[9] At stake in these discussions are questions of value in relation to the abstracting power of commodity exchange, the reciprocity inherent in gift exchange, and the possibility of a logic of exchange that stresses an object's inalienable qualities – social meanings that cannot be reduced to an abstract notion of equivalence, but which are tied to family, kin, ritual, and the sacred. Shell exchanges within European societies were every bit as complex as those of the Pacific cultures described by anthropologists, involving archaic forms of gift exchange within patronage networks and friendship circles, as well as mercantile capitalism's global circuits of commodity exchange. Moreover, shells within European society took on multiple meanings and operated within a range of modalities, from the aesthetic and decorative to the connoisseurial and the scientific. In order to contribute to an understanding of the complex processes by which shells were accumulated, exchanged, and represented within late Georgian society, this book attempts an ethnographic description of the culture of shell collecting, taking as its focus the duchess of Portland's collecting practices.[10]

THE SHELL COLLECTION

How does one study a collection that no longer exists? The duchess of Portland's shell collection was dispersed when its contents were sold in the spring of 1786. Of the thousands of shells that she owned, only three can be traced with any certainty to her collection: one, *Voluta aulica*, is housed in the Natural History Museum in London; the other two, both *Polymita picta*, are at the Hunterian Museum in Glasgow (pls. 5 and 6). This problem – "Writing about a Collection which isn't There" – was addressed by the historian Emma Spary, who states in her study of the long-defunct cabinet of natural and artificial objects belonging to the druggist Pierre Pomet that "vanished collections pose a particular historical and historiographical problem."[11] The most challenging is that textual and visual representations of missing items fail to shed light on the processes by which a collection was assembled and used. Printed catalogues, such as *A Catalogue of the Portland Museum*, for example, do not convey the complex processes involved: naturalists' networks, global exchange, techniques for preserving specimens, practices and aesthetics of display, methods of classification, and regimes of values that were brought to bear on the collected natural object.

In his book on early modern natural history cabinets, Marco Beretta goes so far as to declare that textual sources are a poor substitute for the materiality of the collected object: "The role of writing in natural history is relatively secondary to the naturalist's techniques of handling and preserving specimens, by the care taken when organizing the species in a collection, and by the ways in which the exhibit is displayed to the public." He suggests that we abandon "the study of texts" temporarily and examine "more thoroughly the relationship between naturalists and their natural history specimens";

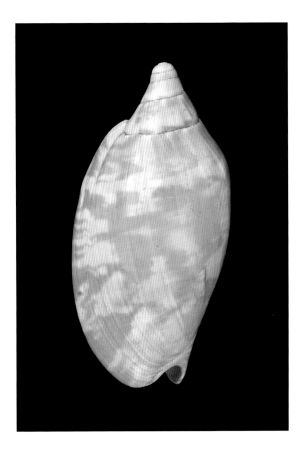

5 *Voluta aulica*. London, Natural History Museum. © Natural History Museum, London

6 *Polymita picta*. Glasgow, Hunterian Museum. © The Hunterian, University of Glasgow 2012

he urges us to think of specimens not as neutral objects, but as offering "precious clues for understanding the events surrounding the birth of natural history museums and the development of the scientific disciplines that grew up around them."[12] But this suggestion makes sense only if there is an extant collection.

According to Beretta, textual representations of collections are unreliable guides to understanding the relationship between collectors and their specimens. A focus on texts distorts the actual way in which natural history collections were used, organized, and displayed: "Collections and specimens tell stories that do not always agree with the texts serving as their framework and introduction."[13] The unreliability of textual representations explains why cultural historians, who have relied on the *Catalogue of the Portland Museum* as a guide to understanding the duchess's collecting practices, have misconstrued her methods and practices as the indiscriminate accumulation of a virtuoso. Furthermore, textual representations, such as a printed catalogue, render the collection static and inert, and cannot convey the "constant flux" that characterized the activities that went into building it. A printed catalogue cannot recover the history of a collection, the ways in which specimens moved in and out of it, or the methods by which a collector manipulated his or her specimens. Textualized representations of collections cannot capture the collecting practices that Spary describes so eloquently as "a choreography of hands moving to bring together, describe, examine, preserve, mount, and of eyes moving between and among specimens and texts."[14]

However, this choice that Beretta posits between a collection and the textual representations of it is a false binary, because there are other textual sources – letters, manuscript catalogues, lists of specimens, notebooks, drawings and illustrations, citations in natural histories, and interleaved and annotated printed catalogues – that can tell us much about just what he and Spary think the focus should be: the social practices and the material culture of natural history collecting. Four lots of shells in the *Catalogue of the Portland Museum* provide examples of how textual sources can illuminate the history of a collection, although these itemized descriptions tell us little about the social lives of shells as collected objects. However, naturalists' letters, notebooks, and specimen lists, as well as voyagers' journals, can provide the contexts for understanding the movement of shells into and out of the duchess's collection, and can reveal the complex world of natural history collecting, shaped by curiosity, desire, and taste.

The *Mya declivis*, lot 918 of the *Catalogue of the Portland Museum*, was one of several specimens that the duchess found at Weymouth: it was a small white bivalve, approximately 33 mm in length.[15] She would spend several weeks at this seaside town every year to indulge in her love of shell hunting, sometimes traveling there with her chaplain, the Rev. John Lightfoot, a talented botanist, and often accompanied on expeditions with a local shell hunter, a Mrs. Le Cocq, of whom she was very fond. The rather dry description of lot 2251 – "A variety of Bivalves and Multivalves, from Weymouth, some of them labelled" – fails to convey the joy with which the duchess would comb the beaches for new shells and go out in rowboats to supervise the dredging of the shallow seas around Weymouth for elusive oysters, scallops, clams, and other mollusks; this is probably how she discovered the *Mya* (now called *Thracia papyracea*), which lives in the intertidal zone "in sand, muddy sand, and sandy gravel." The Portland Museum catalogue contains eight references to the species *Mya declivis*, at least one of which was collected at Weymouth, as is learned from the notebook of the duchess's friend John Timothy Swainson, describing his British shells. His notations record that he received a *Mya declivis* from the duchess as a gift. This fragile little shell had traveled from the shore at Weymouth to London and the duchess's cabinet in Whitehall, and then back out to the coast again, this time to Margate in Kent, where Swainson was a Customs officer. The brief annotation refers, though obliquely, to the complex processes of collection and exchange in which naturalists engaged, much of it not in any sense commercial. In general, shell collectors would give away or trade specimens for which they had duplicates, a practice that will be discussed in Chapter Three, which examines the flow of letters and specimens within the duchess's network of shell collectors.

Shell specimens traveled between naturalists as gifts, trades, and loans, as will be seen from the story behind another shell sold at the Portland auction, the *Venus pensylvanica*. Lot 613 is described as "Various odd Valves of Ostrea, Venus, and other bivalve shells," which admittedly does not sound very exciting, given that these oyster and clam specimens had one valve missing. Unfortunately, the British Library's annotated copy of the *Catalogue of the Portland Museum* does not say who purchased this lot, but the miscellaneous grab bag of imperfect specimens sold for 7s. 6d. What makes these bivalves of interest, however, is that the lot may well have contained Richard Pulteney's specimen of the *Venus pensylvanica* (pl. 7), now named *Venus mercenaria*, a North American edible clam commonly known as quahog. He had sent it to the duchess so that she could compare it to various specimens of the same and similar species that she possessed. In 1779 Pulteney wrote to her: "I have packed up in a Box, and sent by the Coach today a single Valve (for I have not a pair) of that Shell which I have always taken for the *Venus pensylvanica*; and I shall be glad to have it determined from your Grace's Cabinet."[16] He may have waited months for her response, and it is possible that this Venus specimen failed to come back to him: much later in his life he noted that he had lost many of the shell specimens he had sent to other naturalists for classificatory purposes, since these

7   *Venus pensylvanica*, from Linnaeus's shell collection. London, Linnean Society. By permission of the Linnean Society of London

friends had died before returning them. Decades later he complained to a fellow naturalist that the duchess had died while some of his shells were still in her possession and he had had the particularly unpleasant experience of knowing that his own shells were to be sold at the auction. Perhaps Pulteney's *Venus pensylvanica* was sold as part of lot 613.[17]

Here is a shell involved in multiple kinds of exchanges, demonstrating how objects of scientific interest could move from provincial areas to metropolitan centers, where information was amassed and organized and where such objects were also commoditized, sold, and dispersed, perhaps even being absorbed back into private provincial collections. Other shells sold during the Portland auction had social lives that were even more dramatic, moving in much wider arcs than from Dorset to London and across far more distinct regimes of value. They had been gathered in the South Pacific, exchanged for nails and other British manufactured goods by Pacific islanders, then carried by sailors, officers, and shipboard naturalists to England, where they were sold to dealers or given as gifts, making their way to the duchess through various forms of exchange. The abalone shells in lot 380 – "A fine pair of Haliotis Iris, from New Zealand, one its native state, the other cleaned" – had been gathered in New Zealand, where abalones were a source of food, their shells used to provide the raw material for fishhooks and decorative elements in clothing, jewelry, and sacred statuary. Sold at the Portland auction for 8s. to a Mr. Yeats, they were diverted from a Pacific islander economy in which they would have moved through gifting and bartering networks of exchange – the means by which trade in such objects occurred on the beaches and coastal inlets of

8   Thomas Martyn, "*Cypraea Aurora*, or the Orange Cowry," Plate 59 in *The Universal Conchologist* (1784), vol. 2. By permission of University of Glasgow Library, Special Collections

Pacific islands large and small. They were probably brought to London by someone on Cook's voyages, entering the duchess's collection either as a purchase from a dealer or as a gift from a voyager or fellow naturalist. They moved into the commodity phase of their social life when sold at auction, but their short lives as commodities ended when they were bought by Yeats, and they became, once again, natural history specimens. Yeats bought up most of the abalone shells sold at the Portland auction.

Another globetrotter, the orange cowry "from the *Friendly Isles* in the *South Seas*, extremely scarce," lot 3831, would have been worn by elite men in Tongan societies: a small hole was drilled in the shell to accommodate a string of coconut fiber, so that the cowry could be worn around the neck (pl. 8). Highly desirable, these beautiful shells

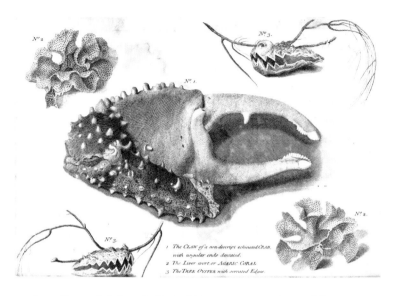

9   George Vertue, shells from the Portland Museum, engraving. Private Collection

circulated widely within Polynesia as emblems of social status, but once bartered for and secured by members of Cook's crew, they entered a monetized economy where their value was measured by the marketplace, some fetching incredible prices ranging up to £10–£15. Beautiful and extremely rare, orange cowries were sought after and displayed by elite members of British society; despite the radical shift in location from Tonga to London, they still functioned as markers of social rank. Their inclusion in Martyn's *Universal Conchologist* testifies to their desirability, for this expensive, gorgeous book of shell illustrations (examined in Chapter Five) served as a guide for cognoscenti eager to possess Pacific shells. With the help of such illustrated natural history books and more ephemeral texts such as Swainson's notebook, it is possible to recover some of the social practices that went into the formation of the duchess of Portland's magnificent shell collection, which in turn can deepen our understanding of how natural history collectors interacted with their collections.

In addition to the texts written by naturalists, collectors, and dealers within the duchess's natural history network, we are fortunate in having visual representations of some of her shells. In addition to the photographs of the three extant shells (see pls. 5 and 6) and George Vertue's engraved illustration of a few shells in the duchess's collection (pl. 9), there are several engravings in shell publications. Images made during her lifetime and with her assistance can be found in the serial *Conchology; or, The Natural History of*

Shells (1770–72), Thomas Pennant's *British Zoology* (vol. 4, 1777), and Thomas Martyn's *The Universal Conchologist* (1784–89); images of the shells drawn after her death and their dispersal at auction appear in Edward Donovan's five-volume *The Natural History of British Shells* (1799–1803). Donovan bought some of these shells second-hand from dealers who had purchased them at the Portland auction, and had engravings made of his drawings of the duchess's actual specimens (pl. 10).[18] Another source of illustration can be found in five volumes held in the Zoology Library of the Natural History Museum, London, entitled *British Conchology; or, Original Drawings of Fresh & Salt Water Shells* (circa 1786–1818) by John William Lewin and J. Agnew, who was the duchess's gardener (pl. 11). Agnew became a professional natural history illustrator after the duchess's death, and drew dozens of watercolors of shells owned by her friend J. T. Swainson, to whom she had also given many shells. Some of Agnew's shell drawings could therefore be of shells once owned by the duchess, and, at the very least, many are of specimens that match those she owned. His drawings, along with the engraved and sometimes hand-colored engravings, bring us closer to the materiality of her shells, giving us a chance to experience their color, texture, form, and size, and allowing us to glimpse those physical qualities that made them so intriguing and aesthetically pleasing.

FOCUSING ON THINGS

One could study the duchess and the culture of natural history collecting during her era using a variety of approaches derived from different disciplinary frameworks. The approach I have chosen is to focus on things, specifically on how her shells moved through various social contexts and geographical spaces. This can help us to understand the ways in which value and meaning were created through exchange, and the ways in which natural objects were transformed into material culture by subjecting them to the various discursive technologies of representation, classification, and display. In choosing to follow the duchess's shells as they moved through different stages in their social lives, I have borrowed from the eighteenth-century "it" novel, in which an object – a penny, a needle, or a lapdog – narrates its journey as it moves from one owner to another, providing descriptive accounts of the environments through which it passes on its way home or to its final destination.[19] Using the episodic structure of the "it" narrative, I have limited anecdotal and biographical detail about the duchess and other collectors to that which can illuminate the social life of the object I am describing. As such, this book is not a biography of the duchess of Portland but of her shell collection.

10 Edward Donovan, "Pinna muricata," Plate x in *The Natural History of British Shells* (London, 1799–1803). SPEC D 756, Arizona State University Libraries

Mytilus rustica

11   J. Agnew, "Mytilus rustica," in *British Conchology; or, Original Drawings of Fresh & Salt Water Shells*, circa 1786–1818, London, Natural History Museum © Natural History Museum, London

Nevertheless, the duchess is crucial to understanding the collection, for a collection is nothing, literally, without its collector. Unlike the "it" narrative's protagonist, an individual object, a collection is a collective entity that is brought into being through the collector's efforts, which involve gathering, organizing, and displaying the collected objects. A collection is an assemblage, to use a term from actor network theory, a hybrid entity, a materiality that comes into being through human interaction with it.[20] An understanding of the duchess's shell collection is therefore dependent on knowledge about her collecting practices and the social milieu in which she collected. To this end, this book has chapters on the duchess herself as a naturalist and a specimen hunter; on the various exchange networks through which she accumulated and exchanged shells and acquired information about them; on what happened to the collection after her death; and how the collection and her efforts were subsequently represented in print culture.

Researchers in anthropology, sociology, and cultural geography who track the movement of objects through various social situations and cultural contexts have found this multi-sited approach useful in analyzing the circuits of exchange and consumption within globalized economies. Tracking an object's movement from one part of the world to another, social anthropologists and cultural geographers have investigated the "social and geographical lives of particular commodities," such as flowers, second-hand clothes, blue-fin tuna, Pokémon cards, and rapeseed.[21] They have studied commodity chains and global flows to see the ways in which commodities move in networks of production, circulation, and consumption; and as these objects move from one social site to another, anthropologists have noted that their value and definition constantly undergo negotiation. Given this dynamic relation between objects and their shifting contexts, a study of "things-in-motion" is a productive way to "illuminate their human and social context" and to explore the various regimes of value through which objects pass over the course of their lives.[22]

Shells from the South Pacific, which were valued by the islanders for their decorative and functional properties, were much sought after by European voyagers, who knew that they could be sold to natural history dealers who had collectors as clients. Pacific shells were collected for a variety of reasons, ranging from their rarity and beauty, to their scientific value, to their association with the celebrated Captain Cook. Following shells from the Pacific to the dealers' auction houses and into the cabinets of collectors, I attend to the vagaries of value by focusing on the moments of exchange, since these reveal the cultural expectations and social codes that inform the transactions, especially when the moment of exchange brings together two very different regimes of value.[23]

The moment of exchange is thus a point of convergence between two symbolic economies. The auction, for example, brings together the collector and the dealer, who regard the object that they exchange as possessing very different kinds of value: for the dealer, the object is a commodity to profit from; for the collector, its value is derived from its aesthetic, scientific, or sentimental appeal. Exchange on the Pacific beaches involved the overlapping and interpenetration of two very different regimes of value – the gift economies of the South Pacific and the commodity driven economies of early modern Europe.[24] Exchange illuminates the structural dynamics of the social context, and, as such, provides an insight into the social mechanisms that constructed meaning and created value.[25]

The story of a shell collection that was assembled in the era of Cook's voyages of discovery must acknowledge the way in which natural history collecting was dependent on exploration, global trade, and the imperial infrastructure that enabled these activities. Wherever British warships and commercial vessels sailed, seamen and officers alike collected natural history specimens with profit in view, either by selling them or giving them to powerful people. Natural history collecting occurred on ships carrying slaves from the Guinea coast to Charleston, South Carolina; on ships transporting cargoes of porcelain and wallpaper from China to London; and on ships taking indigo cakes manufactured in Bengal to the London commodity mart, where they were sold to the rest of Europe. Mary Louise Pratt was among the first to point out natural history's dependence on exploration and colonialism; she described its particular mix of avowal of disinterested curiosity, the dislocating colonial dynamics of transportation, and the removal of objects from their site of origin.[26] In addition, several recent studies focusing on the Atlantic world have sought to deepen our understanding of the way in which knowledge about the natural world was expropriated from its origins and traveled in complex networks made up of actors who included indigenous peoples, settlers, slaves, merchants, and local colonial administrators.[27] Understudied, however, are the ways in which exotic plant and animal matter was assimilated into European systems of signification and cultural practices. In *Entangled Objects* (1991), Nicholas Thomas provided a framework for examining the way in which Europeans appropriated "indigenous things," challenging scholars to examine further "why these objects were acquired and what their collectors thought they were doing. And we must ask what the genealogies of European representation are." Thomas focused on European acts of appropriation within colonial settings, leaving others to explore how exotic objects were domesticated and incorporated into European cultural practices and systems of thought.[28]

The Duchess's Shells aims to help fill the gap in our understanding of natural history collecting in eighteenth-century Britain by reconstructing the methods by which the duchess of Portland acquired foreign and domestic shells and by attending to the movement of specimens, information, and favors within a network of collectors, dealers, and naturalists. In documenting the formation and dispersal of her shell collection, I have attempted to recreate the world she inhabited, the world of conchology with its material culture, exchange networks, and print culture. Chapter One describes the culture of natural history collecting, the way in which it blended art practices with polite science to produce a kind of inquiry into the natural world that partook of an aesthetic sensibility. It argues that the duchess's collecting practices have been misrepresented by cultural historians as an extreme form of acquisitiveness, and to counter this interpretation turns to theories of collecting to help explain these rather unflattering depictions of her collecting practices. Chapter Two provides a detailed description of the methods by which the duchess collected British shells and demonstrates that she herself was a specimen hunter, an assertion that counters the commonly held assumption that her collecting practices were limited to purchases. Chapter Three focuses on the way in which shells from Cook's voyages moved through different kinds of networks involving dealers, collectors, and naturalists, and describes the eagerness with which collectors purchased exotic Pacific specimens. Chapter Four details the methods by which sailors, officers, and other voyagers gathered mollusks on their Pacific journeys with the expectation of selling them to natural history dealers. Chapter Five examines the duchess of Portland's role as a patron of conchology books, focusing on the material practices and publication processes involved. Chapter Six analyzes public interest in the duchess's sudden death and the spectacular Portland auction of the following year. It describes the disassemblage of the duchess's collections, tracing the paths that her shells took when they were auctioned off until the trail goes cold. Chapter Seven examines the way in which the duchess's reputation as a naturalist and collector diminished over time until the third decade of the nineteenth century, when her name and her collection all but disappeared from the print culture of conchology. The focus of Chapters One, Two, Three, and Four is on the material culture of shell collecting and the social practices and discursive technologies that transformed shells into specimens. With their emphasis on print culture, Chapters Five, Six, and Seven explore the role of publication in shaping reputations and producing value.[29]

The means and methods by which the duchess amassed, organized, and displayed her shell collection tell us much about the popularity of natural history within the middling and upper classes of Georgian England, and, in particular, about the operations of polite science as a fashionable pursuit that brought together women, sociability, and natural history. Moreover, tracing the formation of the duchess's shell collection illuminates the global and local natural history networks by which these specimens were accumulated and exchanged, as well as revealing the tensions within the cultures of natural history collecting, tensions generated by the different goals, abilities, and resources available to amateurs, connoisseurs, naturalists, and dealers. Furthermore, this examination of the rise and fall of the collection sheds light on the varied social contexts through which the shells traveled – the ship, the auction house, the connoisseur's cabinet, the publishing house – to reveal how these contexts were instrumental in constructing taste, creating value, and producing meaning.

CHAPTER ONE

12　Thomas Hudson, *Margaret, Duchess of Portland*, oil on canvas, 1744. Private Collection

THE DUCHESS, NATURAL HISTORY, AND CULTURES OF COLLECTING

Among natural history collectors of the last quarter of the eighteenth century, the duchess of Portland reigned supreme (pl. 12).[1] She collected plants, birds, and insects as well as shells, possessing perhaps the largest shell collection in all of Europe. To collect on this scale was to participate in the aristocratic culture of collecting, a tradition that was well established in her family. Born in 1715, Margaret Cavendish Holles Harley inherited her mother's vast fortune and her father's passion for collecting.[2] Her father, Edward Harley, second duke of Oxford, amassed one of the finest collections of medieval manuscripts in Britain, and when he died in 1741 she arranged for Parliament to buy it at a much-reduced price. Comprising more than 7,000 manuscripts, the Harleian Collection, with its psalters and Gospels, European courtly romances, and medical and alchemical treatises, was one of the founding collections of what is now the British Library. Although Margaret, who married William Bentinck, second duke of Portland, in 1734, was also a collector of such fine and decorative arts as Japanese porcelain, Italian cameos, etchings by Dürer and Rembrandt, French snuffboxes, and Roman antiquities, including the Portland Vase now housed in the British Museum, her great love was natural history, and her collections reflect her passionate engagement with the natural world.

Of her collecting practices, her friend Horace Walpole, also a collector, wrote that in her early married life she had collected "old China, particularly of the blue & white with a brown Edge, of which last sort She formed a large Closet at Bulstrode," and later, after a foray into collecting pictures, which Walpole says she "did not understand," she "went deeply into natural history."³ Walpole's overview of the duchess's collecting practices as a shift from acquiring fine and decorative arts to natural history specimens is not entirely accurate. In fact, natural history did not supplant her collecting of china as Walpole suggests, but rather shaped the kind of decorative pieces she acquired for display and purchased for her own use, namely porcelain, silverware, and candelabras decorated with animal and plant figures. The natural world always had intrigued Lady Margaret, who even as a child would fill little handmade notebooks with lists of animals and plants, developing skills that she would use later as she observed, described, and classified the natural world. As a young married woman, she visited Sir Hans Sloane's natural history collection, and let it be known to her friends that she was building her own collection and would appreciate their sending to her whatever botanical or zoological specimens they happened upon.⁴

An actively engaged parent of five children,⁵ she focused her energies on turning Bulstrode, her husband's stiff and formal ducal residence in Buckinghamshire, into a pleasant, sociable space. There she channeled her love of natural history into redesigning the grounds of the estate. A visitor commented in 1753 on the "works and improvements" to Bulstrode: "they are innumerable." The duchess added sweeping lawns, flower gardens, and wooded groves, and she "fitted up the little room out of her conclave that opens into the garden in the Gothic taste, and made it the prettiest cell you can image."⁶ She described her early married life at Bulstrode: "My amusements are all of the Rural Kind – working, Spinning, Knotting, Drawing, Reading, writing, walking & picking Herbs to put into an Herbal."⁷ In the 1730s, 1740s, and 1750s she collected plants, birds, and insects, telling a friend, for instance, "I found to-day a very odd fly – the body black, the legs red, and the tail half-an-inch long, the whole fly rather larger than a gnat."⁸ But it was not until 1762, when her husband died, that she began to collect natural history specimens on a grand scale.

Heiress of the Cavendish fortune and Welbeck Abbey, the duchess was one of the wealthiest women in Britain, and her expenditures at Bulstrode and on her collections were a source of rumors during her lifetime and after her death. Walpole estimated that the duchess had spent "not less than threescore thousand pounds" on improvements at Bulstrode, "indulging her taste for Virtù," which included purchasing every Hollar print, and building a natural history collection, which he reported to have cost her £15,000.⁹ Although Walpole's annotated auction catalogue and commentary provide some information on how she collected objects of *virtù*, noting the prices she paid and the provenance of such items as Raphael's *Holy Family*, he offers little information about the natural history side of this auction. Natural history simply held little interest for him.

This indifference to the duchess of Portland's natural history collecting is not unique. In fact, her massive natural history collections have received scant attention from scholars.[10] Little attempt has been made to describe them or to chronicle the monumental work of constructing them.[11] This indifference to her work as a collector, which is only now beginning to wane, can be explained in part by the fact that natural history collecting does not fall directly within the traditional disciplinary boundaries of either art history or the history of science. Art history's attention has been primarily on the fine arts, as the editors of *Collecting Subjects* (2009) have argued:

> A long-established tradition of art historical writing recording patronage and its system of collecting "high art" objects, namely and exclusively painting and sculpture, has privileged the fine art collector in the cultural and historical matrix. . . . Yet the collector and collections of craft, fashion and other decorative objects have remained in relative obscurity.[12]

Paralleling the way in which collections of fine art have been privileged over the decorative arts in art historical scholarship, natural history collecting has been overlooked as a site of scientific praxis. Experimental science, the laboratory in particular, has been studied as the place where "real" science takes place,[13] and natural history collecting has been perceived as operating "beyond the pale of proper science."[14] As a result, the material culture of natural history collections, as Emma Spary has argued, has yet to be explored fully as a mechanism for acquiring scientific knowledge.[15] Even within the field of museum studies, with its appreciation of collecting as a form of knowledge production, artifacts – "human-made objects of material culture" – are more highly valued as objects of study than natural objects, although, as Susan Pearce points out, once a natural object has been collected, this "natural piece" is transformed "into a humanly defined object, which is to say an artefact."[16] In short, natural history collecting is understudied, undervalued, and poorly theorized. Too easily dismissed as the product of "mindless hoarder's lust,"[17] it has often been slotted by cultural critics and historians into a subset of consumption and has not been taken seriously as a creative and instrumental form of meaning-making.

In order to increase understanding of the duchess's collecting practices, this chapter situates her natural history collecting at the juncture of two distinct but linked activities: collecting and the culture of natural history. It begins with an overview of the popularity of nature-based activities in the second half of the eighteenth century, then describes and attempts to account for the apparent contradiction between the period's love of natural history and its ambivalence toward collecting natural history specimens. To that end, the chapter examines modern-day theories of collecting in the fields of museum studies and art history to elucidate what it is about this practice that distinguishes it from other forms of engagement with the natural world.

POLITE NATURAL HISTORY

Margaret Cavendish Bentinck began collecting shells as scientific specimens in the mid-eighteenth century, when Enlightenment ideals encouraged the free and open exchange of ideas, information, and objects within the Republic of Letters. Natural history was regarded as a form of polite and edifying entertainment.[18] Men and women of the educated classes were avid consumers of books about the Linnaean system of classification, engraved and colored prints of botanical and zoological specimens, and the natural histories of local regions and foreign locales. Natural history publication took many forms, including simple field guides filled with descriptions of plants and animals for the purpose of identification; treatises on economic botany; georgic poetry celebrating agricultural productivity; horticultural manuals and planting calendars such as Philip Miller's *The Gardeners Dictionary* (1731, revised and expanded 1761) and *The Gardeners Kalendar* (1732); elaborately illustrated multi-volume tomes such as Thomas Pennant's *British Zoology* (1768–76); regional guides to local flora and fauna ranging from William Curtis's *Flora Londinensis* (1775?–1798) to Gilbert White's charming meditations on local animal life in the *Natural History and Antiquities of Selborne* (1789); and travel narratives describing the natural history of places such as Jamaica, the coast of Coromandel, and the Carolinas.

Natural history was also consumed in the public arena as a commodity and a spectacle. The culture of curiosity about nature was fueled by visiting places where specimens, both alive and dead, were on display: museums, private collections, public gardens, auction houses, dealer's shops, and country houses with their shell grottoes, menageries, and botanical gardens.[19] Natural history enthusiasts not only bought books on the subject and paid fees to visit the British Museum and Sir Ashton Lever's Museum (pl. 13), but also purchased specimens to display in their own cabinets and drawing rooms. As Elizabeth Montagu, famous for her Bluestocking assemblies, wrote: "It is now grown the fashion to borrow ornaments for cabinets and dressing-rooms of birds and fishes, and vanity and virtuosoship go hand in hand."[20] To oblige those who were eager to display their polite taste for natural history, specimen dealers acquired minerals, fossils, insects, and shells from near and far, selling them to casual and serious collectors alike. In 1762

13 "Sir Ashton Lever's Holophusicon" (1783) and advertisement about opening times. © The Trustees of the British Museum

George Humphrey, father of the George Humphrey who was a part of the duchess's natural history network, ran an advertisement in *Lloyd's Evening Post and British Chronicle* to announce to his "Friends and Customers" that he had "removed from his late House at Hudson's Court, in the Strand, to the SHELL WAREHOUSE next to John's Coffee House in St. Martin's Lane, Strand,"

Where he continues to buy and sell all Sorts of Collection, Grotto, and Flower Shells, and has the greatest Variety of the latter in England; also Corals, Fossils, and other Natural Curiosities, &c. At the said Place, Ladies may be furnished with the Implements used in making Shell-Work, and may be taught that same at Home or Abroad. Any person having any of the above Curiosities to dispose of, by sending a Line shall be waited on.[21]

Humphrey advertised regularly in several newspapers, offering the services of his daughters as instructors for shellwork, including designing grottoes. Although his main focus seems to be shells, he also sold at his shop "a neat assortment of Useful and ornamental China, lacquered Boxes, Fans, Pearl Necklaces, &c., some curious foreign dead Animals, Birds, Fishes, and Insects, dried or preserved in Spirits, &c."[22] After he died in 1776, his "Stock in Trade of Natural Curiosities" was sold by auction. To be sold were his "extensive and beautiful variety of Shells, Corals, and other Marine subjects; uncommon Beetles, Butterflies, and other Insects, Birds, Fishes, Reptiles, and various animals, dried and preserved in Spirits."[23] Natural history specimens were big business, and nature, as a source of entertainment, commercial activity, decorative display, and scientific inquiry, was omnipresent in the lives of the polite classes, appearing in material, visual, and print culture.

Men and women of ordinary abilities also deepened their knowledge of nature by making excursions to urban green spaces or to the countryside. Here they "botanized," an activity that included identifying plants and gathering the flowers. They also caught insects, putting flies, spiders, and beetles in little boxes; and they collected mollusks in the form of snails and freshwater bivalves that inhabited streams and marshy areas. These amateur botanists and zoologists gathered specimens; studied, described, and drew them; and attached names to them, some employing Linnaean taxonomy. Seashells gathered on the beach were cleaned, dried, polished, and housed in cabinets; flowers, mosses, ferns, and seaweed were dried and pressed onto paper and bound into volumes; and butterflies and moths were captured with nets, killed carefully, and pinned onto cork-lined wooden boxes and drawers. The gathering and organizing of plant and animal matter were specific material-based activities located at the intersection of touch and sight, where the tangible was as central as the visible in producing knowledge about the natural world.[24] Natural history collecting – and shell collecting in particular – were embodied forms of knowledge-making that belonged to a period and ethos dominated by the talented amateur.[25]

~

ARTWORK, CRAFTWORK, AND NATURAL HISTORY

In this era, before hard lines were drawn between art and science, and before the fragmentation of knowledge about the natural world into specialized disciplines, inquiry into nature was not sequestered in spaces designed explicitly for scientific activity, such as the laboratory and the university. Instead, it took place in a wide variety of settings, from the domestic interiors of drawing rooms, closets, and kitchens to flower gardens, estate parkland, and uncultivated tracts of wasteland. Engagement with nature took diverse forms, from decorative art practices such as shellwork and needlework; to drawings in pencil or ink or watercolors of plants and animals; to reading and using recipes in cookery books for varnish and ink, and referring to herbals for plant descriptions and remedies to cure illnesses. Botany, zoology, chemistry, and medicine were studied and practiced in proximity to domestic crafts and amateur arts. As the historian Amanda Vickery has observed: "At its peak, shellwork, like floral embroidery and botanical drawing, sat easily with scientific classification and cabinets of curiosity, a domestic performance that signaled modern intellectual horizons, an expression of polite natural science as at home in the drawing room as in the laboratory" (pl. 14).[26]

Enthusiasm for studying the natural world was supported by a vibrant visual and material culture that included a range of activities that appealed to all ages and both sexes. Enlightenment ideals encouraged individuals to observe nature directly and to document their observations with writing, drawing, and collecting. As many scholars have demonstrated, art and science were often practiced together as forms of inquiry into the natural world.[27] Drawing was central to apprehending the complexity and variety of nature, and was crucial to the construction of knowledge about it. Many naturalists were amateur artists, who used their skills to draw their specimens. Henry Seymer (1714–1785), a country gentleman who was an amateur entomologist, conchologist, landscape designer, and artist, made beautiful watercolors of his butterfly and moth collections. He also colored in the engraved illustrations in his natural history books, even going as far as to collect specimens of the insect species represented in the second volume of Dru Drury's *Illustrations of . . . Exotic Insects* (1773) so that he could color them correctly, not trusting professional colorists to do a proper job (pl. 15). The disciplines of art and science, now regarded as separate and distinct, were in this period coeval and mutually productive.[28]

In addition to illustration, the kinds of amateur art practices that flourished in concert with the study of the natural world ranged from two-dimensional work on paper and on textiles to three-dimensional work involving featherwork, shellwork, and taxidermy.

14   Mrs. Beal Bonnell and Miss Harvey Bonnell, shellwork vase, *circa* 1779–81. London, Victoria and Albert Museum, w.70–1981. Photo © Victoria and Albert Museum, London

Gathering flowers, drying them, and pressing them into albums were activities that had wide appeal since they required no special equipment beyond glue and paper. Featherwork and shellwork needed huge amounts of feathers and shells, their practitioners often gathering their own supplies and asking friends to send them whatever they could. When Elizabeth Montagu was making her huge featherwork panels of pastoral themes to decorate her London townhouse, she would write to friends requesting them to get their cooks to save pheasant and goose feathers for her use. She often asked for feathers from exotic birds – "Pray has the macaw dropt some small blue or yellow feathers?" – and noted that some feathers were more useful than others: "My great piece of feather-work

15   Moses Harris, delineator and engraver; coloring by Henry Seymer. Plate V from Dru Drury, *Illustrations of Natural History*, 3 vols. (London, 1770–82). By permission of the Linnean Society of London

is not yet completed; so, if you have an opportunity of getting me any feathers, they will be very acceptable. The brown tails of partridges are very useful, tho' not so brilliant as some others" and

> I am obliged to you for your kind attention to my feather-work. The neck and breast feathers of the stubble goose are very useful, and I wish your cook would save those of the Michaelmas goose for us. Things homely and vulgar are sometimes more useful than the elegant, and the feathers of the goose may be better adapted to some occasions than the plumes of the phoenix.[29]

Shells could be bought by the barrelful, but shell workers could often not afford them, so they would rely instead on their own collecting efforts, as well as depending on friends to send them shells from coastal excursions. Snail shells were more easily acquired; they could be gathered on damp mornings in the countryside or even in London parks and undeveloped areas, such as Battersea Fields south of the Thames. The snails were boiled and removed from their shells, which were then cleaned and polished, ready to be used as design elements in sculptures or to decorate frames, mirrors, and candelabras. Taxidermy skills could be required for more challenging activities, such as bird collecting. This involved eviscerating the creature, stuffing it with a mixture of arsenic, alum, and spices, sewing it up, and then arranging its body to evoke the posture of perching on a branch. Women stuffed and preserved birds for decorative purposes, placing them under bell jars in parlors or incorporating them into scenes in glass boxes with other birds or even animals to create small dioramas.[30] The continuum of natural history activities ran from the purely decorative to the scientific, with much overlapping along the way.

AESTHETICS AND THE NATURAL WORLD

Within polite culture, art and natural history were deeply intertwined, and an appreciation of the aesthetic aspects of nature was central to amateur art practices.[31] It was the beauty they saw in insects, birds, shells, and, of course, flowers that often attracted people to the study of natural history. Linking natural history with art production and an appreciation of the aesthetic, William Curtis, author of *Flora Londinensis*, announced that the goal for his periodical, the *Botanical Magazine*, was "to unite Systematic Knowledge with the Pleasures of the Flower-Garden." He dedicated the first issue to Mrs. Montagu Burgoyne: "not less esteemed for her Social and Domestic Virtues than admired for the accuracy with which she draws and the delicacy with which she paints the Beauties of Flora."[32] Many fine artists and practitioners of the decorative arts were inspired by the shapes, colors, and textures of the natural world. Mary Delany (1700–1788), for instance, used natural forms in her needlework, filling the designs with botanically and zoologically correct images of butterflies, flowers, and ferns. Her cut-paper mosaics of flowers, which she called her *hortus siccus*, were pronounced by Sir Joseph Banks, President of the Royal Society, to be botanically accurate representations of specific species of exotic and domestic flowers (pl. 16).[33] Of her paper mosaics, William Gilpin wrote: "these flowers have both beauty of painting and the exactness of botany." He described

16 Mary Delany, "Magnolia grandiflora," collage of colored papers with bodycolor and watercolor, on black ink background, 1776. London, British Museum. © The Trustees of the British Museum

the process she used, noting her close examination of the flower that she was about to copy: "In the progress of her work, she pulls the flower in pieces, examines anatomically the structure of its leaves, stems, and buds; and having cut her papers to the shape of the several parts, she puts them together, giving them a richness and consistence, by laying one piece over another."[34]

Many naturalists were unable to separate their aesthetic response to nature from the science they practiced as amateur entomologists, conchologists, and botanists. Dru Drury (1725–1804), jeweler and amateur entomologist, wrote and published a three-volume tome, *Illustrations of . . . Exotic Insects* (1770–82), which contained pictures of his own insect collection with descriptions. His aesthetic appreciation of the variety of hues and tones that insect bodies and wings displayed was central to his collecting practices and publishing efforts. One of the reasons he decided to make this book was that he thought it would be the best way of preserving the beauty of the insects, especially the moths and butterflies, from "oblivion, by thus delineating them on paper." Because moths and butterflies were "of such tender and delicate natures," sunlight destroyed "their colours," and air would "totally consume every part of them, leaving nothing behind but a little dust." Drury explained: "however pleasing and agreeable they may be to our sight, they are not easily preserved in all their gay and striking plumage." To achieve these ends, he employed Moses Harris to draw, engrave, and color the insect illustrations in his book. He and Harris were members of the Society of Aurelians, a club of amateur entomologists; Harris proved an excellent choice for illustrator and engraver because he understood and sympathized with Drury's concerns to convey the beauty of the insects accurately. Drury's verbal descriptions of his insects, meant to guide readers through the process of classification, were also exercises in describing the colors and textures of insect bodies. Of one figure on Plate v of Volume i, he wrote:

> *Upperside*. The *Antennae*, are brown, outwardly, and white underneath; the ends being yellow. – The *Head*, *Thorax*, *Abdomen* and *Bases* of the wings are tawny orange. – The *Superior Wings*, are dentated; the *Tips*, and external edges, are dark brown, nearly black; on which five white spots near the tips, the largest being round. The remainder of the wings is a fine orange brown, with several black marks thereon, near the anterior edges.

These descriptions, full of references to shapes, forms, colors, and textures, are vivid testaments to the centrality of visual culture within the world of polite science and the study of natural history.[35]

Equally attentive to the visual aspect of natural history specimens were the descriptions written by Emanuel Mendes da Costa for the shells in *The British Conchology* (1778). He describes the appearance of a mytilus by attending to its shape, color, and texture:

> This is a very small species, of the size of a kidney-bean, extremely thin, semitransparent or horny, and brittle, of a light greenish colour, with a faint tint of brown or

rosy; it is a broad shell, of a squarish shape, and the valves are extremely deep or concave. . . . Inside, smooth and glossy, sometimes pearly, at other times same as the outside, with a more rosy blush. The margins finely crenated; and, by the thinness of the shell, the outside *striae* are seen. . . . The discovery of this pretty and rare species on our coasts is owing to the unwearied pursuits of Dr. Richard Pulteney.[36]

The British Conchology catalogues all the known mollusks found in Britain, offering itself as a scientific treatise, based on the careful and painstaking work of gathering information, specimens, and expert opinions. Aesthetics, however, were never very far from da Costa's agenda, for, as he argues in the preface:

Shells are certainly a beautiful part of the works of Nature: we observe a vast variety in the *genera*, and those branching into an innumerable variety of *species*; we see in them the most *splendid colours*, the most perfect symmetry, and cannot but wonder at all this seeming waste of beauty lavish'd on the depths of the sea, and only accidentally brought to the face of day. . . . – What would be our astonishment, could the beds and caverns of the ocean be thrown open to our view!

The variety, complexity, and beauty of shells cause the "Naturalist" to be "enraptured of the objects to which he devotes his attention." "A *cabinet of shells*," da Costa argues, "not only gives *pleasure*," but "*enlarges* the *mind*":

The *elegance* of their *forms* may inforce our *notice*, the *beauty* of their *colors* may solicit our *admiration*; but the *philosophic eye* penetrates much farther; it *discriminates* every *minute variation*, not only *surveys* the *texture*, but *explores* the *anatomy*, and charm'd with the effect, *darts* forward to the *cause*, and, as Mr. *Pope* has elegantly expres'd it, "Looks thro' Nature up to Nature's God."[37]

Just as collectors were drawn to the beauty of their specimens, so visitors to natural history museums found the displays of shells and stuffed birds aesthetically pleasing. An example of such an aesthetic response to natural history specimens can be found in a late eighteenth-century novel, *The East Indian; or, Clifford Priory* by Mary Julia Young, in which Elinor, a girl in her teens, visits the Leverian Museum, where "she wished to examine minutely the wonderful variety of the feathered race, from the ostrich to the humming bird, and the no less wonderful productions of the ocean" (see pl. 13). Elinor waxes philosophical when contemplating these "wonderful productions" of nature:

"Oh!" she exclaimed, "with what inexpressible delight and astonishment do I behold this choice selection! What a rich display of nature, in her most delicate, yet most permanent charms! Death, who renders the loveliest human exterior disgusting, has no power over the glowing plumage of the birds, or the polished shells of the fishes – they retain their exquisite beauties, when the bodies they so splendidly adorned are crumbled to dust. . . . [T]hese shells and feathers, so judiciously arranged, afford the eye a lasting entertainment, and must certainly fill every thinking mind with love for a Divinity, whose minutest work they behold with admiration.[38]

Elinor's enthusiastic speech captures the complexity of her aesthetic and philosophical response to the specimens, a response that was shared by those immersed in polite culture's embrace of natural history. Mary Delany voiced similar sentiments, though in a more sophisticated manner, in reference to the shells and fossils she collected: "It is impossible to consider their wonderful construction of form and colour, from the largest to the most minute, without admiration and adoration of the great Author of nature," and "Can we view the wonderful texture of every leaf and flower, the dazzling and varied plumage of birds, the glowing colours of flies, &c. &c., and their infinite variety, without saying, '*Wonderful and marvelous art thou in all thy works!*'"[39] Such a view was commonly held by naturalists, amateurs and professionals alike, who regarded their study of nature as a way to appreciate the world that God had made. For some, the close study of nature was akin to a devotional practice. Robert Thornton, author of the elaborately illustrated *Temple of Flora* (1799–1807), wrote in *The Religious Use of Botany* (1824) that the botanist's observation of nature turned a morning walk into a devotional act: "everything in Nature combined to fill our minds with the sweetest and purest delights, and to lead our hearts towards God."[40] Nature, in this era, was a source of entertainment, edification, and inspiration for philosophical reflection and artistic practice.

NATURAL HISTORY AT BULSTRODE

As a fashionable pursuit, the study of natural history brought people together to observe, draw, collect, classify, and discuss the natural world. Gathering botanical specimens to make floras and herbaria, collecting butterflies, spiders, and shells, making decorative objects from natural materials such as feathers and shells, and sketching and painting images from nature were social activities that men and women of the middling and upper classes did together. Indeed, sociability, collegiality, and collaboration were the hallmarks of the study of natural history and the art practices derived from observing the natural world and manipulating natural objects. An active participant in this polite culture of natural history, Margaret Cavendish Bentinck set the standard for such intellectual and

17 Richard Corbould, *Bulstrode Park, Buckinghamshire*, watercolor with pen and ink, *circa* 1790(?). Engraved by Walker, Plate 56 of the *Copper Plate Magazine*, 1794; also appeared in Harrison's *Picturesque Views* (1788). London, British Museum © The Trustees of the British Museum

artistic engagement with the natural world. Under her direction Bulstrode (pl. 17) became a major center for the study of natural history; as Mary Delany noted, it was "a *noble school* for such contemplations."[41] Rebecca Stott suggests that "if Bulstrode had survived it would rank with the Ashmolean Museum in Oxford and the British Museum in London as collections embodying the spirit of learning and hunger for knowledge which is at the heart of the Enlightenment."[42] Bulstrode, "something between a museum and a university" as Amanda Vickery asserts, attracted visitors who shared the duchess's enthusiasm.[43] Visitors included such famous naturalists as Sir Joseph Banks (1743–1820) (pl. 18), Daniel Solander (1733–1782), premier taxonomist and British Museum curator, William Curtis (1746–1799) of the Chelsea Physic Garden, and Thomas Pennant (1728–1798), author of *British Zoology*, as well as ordinary people who, because of their interest in natural history, were just as welcome as the more powerful and prominent guests. The Customs officer John Timothy Swainson (1757–1824), for instance, was

18 William Dickinson after Sir Joshua Reynolds, "Joseph Banks, Esqr." Mezzotint, 1774. London, British Museum. © The Trustees of the British Museum

invited to Bulstrode to participate in a shell collecting expedition, on which he found a snail "in a canal at Bulstrode," a very rare *Helix polita*.[44]

Surrounding herself with artists and naturalists, the duchess created an atmosphere that was very productive for those who had the good fortune to reside at Bulstrode. In residence there in the late 1760s was the great botanical illustrator Georg Dionysis Ehret (1708–1770), who taught the duchess's daughters how to draw. During his tenure, he made hundreds of botanical illustrations for the duchess. The Rev. John Lightfoot (1735–1788), an accomplished botanist and author of *Flora Scotica* (1777), was chaplain from 1767 until the duchess's death in 1785. Although he was appointed the curate at

19 John Opie, *Mary Delany*, oil on canvas, 1782. London, National Portrait Gallery. © National Portrait Gallery, London

Colnbrook, Middlesex, which included a "lectureship" at Uxbridge, he stayed at Bulstrode from Wednesday until Saturday each week, and participated in numerous natural history projects.[45] It was not only Lightfoot who spent a great deal of time at Bulstrode: Mary Delany (pl. 19) lived there as the duchess's guest for nearly two decades, most of her second widowhood, and it was at Bulstrode that she did most of her exquisitely crafted and botanically correct cut-paper flower mosaics.[46] Until recently, Delany was known to literary historians and scholars of women's history primarily for her voluminous correspondence, a source of much information about the world of educated, socially prominent Georgian women, in which she circulated as a member of an aristocratic family but without the advantages of wealth or financial security. Twice married, once disastrously in her youth to a much older man, Alexander Pendarves, who left her an impoverished widow in 1725, she remarried in her early forties one of Jonathan Swift's friends, the Rev. Patrick Delany, an Anglican clergyman, with whom she lived happily in Ireland until his death in 1768. Returning to England, she settled into a life divided between Bulstrode, living in London near the duchess's Whitehall residence, and visiting her many friends. Her artistic accomplishments have finally caught the attention of scholars, who now recognize her needlework creations, drawings, paintings, and paper mosaics as art rather than quaint craft projects, to which they had been relegated for two centuries. Delany, no doubt, benefited as an artist from her residence at Bulstrode. Delany's affection and gratitude are evident when she says that the duchess is

> a blessing to *every creature* within her possessions. But what makes her so, my dearest child? *Not* her great fortune, – *not* her high station: but the *goodness of her heart*, the excellence of her principles, the sweetness of her manners, an understanding improved by reading and observation and her many *ingenious pursuits*, which are a constant source of entertainment to herself and those she honours with her conversation. How happy must I be in such a friend![47]

She repeatedly proclaimed her debt to the duchess for inspiring her to make her *hortus siccus*:

> To *her* I owe the spirit of pursuing it with diligence and pleasure. To *her* I owe more than I dare express, but my heart will ever feel with the utmost gratitude, and tenderest affection, the honour, and delight I have enjoy'd in her most generous, steady, and delicate friendship, for above forty years.[48]

In addition to supporting Lightfoot's botanical research trips to Wales and Scotland and Delany's nature-based artwork, the duchess encouraged her head gardener, Agnew, to become a natural history illustrator.[49] Some of Agnew's work from his days at Bulstrode during the last years of the duchess's life survives in the Botany Library at the Natural History Museum in London. These strange but lovely drawings of fungi are annotated and dated, and the notations are poignantly revealing of the kind of attention

20   J. Agnew, "Perira arugovirens," watercolor. Written at bottom of drawing: "Found at Bullstrode Gardens. Amongst the Yew and Holly Bushes by the side of the Lime walk. Decr 1783." London, Natural History Museum. © Natural History Museum, London

to nature that the duchess no doubt cultivated in the people who surrounded her. At the bottom of his drawing of *Perira arugovirens* Agnew wrote: "Found at Bulstrode Gardens. Amongst the Yew and Holly Bushes by the side of the Lime Walk. Decr. 1783" (pl. 20). Of *Perira flexicaulus*, he noted: "This little plant was found 17th of Dec 1781 In Druid's grove amongst the Beech leaves," and of *Perira dentate*: "Found in the Orchard Oct 1782 Growing out of the underside of a piece of board which lay flat against the ground." These comments give glimpses into Bulstrode's unique milieu where Agnew, in his duties as a gardener, would lift up an old board in the orchard carefully to see what might be growing underneath and on its underside, and then gather up the tiny bits of fungi and bring them back to the house, where he had access to a microscope. He could then examine the specimens further, drawing them and then coloring in the pencil sketches with watercolors. This loving appreciation of the beauty and wondrous intricacy of the natural world was a product of the reigning ethos of Bulstrode.

Agnew's notations are also poignant in that they reveal much about the landscape at Bulstrode that was irrevocably altered in the 1790s by Humphrey Repton at the behest of the third duke of Portland, Margaret's son and heir. From Agnew we learn of the holly bushes, yew trees, beech trees, a row of lime trees, an orchard, a dairy, the Druid's Grove, Wasp's Wood, and something called Cain and Abel, where an echina was "Found upon the Right-hand edge of the walk, going from Cain and Abel to the Dairy. 19 of August 1782." After the duchess's death, Agnew joined forces with John William Lewin to produce more than 500 drawings of shells for J. T. Swainson, who may have had plans to have these illustrations engraved and published (pl. 21). Now housed safely at the Zoology Library of the Natural History Museum, they illustrate Swainson's own collection of British shells, several of which were given to him by the duchess, as will be seen in the next chapter.

ELITE CULTURES OF COLLECTING

In this environment, which fostered creativity in the service of contemplating the natural world, the duchess of Portland engaged in various forms of natural history collecting and oversaw the construction of a botanical garden, a hothouse, a menagerie, and an aviary at Bulstrode.[50] Such an assembly of live specimens was derived from older aristocratic forms of collecting related to the princely *Kunst und Wunderkammern* of the Renaissance and the early modern cabinets of curiosity, in which the collected object, a rarity, represented a region or a people and displayed the owner's global reach and figurative dominion over the world's resources.[51] As Krzysztof Pomian has argued in his study of collecting in Europe, curiosity is the "desire to see, learn or possess rare, new, secret or remarkable things, in other words those things which have a special relationship with totality and consequently provide a means for attaining it."[52] The English gentleman's cabinet of curiosities, as Katie Whitaker has suggested, was filled with rarities, objects that aroused a sense of wonder, admiration, and fascination, and excited viewers

21 J. Agnew, "Ostrea opercularis," in *British Conchology or Original Drawings of Fresh & Salt Water Shells*, circa 1786–1818. London, Natural History Museum. © Natural History Museum, London

to respond with reverence and to question the curious object's natural or artificial origins. These seventeenth-century virtuosi enjoyed "displaying rarities in [their] houses, gardens, and estates, and visiting each other to view and discuss these rarities."[53] This form of collecting continued to be practiced in the eighteenth century and was understood as a sign of elite status. Several of the duchess's peers imported exotic species of plants and animals to enhance the beauty and prestige of their estates. The Hon. Jane Barrington, Lady Amelia Hume, the Hon. Dowager Lady de Clifford, and Lady Shelburne collected exotic plants to grow in their hothouses and "stoves," while the duchess's friends Lord and Lady Bute took active interest in raising North American trees and shrubs on the grounds of their princely estate, Luton Park, Bedfordshire; Lady Bute also enjoyed collecting shells.[54]

In keeping with these aristocratic traditions, the duchess owned exotic animals and a vast array of live birds, some of which roamed freely on the Bulstrode lawns. Mary Delany, when describing in July 1778 this ducal seat, grew rapturous, calling it "Paradise":

> This place is now in its full beauty, and if any situation can bear a resemblance to Paradise it is this. The *variety* of *creatures* (in perfect agreement) *and vegetables* are a constant scene of delight and amusement, besides the good taste in which all the improvements are laid out. It has as much magnificence as is necessary, with every elegance and comfort that can be wish'd for, such as everybody must approve and enjoy . . . A curious and enquiring mind can't fail of being gratified . . . at Bulstrode, wth *every branch* of virtû.

The following summer she wrote:

> [B]eautiful deer, oxen, cows, sheep of all countrys, bufalos, mouflons, horses, asses, all in their proper places. Then, *hares* and *squirrels at every step you take*, so confident of their security that they hardly run away! The great lawn before the house is the nursery of all sorts of pheasants, pea fowl and Guinea fowl, besides interlopers of Bantam pidgeons; and not withstanding these numerous *familys*, the lawn is kept with as much neatness as the drawing room; such is the diligence of the attendants and the diligent eye of their sovereign lady, who delights in having everything in the best order.[55]

According to Mrs. Powys, who visited Bulstrode in July 1769, the menagerie contained many beautiful larger birds – storks, pheasants, bustards, and goons – while the aviary housed

22   John Raphael Smith, after Sir Joshua Reynolds, *Elizabeth Montagu*, 1776, mezzotint. London, National Portrait Gallery. © National Portrait Gallery, London

a most beautiful collection of smaller birds – tumblers, waxbills, yellow and bloom paraquets, Java sparrows, Loretta bluebirds, Virginia nightingales, and two widow-birds, or, as Edward calls them "red-breasted long-twit'd finches." Besides all of the above mention'd, her Grace is exceedingly fond of gardening, is a very learned botanist, and has every English plant in a separate garden by themselves. Upon the whole, I never was more entertain'd than at Bulstrode.[56]

Delany sketched some of these exotic animals: there is a drawing of a zebu bull and one of a Java hare, as well as drawings of the birds, in private collections. The duchess's fondness for birds was clearly shared by many, though it was mocked by her friend Elizabeth Montagu (pl. 22), who teased her for liking birds better than her friends. "It is a hard case that your Grace forgets your correspondents for your Bantam fowl.

Though I have not my head so well curled as your Friesland hen, nor hold up my head like your upright duck, do you think I consent to be laid aside for them?"[57]

The duchess of Portland's collecting activities can be explained as a product of her class position, aristocratic heritage, and family traditions. Ann Shteir has labeled the duchess as "the paradigmatic aristocratic woman collector of the eighteenth century."[58] Her interest in collecting rare birds and plants can be understood as enacting her cultural heritage, which was one of aristocratic expenditure and display, and as a way of enhancing the visual splendor of her ducal residence and to arouse the curiosity and interest of visitors, some of whom were her friends and acquaintances while others, like the Powys family, were country house tourists. In the latter context, Elizabeth Montagu suggested to her cousin that he visit Bulstrode while touring Buckinghamshire: "I believe the menagerie at Bulstrode is exceedingly well worth seeing, for the Duchess of Portland is as eager in collecting animals, as if she foresaw another deluge, and was assembling every creature after its kind, to preserve the species."[59] Montagu's playful and hyperbolic portrait of Bulstrode's animal life, though meant ironically, conveys the impression that the duchess had gone too far, and was excessive in her quest for botanical and zoological rarities. Still, the letter acknowledges that the stocking of Bulstrode with exotic plant and animal species enhanced the edifying and entertainment value of this country estate. The duchess's interest, however, in botany, zoology, and especially conchology, went well beyond the aristocratic culture of collecting. With her mastery of the Linnaean system of classification, her natural history collecting took on the seriousness of a scientific endeavor with the specific aim of contributing to knowledge about the natural world. It was this seriousness of purpose that Montagu mocked as excessive, as taking the duchess beyond the bounds of polite natural history.

NATURAL HISTORY COLLECTORS

Natural history collectors were often the objects of ridicule. Emanuel Mendes da Costa, an Anglo-Portuguese Jew living in London, wrote in 1776 to a fellow naturalist to express his anxiety over the negative reception he was anticipating for his soon-to-be published book on British shells. "I am almost frightened at my undertaking," he wrote, because the book's sole focus was the "local or the meer English shell." He confided that he almost wished he had stuck to his correspondent's "maxim of not talking about these subjects in most companies," for collectors of British shells were regarded as odd or worse, and as a result, "we generally are the but[t] of ridicule."[60] Da Costa was not being paranoid; even within his world of conchologists, domestic shells, unlike exotic shells, especially those from the Pacific, were not considered appropriate objects of inquiry or desire. Not until the Napoleonic Wars did the study of native shells assume importance as a legitimate field of inquiry within Britain, and da Costa's book *Historia Naturalis Testaceorum Britanniae; or, The British Conchology* (1778), the first in nearly a century since Martin Lister's *Historiae conchyliorum* (1685–92) to focus on domestic shells, paved the way for the early nineteenth-century fascination with British shells.

Da Costa's initial anxieties about the reception of his book, his fear of mockery and ridicule, were not uncommon concerns for collectors of natural history specimens, who were well aware of the satirical stereotypes that circulated in essays, prints, poems, and plays. They were depicted in popular culture as dull-witted, dry-as-dust, and methodical accumulators of nature's detritus; as anti-social, miserly, prone to hoarding and secrecy, and bizarre in their passionate engagement with dried flowers, dead leaves, skins, feathers, fossils, bones, and shells. A long literary tradition beginning with Thomas Shadwell's play *The Virtuoso* (1676), extending through the eighteenth century into the nineteenth, portrayed the collector of natural curiosities as a ridiculous figure, who was represented either as an anti-social miser more attached to things than people or as the petty-minded, fussy, scatterbrained, and ultimately non-productive type incapable of understanding the significance of the very objects he had collected. The latter type, the fussy old fool, is captured in Mr. Woodhouse, a character in Jane Austen's *Emma* (1815), who happily spends an afternoon at Mr. Knightley's Donwell Abbey, going through "drawers of medals, cameos, corals, shells, and every other family collection with his cabinets." Mr. Woodhouse, with his courtly old ways and gentlemanly refusal to engage in useful activities, is delighted with these collections because they suit his temperament perfectly: "he was slow, constant, and methodical," and, like "a child," "he had a total want of taste for what he saw."[61]

In contrast to the fussy old fool was the virtuoso, a figure that had been derided repeatedly since the founding of the Royal Society (1662). A late seventeenth-century lampoon on Dr. John Woodward, a fossil collector who eventually donated his collections to Cambridge University, defined a virtuoso as:

> one that has sold an Estate in Land to purchase one in Scallop, Conch, Muscle, cockle Shells, Periwinkles, Sea Shrubs, Weeds, Mosses, Sponges, Coralls, Corallines, Sea Fans, Pebbles, Marchasite and Flint Stones; and has abandon'd the Acquaintance and Society of Men for that of Insects, Worms, Grubbs, Maggots, Flies, Moths, Locusts, Bettles, Spiders, Grashoppers, Snails, Lizards and Tortoises.[62]

An essay in *The Tatler* in 1710, entitled "The Will of the Virtuoso," takes up Shadwell's character Sir Nicholas Gimcrack, depicting him as someone who is *so* cut off from normal social ties that he bequeaths his most valued possessions – a hummingbird's nest and a rat's testicles – to his family, not realizing that what he values possesses no value for them.[63] Often portrayed as anti-social, the virtuoso was mocked for substituting things for people. In Hannah Cowley's *The Belle's Stratagem* (1780), Lady Frances's father, a virtuoso, "kept her locked up with his caterpillars and shells and loved her beyond anything – but a blue butterfly and a petrified frog."[64] Hinted at in these portraits is the idea that collectors were engaged in a perverted and deviant economy of accumulation, perverse because they collected objects rather than money or land, and deviant because they took objects out of economic circulation, diverting them from the so-called rational world of commodity exchange, and inserted them into the affective realm of curiosity.[65] Da Costa observed:

> The World is but too much inclined to treat with levity those *studies* which do not lead to its riches or preferment; and, while the generality of mankind have only these objects in view, it is no wonder that the *enthusiasm* of the *Naturalist* is so often ridiculed, who can desert the beaten road . . . and turn aside into the silent paths of life, in quest of nothing better than *contemplative pleasure*.[66]

Collectors of natural history objects were ridiculed for having refused the category of the useful and the productive to dwell in an alternative economy, one that placed value on natural objects beyond or outside market value. They were viewed even more negatively than art collectors or antiquarians, whose objects possessed either some intrinsic aesthetic value or could be linked to patriotic narratives of national origins. Natural history collectors were portrayed as male and privileged, engaged in useless activities that only a leisured lifestyle could support. They appeared in print and popular culture as driven by curiosity and desire rather than motivated by rational principles, taste, or erudition.[67]

This stereotype of the virtuoso has found its way into present-day depictions of eighteenth-century natural history collectors. As Craig Hanson has observed, "all too often, in fact, twentieth-century commentators simply followed the scripts established by the early modern critics of the virtuosi, with little judicious evaluation of their own."[68] Historians of science and cultural critics have given short shrift to natural history collecting. Other than a few popular histories that take the "freaks and geeks" approach to their material, focusing on what is odd or bizarre about collectors, their fetishistic desires, and their bizarre collections of pickled body parts and nature's mistakes, there are only a few sustained scholarly treatments of natural history collecting as a social practice.⁶⁹ Only a handful of historians of science have attended to natural history collections; and shell collecting, perhaps because of its conflation with the aesthetic presentation of objects and its engagement with commercial culture, has been largely overlooked as scientific praxis and as a form of rational inquiry into nature.⁷⁰

THE DUCHESS AS A COLLECTOR

Outside art history and museum studies, collecting as a topic of historical inquiry and cultural critique is burdened by a hint of disgust or condescension derived, in part, from Freudian innuendos about accumulation. Baudrillard's system of objects depicts the collector as male and as a sexually impotent fetishist, since he followed Freud's formulations about the fetish as a way for men to deal with castration anxiety.⁷¹ According to classic Freudian sexual theories, men only are psychologically capable of fetishizing objects. Women collectors tend to be portrayed instead as consumers and accumulators, which places them at an even more primitive stage in Freud's schema of sexual development, at the oral stage of incorporation, when a baby puts everything into its mouth as a way of interacting with the world beyond the body. Collecting has also been linked to the anal stage, in which a fascination with excrement is replaced with the hoarding of possessions. I suggest that these ideas, however unconsciously marshaled in analyses of collecting, haunt current scholarship on the duchess. Nor has collecting fared well with Marxian sociologists and cultural critics, who, viewing collecting as a form of commodity consumption, tend to portray it in their most generous moments as a form of self-fashioning and as a display of cultural capital.⁷² Too easily dismissed as mindless and compulsive, natural history collecting has often been slotted by cultural critics and historians into a subset of consumption and has not been taken seriously as a creative and instrumental form of meaning-making.

It is not surprising, then, that the duchess and her collecting practices have been portrayed in a less-than-flattering light. Her position as a woman and an aristocrat, two categories too easily dismissed as incapable of reasoned inquiry and sustained effort, have contributed to the portrayal of her collecting activities as those of a "magpie" or a "bowerbird."⁷³ These unfortunate bird metaphors convey the impression that she was an indiscriminate collector of everything and anything, that her collecting practices were those of a giddy, thoughtless, scatterbrained, and bird-brained woman who was feathering her nest with bright, shiny objects. In addition, scholars have portrayed the duchess's collecting as compulsive and out of control, claiming that she accumulated much more

23   Charles Grignion after E. F. Burney, frontispiece to *A Catalogue of the Portland Museum*, 1786. Courtesy of The Lewis Walpole Library, Yale University

than she could deal with, and that Bulstrode and Whitehall, her two main residences, were in a state of chaos because she was simply unable to impose order on the plethora of objects she had collected.[74]

The origin of this trope of disorder can be traced to the engraved frontispiece of the Portland Museum auction catalogue, which pictures the duchess's objects piled haphazardly in stacks and lying on the floor of one of her Whitehall apartments (pl. 23). Drawn by Edward Francis Burney (Frances Burney's cousin), the illustration portrays the room as a site of abundance and disorder, which at this point, several months after the duchess's death, it well may have been, for the Whitehall residence was the site of the auction. Along with some of her china, the shell collection was housed at Whitehall during her lifetime, and after her death other items were taken there from Bulstrode for sale – porcelain, cameos, snuffboxes, Roman sculpture and other antiquities, and natural history specimens such as insects, fossils, birds, and minerals. The illustration owes its design and subject matter to a specific genre used to advertise auctions and other kinds of merchandizing. Following in the French tradition of illustrations accompanying auction catalogues, a genre perfected by François Boucher, Burney's image employs the visual trope of the cornucopia, a spilling forth of abundance, which is heightened by its disarray. It signals the dissolution of the collector's imposition of order on the objects and their readiness for insertion into someone else's collection. The image is designed to invite customers to fantasize about rescuing some of these precious objects from the chaos to which they have been consigned either by carelessness or by the disorder that death brings to possessions. It does exactly what Cynthia Wall has so precisely observed in her study of auction catalogues: it pictures the disassemblage of a collection in a way that invites viewers to impose their fantasies on the soon-to-be-auctioned objects. If the illustration had shown the duchess's massive cabinets with objects lined up in order, it would have failed to entice prospective purchasers to bid on the individual items.

Mistakenly assuming that the Burney illustration captured the "logic of the collection itself," cultural historians have portrayed the duchess's collections as being in a constant state of disorder, "a chaotic jumble" and "a clutter of shells in a jumble of unsorted boxes," scattered higgledy-piggledy all over the apartments at Bulstrode and Whitehall.[75] A comment made by Mary Delany has been interpreted as evidence of the disordered state of the collections. She describes how the duchess's passion for natural history transformed the spaces at Bulstrode, turning "her Grace's breakfast room" into a

> repository of sieves, pans, platters, and filled with all the productions of *that nature*, [which] are spread on tables, windows, chairs, which with books of all kinds, (opened

to their useful places), make an agreeable confusion; sometimes, not withstanding twelve chairs and a couch, *it is* indeed a little *difficult* to find a *seat*.⁷⁶

Historians have interpreted this scene as chaotic, overlooking what the duchess was actually doing here. She had commandeered domestic space, transforming it into a site where she could classify specimens, which, after all, was an act of "the setting in order (out of 'agreeable confusion') of God's creations."⁷⁷ Although the breakfast room may have appeared to Mary Delany as a site of "confusion," the process of cataloguing specimens – shells, insects, fungi, mosses, mushrooms, and ferns – required space in which to lay out and sort through them; it necessitated containers (hence the sieves, pans, and platters) to hold the sorted objects, as well as reference books with illustrations and descriptions for comparison purposes. Though the scene she describes is visually one of disorder, the opposite – the imposition of order – is what was actually going on. The duchess was doing what every naturalist does when trying to identify specimens; she was in the act of imposing systematic order in the form of Linnaean taxonomy on the diversity of the natural world.⁷⁸

Even the great twentieth-century collector of eighteenth-century manuscripts, letters, and prints, W. S. Lewis, colored his mostly positive portrait of the duchess's collecting with a slightly sarcastic remark. "Few men," he wrote in 1936, "have equaled Margaret Cavendish Holles Harley, Duchess of Portland, in mania of collecting, and perhaps, no woman. In an age of great collectors, she rivalled the greatest."⁷⁹ His use of the word mania is surprising considering that he himself was a famous bibliophile; one would assume that he might have shown more sympathy for the duchess's activities. Clearly, there is more than a hint of admiration in Lewis's statement for the scale on which she operated, and his linking of insanity with collecting may have more to do with the object of her "mania" – natural history specimens – than with the practice of collecting itself. Also, he may have been channeling opinions expressed by Horace Walpole, the object of Lewis's own collecting zeal, for Walpole, though the duchess's friend, was dismissive of her interest in natural history since his own interests lay in the decorative arts. This linking of madness and collecting has the effect of dismissing natural history collecting as a meaningful activity, one that could be productive of knowledge. Following Lewis's lead, the duchess's activities have been understood by most cultural critics as that of a virtuoso, collecting practices that were open to ridicule because they were associated with an indiscriminate desire for objects regarded as curious, a term that could be applied equally to exquisitely crafted handmade objects as well as to beautiful, rare, or exotic objects from the natural world.⁸⁰

The eighteenth-century collections of virtuosi are vestiges of older forms of collecting. In the sixteenth and seventeenth centuries princely collections, *Wunderkammern*, and cabinets of curiosity were built upon the accumulation of natural objects from early modern Europe's expansionist travel and empire building. With the introduction of Linnaean taxonomy, and the publication in the 1750s of Linnaeus's *Species Plantarum* and *Systema Naturae*, natural history collecting became less idiosyncratic and more systematic, though this did not happen overnight. The power of the curiosity cabinet, designed to arouse the viewer's wonder at the strangeness and beauty of God's creation, continued to be part of genteel life, residing comfortably in the townhouses or country seats of the gentry, as demonstrated by Austen's depiction of Mr. Knightley's family collection. By the third quarter of the eighteenth century collectors of natural history objects were participating concurrently in two very different practices: collecting curiosities and "scientific" collecting. For example, a "scientific" collector, intent on using Linnaean systematics to arrange his collection, would often attend the same natural history auctions as the dilettante and curio collector. Also, they could even be the same person, depending on the moment and the collected object, for someone who had begun to collect shells as objects of rarity and beauty could gradually turn from connoisseurial delight to the Linnaean classificatory system to order and arrange his collection.

To clarify the distinction between connoisseurial and scientific collecting, it will be helpful to locate natural history collecting within the larger social field of collecting. I turn to theories of collecting as formulated within museum studies to suggest that understanding the social significance of the collected natural object, a bone or a shell, requires that one considers its structural similarities to other collected objects, even those, like a weapon or a piece of jewelry, that are made by humans. Susan Pearce argues that natural history specimens should be classified as material culture, since these natural objects, when collected, are submitted, like human-made artifacts, to the same social processes that define the act of collecting: "selection according to contemporary principles, detachment from the natural context, and organization into some kind of relationship (many are possible) with other, or different material."[81]

COLLECTING AND CONSUMPTION

Pearce's impressive body of work suggests that in general there are three modes of collecting. One is to choose objects that function as an extension of oneself: souvenirs are an example, since, in stressing narratives of the self, they locate the object within the personal realm of the collector. The souvenir, according to Susan Stewart, invites narratives of the possessing self, "not a narrative of the object": "the souvenir displaces the point of authenticity as it itself becomes the point of origin for the narrative. Such a narrative cannot be generalized to encompass the experience of anyone; it pertains only to the possessor of the object." For example, the story of a miniature Eiffel Tower bought by a tourist to take its place in a collection of European souvenirs begins with the purchase; the tourist then locates this object in his or her own narrative of travel and collecting, which might go something like this: "I bought this in Paris just after seeing the view from the top of the Eiffel Tower. I have a miniature of St. Peter's Basilica, which I got in Rome, and one of London's Big Ben." This kind of collecting reinscribes the collecting self as subject and agent through a process that removes the collected object from its origins, retaining the barest narratives of origins, only those necessary to turn the object into an entity constitutive of the "possessive self."[82]

Pearce's second mode of collecting, one that resembles the way in which the fictional Sir Nicholas Gimcrack collected natural curiosities, also functions as a way to shore up an individual's identity. These collections are amassed without "intellectual rationale," for the principles ordering them are usually related to events in an individual's life. Pearce's examples range from John Tradescant's and Sir Hans Sloane's natural and artificial curiosities to Freud's collection of antiquities, mostly small figurines dating from Roman and pre-Roman times, and Walter Benjamin's book collection. She suggests that "this kind of collection is made by people whose imaginations identify with the objects they desire. . . . The whole process is a deployment of the possessive self, a strategy of desire, and this is part of the reason why this mode of collecting is described as fetishistic." As has been seen, this type is easily parodied, and yet, as Pearce attests, "much of the material in all of these collections has always been valued for its perceived intrinsic, and therefore financial quality, and has been taken correspondingly seriously by the museums which hold it."[83] The recipients of centuries of this kind of collecting are the great natural history museums of today, the British Museum, the Natural History Museum in London (formerly part of the British Museum), the American Museum of Natural History in New York, and the Field Museum in Chicago, as well as the smaller museums and historical societies that store and display natural history collections. A variation on this model of collecting, what has been called "romantic collecting," an attempt to address time's destructive powers and to suture the fragmentation of the self through the possession of "the detritus of history,"[84] can also be understood as fetishistic, for, as Pearce suggests, when a collector "maintains a worshipful attitude towards his objects," the objects are perceived as a path through which the subject can create "a romantic wholeness."[85]

Pearce's third mode of collecting, one that I claim for the duchess, is systematic, in that it functions to illustrate a system of thought, in this case, the Linnaean taxonomic system, which hinges on the "ability to compare and contrast collected specimens in order to distinguish the fine detail which divides one species from another, and so carry out identifications."[86] The key to taxonomy is seriality, which is dependent on extracting one specimen from its context and placing it in a relationship to another specimen. Seriality, as Baudrillard argues, is what turns an accumulation of objects into a collection. In "The System of Collecting," he distinguishes collecting from accumulating objects:

> It should be stressed that the concept of collecting (from the Latin *colligere*, to select and assemble) is distinct from that of accumulating. The latter – the piling up of old papers, the stockpiling of items of food – is an inferior stage of collecting, and lies midway between oral introjection and anal retention. The next stage is that of the serial accumulation of identical objects. Collecting proper emerges at first with an orientation to the cultural: it aspires to discriminate *between* objects, privileging those which have some exchange value or which are also "objects" of conservation, of commerce, of social ritual, of display – possibly which are even a source of profit. While ceaselessly referring to one another, they admit within their orbit the external dimension of social and human intercourse.[87]

Using Baudrillard's distinction, it can be seen that Pearce's first two kinds of collections, the souvenir and the fetish, are forms of accumulation, because the collections of souvenirs and mementos – and even those of gentlemen collectors such as the architect Sir John Soane, whose London townhouse was stuffed with antiquities – are without seriality, the systematic "serial accumulation of identical objects" with the goal of discriminating between them. Items in collections that lack seriality can be arranged in any way that pleases the possessor: "the personality of the collector, in a very particular sense, is the mainspring of this kind of collecting activity and runs beneath much collection-forming," even though collections such as Freud's and Soane's antiquities and Walter Benjamin's books are "presented in a more intellectual, dignified and objective light."[88] In contrast to these dignified forms of accumulation, the hallmarks of a systematic collection are seriality and recontextualization, with objects organized in relation to each other, arranged along some scheme that exists outside the individual collector's mind, usually illustrating an agreed-upon conceptual system and "designed to demonstrate a point."[89] For Susan Stewart, collecting based on classification is distinct from other forms of object accumulation, notably the acquisition of souvenirs, mementos, and antiquarian artifacts, for these latter types of objects function to elicit narratives about the past, albeit an "imagined past," which, unlike the "actual past," is "available for consumption." Classification, on the other hand, seeks to sever ties with history and the originating context in order to ensure a spacialized order that is "beyond the realm of temporality."[90] Drawing on these arguments about accumulation versus collecting, I argue in subsequent chapters that in organizing her shell collection according to Linnaean systems of classification, attending to the differences between similar objects, the duchess of Portland's engagement with collecting natural objects cannot be dismissed as belonging to those virtuoso practices of curiosity collecting mocked by social critics in the eighteenth century. Her collecting practices stand in distinction to those of the virtuosi with their fetishistic accumulation of objects, which is, according to Baudrillard, Stewart, and Pearce, about extending and consolidating the possessive self.[91]

The relationship between collecting and consumption is a contested subject within a variety of academic disciplines. Some sociologists, cultural critics, and social historians have sought to link the rise of early modern collecting with the new consumer society, suggesting that collecting is a form of primitive accumulation typical of the commercial society of early modern Europe. James H. Bunn sees curiosity collecting as growing out of the "aesthetics of mercantilism," for with the removal of the object from its context, collecting mimics the exchange economy's abstracting dynamics.[92] Although Susan Stewart is less explicitly Marxian than Bunn in her explanation of the connection between collecting and consumption, she suggests that collecting is a response to an "exchange economy," in short, to living in a capitalist society. She argues that the desire for objects that operate within the categories of the souvenir, the antique, and the exotic derives from the search for the authentic within a commoditized economy. She contends that this search for authentic experiences, "the lived relation of the body to the phenomenological world," gets displaced onto a search for authentic objects, which appear as a trace of the lived experience. For Stewart, the accumulation of objects is compensatory for the loss of authenticity, a loss created by a system of "mechanical modes of production," that in mediating experience distances us from fully inhabiting a lived, bodily presence.[93] Taking Stewart's analysis a bit further, one could argue that collecting compensates for a lost relation to artisanal labor, that lived experience of work where the artisan's skilled hands, emblematic of the embodied knowledge of his craft, encounter materiality.

While it is not clear to me that collecting appears only in capitalist societies or even more broadly defined commercial societies (which would encompass, for instance, the Roman and Ottoman empires, and the Chinese dynasties of the sixteenth, seventeenth, and eighteenth centuries), I would not categorize collecting as simply a variant form of commodity consumption.[94] Although I find Stewart's argument persuasive and agree that capitalism's abstracting power can strip objects of their authenticity, I would like to entertain the idea that collecting is more dynamic than her compensatory model of collecting. For Stewart, collecting is a substitution for production, but is it not possible that collecting is a form of production? What a collector does is to enclave objects (to borrow a term from Arjun Appadurai), removing them from commodity status and placing them in an affective, social, or discursive realm where they are not alienable but endowed with attributes that are derived from a sense of self, family, nation, ideology, ritual, or the sacred. Collecting, though a reaction to commoditization, is not only or merely compensatory but also an active and inventive response to the absence of affect and narrative that capitalism demands of exchange, where objects must be stripped of sentimental and social ties to circulate freely within the commodity exchange system. The only value that can be tolerated by the "free" market is exchange value. Everything must be convertible into the abstractly expressed value of currency; in short, everything has its price, nothing can be left outside this system, not even sacred objects, family heirlooms, national memorials, or objects of scientific scrutiny.

Although collecting mirrors the dynamics of the marketplace with its operating principles of alienation, abstraction, and decontextualization, it also offers an alternative to the very system it seems to mimic. Collectors' high level of attachment to the collected object as well as the process of collecting, though easily dismissed as excessive, aberrant, or fetishistic, speaks to a relationship with the material world that cannot simply be subsumed under the category of commodity consumption. This desire for something other than exchange value is what collecting is about. At war and in league with the commercial world, the collector engages in commodity exchange but then tries to undo its alienating effects by endowing his or her collected objects with a value that is not determined by the market – a value derived from such sources as the aesthetic response to an object's attractive qualities, or the sentiment fostered by personal, familial, regional, and national narratives, or the value bestowed by scientific principles, nomenclature, and categories of description and analysis. In seeking to reinstate some kind of value beyond market price, collecting tries to bind the object to sentiment, or to intellectual curiosity, or to notions of family, heritage, or nation, to enrich the collected object with affect, narratives, and/or ideas in order to complicate one's relationship with materiality.

THE DUCHESS AS A NATURALIST

The duchess of Portland's collecting has been portrayed by scholars as a form of primitive accumulation, a distracted form of consumption driven by irrational desires, and as an acquisitiveness born of the commercial society she inhabited. Indeed, if her collecting of natural history specimens had been limited to purchases from dealers, bidding at auctions, or paying someone to gather them for her, then there might be some validity in these views. Her collecting practices, however, were far more complex than mere consumption.

Recent scholarly work on Mary Delany characterizes Bulstrode as a site where naturalists gathered to discuss natural history. Mark Laird affirms what David Allen suggested more than two decades ago: "the role the duchess played in offering private patronage to a wide circle of naturalists . . . is indisputable."[95] Bulstrode is figured as a precursor to the Bluestocking assemblies, some scholars even suggesting that Montagu "considered the duchess of Portland's estate at Bulstrode an ideal of social and intellectual life," and that she modeled her London assemblies on it.[96] Though meant to be positive in depicting the duchess's botanical and zoological activities, such commentary, in portraying her as a gracious and intelligent hostess, reduces her participation in natural history circles to that of a facilitator who provided the space and resources for others, primarily men, to study nature. She is at best in these descriptions a patron of natural history, but nothing more.

Margaret Cavendish Bentinck was herself an accomplished naturalist as well as a specimen hunter who went out into the field to gather plants, fungi, insects, and mollusks.[97] Often in the company of friends and fellow amateur naturalists, she wandered about the countryside, carrying nets, shovels, baskets, and boxes to gather specimens. She "botanized" with Jean-Jacques Rousseau in the Peak District in 1766,[98] and traveled to the southern coast of England with the Rev. Lightfoot to search for butterflies, mosses, and seashells. While he combed the high ground above the sea, she walked along the beach looking for shells, digging in the sand, wading in tide pools, clambering over rocks, and taking out small boats a short distance to trawl for live specimens. At Weymouth she found several new species of mollusks, some of which she most generously gave to her fellow conchologists. She also scoured Buckinghamshire for specimens, finding in the grounds of Bulstrode land snails that crept about in the early, dewy mornings and rainy afternoons. In the company of her gardener, visitors, and fellow conchologists, she gathered freshwater snails that clung to reeds in marshy areas and bivalves that lived in the muddy bottoms of ponds and in the silt of shallow meandering streams. For example, in a river near Bulstrode she found a new species of *Tellina*, a small bivalve that had not yet been named.

Although the duchess possessed one of the largest collections of shells in Europe, rivaling and surpassing that of Queen Louisa Ulrika of Sweden, aristocratic women were not the only female shell collectors in the late eighteenth and early nineteenth centuries. Ordinary middle-class women collected shells. The sisters, daughters, and wives of dealers and collectors participated in gathering, cleaning, classifying, drawing, and displaying shells; for instance, George Humphrey's sister, Elizabeth Humphrey Forster (wife of the mineralogist and natural history dealer Jacob Forster), found rare snail shells in Battersea; Margaret Fordyce, who accompanied her father, Dr. George Fordyce, to the Portland auction, purchased some of the duchess's shells; and George Montagu's life companion, Elizabeth Dorville, drew most of the dozens of shell illustrations for his *Testacea Britannica* (1803) and its supplemtent (1808), and engraved all of them. Montagu describes her as "a friend of science" whose goals as an illustrator and a self-taught engraver were "to further science by a correct representation of the original drawings, taken by the same hand." He insists that though her drawings and engravings may not please the "critical artist," they serve science well: "we trust science will be considered as having reaped more advantage from such, than from highly finished engravings devoid of correctness and character."[99] In the five-volume *Natural History of British Shells* (1799–1803), Edward Donovan repeatedly thanks a Miss Pocock for shells she found along the Welsh coast. Captain Bligh's wife, Elizabeth, possessed a very large shell collection, which was purchased and then sold at auction by John Mawe and his wife, Sarah, both of whom were dealers in natural history specimens. According to Hugh Torrens, Sarah Mawe "became a highly competent mineral appraiser, purchaser, and identifier in her own right, and supplier of shells and fossils to naturalists in Europe and North America." On her husband's death Mrs. Mawe inherited his natural history shop and eventually became mineralogist to Queen Victoria.[100] Thomas Martyn's *The Universal Conchologist* (1784–89) lists in its introduction several names of women collectors: the countess of Bute (Lady Mary Wortley Montagu's daughter), Mrs. Thomas Heron, Mrs. Isaac Walker, and Mrs. John Barclay.[101] Clearly, these names are only the tip of the iceberg. Much more work, along the lines pursued by Margaret Jacob and Dorothée Sturkenboom with their focus on "ordinary women in technical and scientific settings," is necessary to discover and document women's engagement in natural history activities and their contribution to natural knowledge.[102] The attention I give to the duchess of Portland's engagement with natural history is not to deny the richness and complexity of middle- and working-class women's participation in natural history collecting; I merely seek to rectify the refusal to take her activities seriously.

In addition to the duchess's activities as a field naturalist, she was also actively engaged in shell classification. The goals for eighteenth-century naturalists, amateurs and professionals alike, were the description and classification of nature. Central to these activities were the physicality of the specimen, its visual and verbal representation in a book, and the process of moving back and forth between the specimen in one hand and the reference book in the other; this was the period's *techné* – the triangulation of eye, book, and object – that produced natural knowledge and continues to be a method currently employed by taxonomists.[103]

Having mastered the principles of Linnaean taxonomy, the duchess was adept at identifying the specimens' taxa, a practice that involved comparing the collected object with verbal and visual representations of it in various reference books, some of which were organized along Linnaean categories. Classification for shell collectors usually meant working indoors, since shells were often delicate and easy to lose; some were so small that they had to be viewed through a microscope, and the reference books were often large-format catalogues, too expensive and heavy to carry outdoors, as one might a field guide. Classification took place in a variety of indoor spaces, including libraries, drawing rooms, breakfast rooms, sitting rooms, and closets; naturalists were surrounded by shells on table tops, desk tops, and other available surfaces, and on sorting trays and in little boxes and patty pans (cupcake tins). This was often a social activity that involved conferring with friends, exchanging information and guesses about the species of new and unnamed specimens. The duchess performed these activities in the presence of visiting naturalists and interested members of her household, including Mary Delany, the Rev. Lightfoot, the artist Ehret, and even her gardener Agnew.

The duchess of Portland also established and maintained a correspondence with naturalists, writing with regularity to Dru Drury, the London-based author of *Illustrations of . . . Exotic Insects* (1770–82); John Ellis, author of the *Natural History of Corallines* (1755); and Dr. John Fothergill and Peter Collinson, two Quaker natural history collectors who had strong ties to North America and the West Indies. She conferred with them about specimens and collaborated on projects that ranged from supporting book projects to funding specimen-hunting expeditions. She contributed quite a bit of money, at Dru Drury's invitation, towards Henry Smeathman's expedition to Sierra Leone (1771–75), from whence he sent specimens, including shells and insects for the duchess, insects for Drury, seeds for John Fothergill, insects and birds for Marmaduke Tunstall,[104] and plants for Sir Joseph Banks, all of whom also contributed, though less than the duchess, to funding this expedition.

A less far-flung natural history network in which the duchess participated was one to which she was devoted: it consisted of two provincial natural history collectors, Dr. Richard Pulteney of Blandford (1730–1801) (pl. 24) and Henry Seymer, who were near neighbors and rarely traveled beyond their Dorset homes. Her friendship with Dr. Pulteney spanned more than two decades, and their correspondence reveals the workings of that particular kind of intimacy based on a shared love of natural history. The affection with which she wrote to Pulteney was based on respect for his talents as a naturalist and manifested itself in the exchange of information, gifts of game, the loan of books, and the circulation of specimens.[105] She gave him exotic shells from Cook's voyages, shells that he could not afford to buy, and also shells she had collected herself at Weymouth, knowing that he could rarely take the time from his duties as a physician to gather specimens from the seashore. Pulteney's specialty was the natural history of Dorset, and his strong suit was botany. He had mastered Linnaean systems of classification and wrote *A General View of the Writings of Linnaeus* (1781), an influential introduction to Linnaeus's life and work, and *The Progress of Botany* (1790), a book that celebrates Linnaeus and argues that the popularity of botany in Britain with amateurs – men, women, and children alike – was due to the simplicity and accessibility of the Linnaean binominal system of nomenclature. Published posthumously was his life's work, *The Natural History of Dorset* (1813), a catalogue of the county's birds, plants, and mollusks. The duchess visited Pulteney on her annual trips to Weymouth, stopping off at Blandford for a few days, and on occasion he and his wife traveled to London to visit her.

Through Pulteney, the duchess met Henry Seymer (pl. 25), a landed gentleman who lived five miles from Pulteney, and who was an enthusiastic amateur naturalist and a talented artist, whose "finished drawings of Birds, Shells, and Insects" were "excelled by very few." His butterfly drawings and watercolors, for instance, are beautifully rendered depictions of more than 100 different species, each illustration containing the various stages of development of the species as well as the plant life to which the butterflies were attracted and dwelt among. Though Seymer's specialty was insects, he also collected fossils and shells, and was particularly interested in shells from Cook's voyages. His son-in-law, writing a brief biographical sketch for the Linnean Society, described Seymer's shell collection as "very rich" because

24   Thomas Beach, *Richard Pulteney, FLS*, oil on canvas. London, Linnean Society. By permission of the Linnean Society of London

he never lost any opportunity of procuring the finest [shells] that came to market; and although he resided at so great a distance from the metropolis he had always agents in town on the look out to secure any collections that might be brought to this country by Voyagers and he procured many from the celebrated voyage of the immortal Cook.

The excellence of Seymer's natural history collections was well known, Thomas Martyn's *Universal Conchologist* stating:

25 Thomas Beach, *Henry Seymer, FLS*, oil on canvas. London, Linnean Society. By permission of the Linnean Society of London

No gentleman out of the limits of the metropolis, is supposed to possess so numerous and well-chosen a collection of Shells, extraneous fossils, and insects, as Henry Seymer, Esq. at Hanford, in Dorsetshire. Endowed by nature with great abilities for the investigation of every branch of natural history, he has added the experience resulting from a long application to these studies; accordingly, his museum as far excels the generality of others, as in the knowledge of the various subjects he is allowed to be superior to most of his contemporaries.[106]

26 William Parry (previously attributed to Johann Zoffany), *Daniel Solander*, circa 1775–76, oil on canvas. London, Linnean Society. By permission of the Linnean Society of London

Seymer's son-in-law described Seymer's relationship with the duchess with pride: "The celebrated Duchess of Portland frequently presented him with rare specimens of shells from her noble cabinet; and for the last twelve years of her life never omitted spending a week at Mr. Seymers on her way to Weymouth."[107] Seymer's abilities as a naturalist were most welcome to the duchess and she happily cultivated a relationship with him, which blossomed into a long-term relationship that lasted until his death.

As will be seen, the duchess's relationships with these two Dorset naturalists was one of friendship, which, in the eighteenth-century sense of the word, often involved complex forms of gift exchange within a hierarchical yet reciprocal dynamic of obligation and deference, all carried out under the rubric of natural history inquiry. Because

Seymer and Pulteney rarely traveled beyond their homes in Dorsetshire, they used the "Blandford waggon car" to exchange shells with the duchess, Pulteney receiving gifts of shells while loaning specimens to her for classification purposes, and Seymer swapping his duplicates for hers. They also shared their expertise as they worked at organizing their shell collections along Linnaean categories. The duchess worked hard at classifying her own collection of shells, and though she could recognize and label known species with the assistance of shell catalogues, she left the naming of new species – new in the sense of never named before within Linnaean systematics – to Daniel Solander (pl. 26), since he was the foremost expert on mollusks in Britain. Beginning in January 1788 Solander spent every Tuesday for more than four years working on her huge collection, naming her non-descript shells and using it as a means to refine and expand on Linnaeus's system as it applied to mollusks. Although the actual practices of classification that the duchess and her network engaged in as they cleaned, sorted, identified, and named their specimens can only be inferred from their written communications, it is fairly clear that she was respected by her fellow amateur conchologists, who consulted her for advice when classifying their specimens.

Far from the secretive, reclusive stereotype of the virtuoso fondling his bits of dead natural history specimens, the duchess and her naturalist friends were active, energetic, and sociable, hunting for specimens and working together to classify them.[108] It is important to recognize that natural history collectors were engaged with foundational questions concerning the organization of the natural world. Classification was dependent on access to large collections and could not proceed without them. Collectors of natural history specimens were ultimately engaged in exploring the tension between order in nature and the variety of life, testing systems based on Linnaean taxonomy against the plentitude and variation of life. These activities were at the heart of the Enlightenment's project of bringing concepts of systematic order to bear on the natural world.[109]

CHAPTER TWO

27 Edward Donovan, "Tellina variabilis," Plate XLI of *The Natural History of British Shells* (London, 1799–1803).
SPEC D 756, Arizona State University Libraries

THE DUCHESS AND SHELL COLLECTING

In 1771 the amateur entomologist Dru Drury put together a consortium of natural history collectors to sponsor Henry Smeathman's voyage to the west coast of Africa. The plan was that Smeathman, a friend and fellow member of the Aurelian Society, would collect the "*Naturalia* of that Country," sending boxes of insects, seeds, shells, and birds to Drury, who would pass them on to the sponsors of the expedition. To fund the trip, Drury had tapped Sir Joseph Banks, Marmaduke Tunstall, and Dr. John Fothergill for £50 each, but once Smeathman arrived in Sierra Leone, he wrote to his benefactor saying that this amount had proved insufficient and that he needed more money for living expenses and to purchase equipment, including a boat. Drury responded, writing that Tunstall would not donate any more money, but Banks had offered £10 more and Fothergill promised another £50. Drury was worried that this would still not be enough. Finding a solution to this problem, he announced to Smeathman in June 1773: "I hope to get the Dutchess soon to subscribe." By August he had successfully solicited the duchess of Portland's support for this venture, having collected £100 from her, by promising that "the things he [Smeathman] will send over in less than a twelve month will be more than sufficient to discharge" the sum she had contributed. He had also sweetened the deal by telling her that a cargo of boxes filled with "exceedingly fine" specimens of insects, birds, and shells from Africa would be arriving soon, and that she would be the first to examine them and to make her choice, since the other sponsors were out of town for the next two months: Fothergill and Tunstall were spending the summer on their country estates and Banks was on an expedition to Iceland. The duchess must have taken up Drury's offer, since her natural history collection, according to the auction catalogue, had many African specimens, including dozens of insects and a few very rare mollusks from West Africa.[1]

Sponsoring expeditions such as Smeathman's was one of the ways that the duchess acquired rare and exotic specimens. She also dealt directly with ship captains, asking Frances Boscawen's husband, a naval officer, if he would keep his eye out for interesting specimens, and, as will be seen in Chapter Three, she received shells as gifts from Captain Cook and Lieutenant Clerke. She also purchased Pacific shells from natural history dealers, George Humphrey selling her very expensive shells that he had acquired from officers and ordinary seamen. Alternatively, she used her connections with other naturalists who traveled abroad, hoping that they would remember to bring her specimens from their journeys. For instance, she wrote to her friend John Ellis, who was to journey to Norway, gently suggesting that she would find Norwegian shells a welcome addition to her collection: "there are many things in the shell and coral tribe on the coast of Norway, I should imagine, must be very curious."[2] She also provided naturalists with funds to defray their expenses while they traveled abroad with the expectation that they would collect specimens on her behalf. She gave £600 to the Oxford professor Dr. Thomas Shaw, a friend of Elizabeth Montagu and a frequent visitor to Bulstrode, to collect shells for her while he traveled in the Ottoman Empire.[3] A similar arrangement was arrived at with Dr. Sibthorpe, who made "a complete Collection of the Land & Water shells of Switzerland" for the duchess.[4]

Her methods of acquisition have been described by David Elliston Allen, a prominent historian of science, as "purely monetary in character, inasmuch as her activities did not extend beyond employing people herself or making payments for specimens."[5] This is true to a large extent, but it is not the whole story. In this chapter I examine what has been overlooked by historians of science: the fact that the duchess was herself a specimen hunter. Her shell hunting expeditions, given the limitations of her position in society, were confined to Britain; she journeyed to the south coast, where she collected several new specimens, which were eagerly sought after by collectors. Several of these British shells were purchased at the Portland auction by collectors, some finding their way into Edward Donovan's collection. Donovan's five-volume catalogue, *The Natural History of British Shells* (1799–1803), contains illustrations of a few of the duchess's domestic shells (pl. 27). This chapter will also argue that her involvement with her collection went well beyond an exclusive concern with acquisition. Far from evincing a "bowerbird mentality," as Allen wrote, the duchess was deeply earnest about organizing and arranging her collection along Linnaean principles.[6]

This chapter provides an overview of the material practices and discursive technologies that transformed shells into objects of scientific interest. In laying out for my readers all the steps involved in gathering, cleaning, and sorting shells, I describe the life cycle of the shell as a collected specimen. From a snail gathered in the morning in wet grasses or a bivalve dug up in seaside mudflats to the cleaning processes that transformed shells into objects worthy of display, these are social mechanisms and physical processes by which natural objects were turned into material culture. Embodied practices, in particular the gathering and preparing of specimens, included: the soggy stockings and shoes damp from walking on the squishy mudflats and sandy stretches left exposed by the ebbing tides while searching for clams buried beneath the surface; trying to keep snails, still alive, from crawling out of the wicker basket into which they had been dropped; the feel of the bristles of the boar's brush while scouring the shell's epidermis; the burn of the acid solution used to remove debris from shells as it splashed on one's hands; the smell of the putrid bodies of mollusks that waft off shells, and the sight of maggots as they crawl out of shells not thoroughly cleaned; cutting and folding cardboard playing cards to make little boxes to house the shells; the careful packing of specimens in boxes, embedding them in sawdust or sand to avoid breakages; and the tricky business of making tiny labels and gluing them onto both the shells and their containers.

In describing these taxidermic processes as they relate to shells, I touch upon the way in which some of the practices of scientific shell collecting mirrored the decorative practices of shellwork, for shellwork involved techniques for killing the animals in the shells, removing them, and using acids to clean and polish the shells, techniques developed in conjunction with the protocols of collecting natural history specimens. When these transformative processes were combined with the rubrics of classification, which involved reference books, consultation, expert advice, and the borrowing of other specimens to help name the object and locate it within a category, shells were transformed from objects of aesthetic interest into ones of scientific value.[7] This chapter outlines the processes by which shells became natural history specimens, charting the physical transformations they underwent and the conceptual categories they traversed as they moved from the out of doors – the seashore, the pond, or some soggy pasture – to the glass-topped drawers of the collector's mahogany cabinet, a process by which they accrued value as objects of beauty, curiosity, and scientific inquiry.

COLLECTING BRITISH SHELLS

Canals, rivers, marshes, ponds, and ditches were popular places for conchologists to search for new specimens, and Bulstrode proved to be productive as a place for mollusks. The duchess of Portland found land snails in the damp grasses of its pastures and parkland (pl. 28), and freshwater snails and other mollusks in the rivers and canals on the estate and surrounding countryside. She performed these activities in the company of visitors and interested members of her household, primarily the Rev. John Lightfoot and J. Agnew, her gardener turned illustrator. Among the many specimens she collected in Buckinghamshire was a *Tellina lacustris*, a freshwater mussel, which she found in a river near Bulstrode. Agnew discovered a *Patella oblonga*, a freshwater limpet, "adhering to the leaves of the *Iris pseudacorus*" in a nearby river (pl. 29);[8] John Lightfoot wrote a paper on this minute shell, publishing it in the *Philosophical Transactions of the Royal Society*.[9] A *Helix spinulosa*, a land snail, according to George Montagu, author of *Testacea Britannica* (1803), was found near Bulstrode "upon old bricks and stones, after rainy weather, in June and July."[10] John Timothy Swainson, father of the naturalist and author William Swainson, was invited to Bulstrode to participate in a collecting expedition; he found a snail "in a canal at Bulstrode," a *Helix polita*, which he treasured because it was very rare.[11]

How did the duchess and her colleagues collect these land and freshwater shells? What kind of tools – nets, shovels, baskets – did they take with them as they rambled over the countryside, stopping at ditches and ponds to look for mollusks? Other than references to bags and boxes in the duchess's correspondence, there is little explicit description of shell gathering tools. Even the pamphlets published by George Humphrey, a natural history dealer, and J. R. Forster, a professional naturalist, do not explain what kind of equipment might be needed for such expeditions; aimed at travelers going abroad, their instructions do not go into great detail about the type of tools to use. Even eighteenth-century guides to building shell collections, such as Emanuel Mendes da Costa's *Elements of Conchology* (1776), fail to elaborate on the tools and containers needed to gather specimens.[12] In the 1820s, however, William Swainson wrote a guide to natural history collecting. He learned where and how to find shells from his father, John Timothy, and from George Humphrey, his father's good friend, of whom he was very fond. He recalled his childhood visits to Humphrey – the "good old man," who would help him classify "all my shells, whose names I knew not" and write "little tickets for each, in one of the most beautiful and legible hands I ever saw" – as the "greatest happiness of my life."[13] He grew up seeped in stories about natural history collecting, stories told by

28   J. Agnew, "Helix lucorum," watercolor. London, Natural History Museum. © Natural History Museum, London

29   Edward Donovan, "Patella oblonga," Plate CL of *The Natural History of British Shells* (London, 1799–1803). SPEC D 756, Arizona State University Libraries

Humphrey and by his father, who had been to Bulstrode on a shell hunting expedition, and young Swainson learned from them how to gather specimens in the field. Although he belonged to a different generation from the duchess and her fellow collectors, his insights into shell collecting give valuable access to the natural history practices of the late eighteenth century. He described what he regarded as the proper implements for collecting freshwater mollusks:

> The principal of these [implements] is a ladle or spoon, made of tin or thin iron, 5 inches long and 3½ wide, with a rim about an inch in height: it should have a short hollow handle, by which it may be fixed to the end of a long walking stick; the middle should be perforated with holes, no larger than is sufficient for the passage of water.

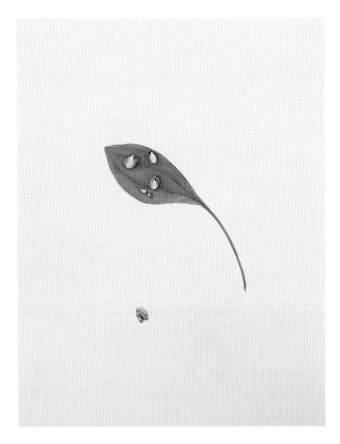

30   Edward Donovan, "Patella lacustris," Plate CXLVII of *The Natural History of British Shells* (London, 1799–1803). Author's photo

> This instrument is very useful in fishing for small river shells. . . . In searching for the larger freshwater bivalves, a landing net, with very small meshes, is of great service, and it may be made to fit upon the same stick as that which receives the spoon already described.[14]

Swainson's advice is mirrored in a pamphlet from the early nineteenth century entitled "Short Instructions for Collecting Shells." According to this, early mornings in rainy weather are best for land snails, and rivers, lakes, brooks, and canals for freshwater snails, often found under vegetation. To gather snails and limpets, which might cling to water plants (pl. 30), the author recommends using a net, "such as used to land fish with," offering advice on ways to make this net transportable: "by having four joints to the hoop of this net, it may be folded up, so as to go into the pocket; and by a stout screw, may be fixed to the end of a thick walking stick." As for bivalves, "muscles and such like," they can be found "under the mud." To scoop up freshwater bivalves, "it will be necessary to employ a boy, or other person, to wade for them, as they are often a foot or more under the mud."[15] In addition to nets and spoon-like shovels, also necessary on any shell gathering expedition were containers: small boxes, bags, and baskets. The pamphlet suggests that "[w]hen collecting, it will be proper to have with you a covered basket to put the shells in, to prevent their getting broke, and some of them, particularly the land shells, from creeping away."[16]

The duchess also spent time at the seashore collecting specimens. The enthusiastic field collector J. T. Swainson recorded the provenances of his British shells in a small notebook, including those he collected himself and those that were given to him by the duchess (pl. 31).[17] Of the several shells received as gifts from her, many were from Weymouth on the south coast. He noted that his specimen of the *Trochus papillosus* was "found by the D: of Portland at Weymouth & had from her"; the *Strombus pespelecani* was found "at Weym: by the D. of P."; the *Bulla patula* was "given me by the D of Pd found

31   J. T. Swainson's notebook of British shells. London, Natural History Museum © Natural History Museum, London

32 John Constable, *Weymouth Bay: Bowleaze Cove and Jordon Hill*, 1816–17, oil on canvas, 53 × 75 cm. Salting Bequest, 1910 (NG2652). London, National Gallery. © The National Gallery, London/Art Resource, NY

by her at Weym"; the *Venus undata* was "found by Dss. Portland off Weymouth"; the *Patella groeca* was "found by the D of P off Weymouth"; the *Mya declivis* was "found by the D of P off Weymouth"; and of his *Cardium tuberculatum*, he wrote: "my large specimen given me by GH/ found at Weym by the D. of Portland, escaped Dr. Solr. most probably found after his disease." The last entry conveys the complexity of the shell exchange circuit with Humphrey either giving or selling this shell to Swainson, a shell that he had most probably bought at the Portland auction, since it had been found by the duchess at Weymouth sometime between 1782, the year of Solander's death, and 1785, the year she died.

Most of the new molluscan species the duchess of Portland discovered came from Weymouth (pl. 32), a place she used to visit often because it was near her married daughter's residence (Elizabeth's husband was Thomas Thynne, third Viscount Weymouth, later first marquess of Bath). The duchess was one of several naturalists who frequented the region, which was rich in shell and fossil specimens.[18] Pulteney went there as part of his project to write the natural history of Dorset, and so did George Montagu (1753–1815), who in the early 1800s published a book on British shells based on his own collection, which he gathered himself in the field. Montagu notes with pride how most of the shells he describes in the book are of his own collecting. Using the plural, or "royal we," to express his endeavors, he wrote: "The species hereafter described, with a few exceptions, are in our own cabinet, and have chiefly been collected from their native places by ourselves, or by the hands of a few friends, whose conchological knowledge, and scientific researches are too well known to be doubted."[19] In one of the few statements issued by the duchess about her own collecting practices, in a letter to the naturalist Thomas Pennant about a Weymouth shell she had found, she is as usual polite and even deferential:

> I am always glad of an opportunity of Obeying your Commands as my stay in Town will be very short I sent to Mr Brown & gave him the two Shells you desired to have drawn & a new *Solen* or *mytilus* which I found last year at Weymouth which they tell me has not been described but if you don't chuse to take the drawing of that shell I told Mr Brown I wou'd have it.[20]

Pierre Brown was illustrating Pennant's *British Zoology*, and the duchess had given him a few of her very rare shells to draw and engrave for the volume on shells.

The shells that the duchess found at Weymouth were gathered in a variety of ways. Sometimes she found them by walking along the beach, often accompanied by Mrs. Le Cocq, an expert specimen hunter who resided there. Describing an excursion to look for specimens, the duchess wrote to Mrs. Delany:

> The other day I was on the beach, the wind blew a brisk gale; I got into a little sand grotto, and read your charming letter over and over while Mrs. Le Cocq was travelling about in search of shells, butterflys, and plants. Mr. Lightfoot desires his best comp[ts]; he goes away on Monday, and I hope to set out next Thursday. He has traversed all the island of Portland in search of plants, but has met with nothing new, neither animals nor vegetables – and has been very seasick, but he will give you a better history of his adventures than I can do, for I have no time as I expect visitors . . .[21]

The duchess's visits to Weymouth gave her the opportunity to indulge her love of natural history, to collect specimens, and to experience freedom from the demands and restraints of her social rank as she wandered for hours along the shore. While visiting the town in

1771, she received a letter from the Rev. Lightfoot, stating how delighted he was to hear that she was in good health and

> that Weymouth had wonderfully recovered your Grace's Health & Spirits. . . . Three or four hours daily Exercise or Motion in the open Air, your wholesome sea Breezes, your Sand Grottos, but above all the odoriferous Fragrancy of sea . . . & sea shells, cannot fail to brace up the languid System . . .[22]

The duchess's love of the seashore continued well into her late sixties, though the pleasures of Weymouth diminished with the death of Mrs. Le Cocq. In 1782 she wrote: "I propose setting out for Margate next week to try if the sea air has still the same Efficacy, for alas, I can never think of Weymouth again since the loss of my kind affectionate Friend Mrs. LeCoq."[23] Even with this shift in location to Margate in Kent, the duchess continued her practice of walking along the shoreline in search of shells in the summers of 1782 and 1784. Here she was sometimes less successful than she wished: "This air is very good & I have found great benefit from it but by way of amusement[,] except very pleasant airings there is none[,] not a shell to be met with tho' I have been very diligent in my searches."[24] Perhaps such disappointments were offset by meeting J. T. Swainson in Margate, who, though her junior by some forty years, shared her passion for collecting shells. She gave him shells and went to his apartments to examine his collection; it is possible that they went out together to comb the shoreline for new specimens, the duchess writing to Mrs. Delany that "he will be a good acquisition" to her community of natural history enthusiasts.[25]

The sociability of shell collecting glimpsed in the duchess's correspondence is conveyed in greater detail in Mary Delany's letters from the 1750s, when she was an avid shell worker in Ireland. Gathering shells with friends was a great source of pleasure, combining rambling with conviviality:

> Yesterday at five o'clock in the afternoon we took a boat and went to a shore about a mile off to gather shells, where we found a vast variety of beauties. We were very merry in our work, but much merrier in our return home, for five of us, viz., Phill, Mrs. Don, Mr Lloyd, and a young clergyman (who is here very often, one Mr Langton), and Penelope all mounted a cart, and home we drove as jocund as ever five people were.[26]

She traveled modest distances to gather shells: "The coach is at the door, and we are going to Burdoyl, a strand about six miles off, in search of shells" for making "festoons of shell flowers in their *natural colours*, that are to go over the bow window."[27] The pleasures of travel to the seashore and of walking along the beach in search of seashells were intensified by sharing the experience with others who were equally passionate about collecting natural history specimens, a practice that appealed to fancy workers as well as collectors.

Walking on the beach at low tide and after a storm were popular ways of finding shells. The "best time" to find marine shells, according to William Swainson, was at "the lowest ebb of the tide":

> The rocks, corals and stones, which are then left exposed should be carefully examined for chitons, limpets, ear-shells, and other adhesive tribes, which are fixed upon the surface, or shelter themselves in the crevices: they are detached by suddenly passing a knife between them and the substance they are upon. Muscles, and other gregarious bivalves, furnished with a byssus, likewise occur in such situations. Wherever the rock, mud, or sand is pierced with round holes, the collector may be tolerably sure of finding bivalves: they are procured either by breaking the rock with a hammer, or digging into the sand or mud with a spade. The little puddles of salt water, left by the tide, are the habitations of many univalve shells; and others will be found beneath loose stones and sea-weeds.[28]

At low tide, those collectors and their assistants who were inclined to dig in the sand and mud would have found at "the depth of two or three feet . . . many various tribes of Cockles, Clams, Razors, Purrs, and other double shells, many of which spirit up the water, or sink the surface of their beds, by which they may be discovered." In addition to examining tide pools and the mudflats that were exposed by the receding tide, collectors could explore in the shallow water the sandbanks that were submerged at high tide and which were full of "thorny and other oysters, worms, Suns, Purples, and various other kinds of Shells."[29] Also recommended was the period after a storm, for

> [a]fter a gale of wind, or violent storm, the shore should be immediately visited, as fine shells are frequently to be met with: if the line of coast is extensive, a few boys should be engaged to assist in the search. This must be done quickly; for it not infrequently happens, that the next flow of the tide takes away every shell.[30]

Another reason for picking up shells quickly was that the sun could fade their colors and ruin their beauty. Da Costa warns that "shells that have lain for some time on the shores, known by the name of *Dead Shells* among collectors, they are seldom of good colour, by being exposed to the sun, and are also often imperfect, from being bowled to and fro, and by that means worn and broken."[31]

The duchess must also have employed others, fishermen most probably, to dredge for shells just off the coast of Weymouth. Several of the shells she discovered there were found with dredging, for, as J. T. Swainson's notebook testifies, his *Tellina fausta* was "dredged up at Weymouth, under the inspection of the late duchess dowager of Portland" and his *Mya pubescens* was "first noticed by the last duchess-dowager of Portland" and "Dredged up at Waymouth [*sic*]." His *Venus undata*, *Mya declivis pubescent*, and *Patella groeca* were found "by Dss. Portland off Weymouth," the word "off" suggesting dredging, a technique adapted from the way in which fishermen would use a dragnet or trawling net to gather oysters and scallops (pl. 33).[32] A dredge, basically a net on an iron frame or bucket, would be lowered to the sea floor and dragged to gather shellfish that rarely made it to the shore alive or intact. Using a dredge fashioned with a net, larger shells could be captured, the sand and other small debris sifting back down to the seabed. Dredging, or trawling as some called it, was a way to find rare specimens: "Trawling for fish in deep water, affords the best opportunity for collecting rarities, as much the rubbish, sea-weeds, &c. which come up in the net, many curious sorts of Scollops and other shells are to be got, which are not to be found near the land."[33] Swainson dredged for shells, noting this activity in his notebook and stating, for instance, that his *Patella ungarica* was "got by dredging off Fowey" and his *Murex dispectus* was "dredged up off Margate in 1783." Although his biographer Nora McMillan suggested that Swainson may have pioneered the use of the dredge to obtain specimens,[34] dredging was a technique that the duchess had employed with great skill and success much earlier. Da Costa also recommended it for collecting shellfish such as oysters, scallops, and other bivalves, writing to Pulteney in October 1776: "the surest & best way to collect Shells" is to employ "fishermen or dredgers . . . as they not only get them alive & perfect, but also get some sorts that the tides never fling upon the shores." Though he had been duped by fishermen in the past who had, it seems, sent him barrels of worthless shells, he asked Pulteney to make arrangements to have shells dredged along the coast of Dorset by fishermen and sent to him in London: "for though I became their Dupe, yet certainly I should acquire some Shells worth my cognizance, in a preservation I could not hope to get by other means."[35] Dredging was an effective method for harvesting live shellfish and could yield high-quality specimens for collectors, though today it is considered to be a destructive environmental practice and is in some waters illegal.[36]

Emanuel Mendes da Costa, Edward Donovan, John Mawe, William Swainson, and other authors of instructional pamphlets and collecting guidebooks cautioned their readers that most shells when found on the beach were "dead shells," often discolored by the sun, fragmented or distorted, having been pecked by birds and agitated by waves.

33   J. Agnew, "Pecten maxima," in *British Conchology or Original Drawings of Fresh & Salt Water Shells*, circa 1786–1818 © Natural History Museum, London

They suggested dredging as a remedy and as the best way to ensure perfect specimens and – even better – shells with live animals in them. Da Costa explained the importance of the live animals, not only for classifying the shell but also for keeping it at its best:

> It is always necessary, if possible, to get them alive, or with the animals in them. It not only instructs us in the natural history of the very animals, a part extremely useful to a thorough knowledge of Shells . . . but it also preserves the Shell in its perfect nature and beauty; for only live Shells bear the full glow of their colours.[37]

The younger Swainson advises prospective shell hunters to use "the *trawl*, or dragging net, upon a productive coast," because it will "generally bring up a variety of living shellfish." Before trawling, however, it is important to ascertain whether the seabed contains live shellfish, and the best way to tell if a coast is productive is to examine the beach for dead and broken shells, which will indicate what species might be found beyond the shoreline: "The trawl should be tried in every direction, both in deep and shallow water; and when once the shelly ground has been discovered, the collector may calculate upon procuring a variety of species peculiar to such waters."[38] The process required the collector to be able to read the dead shells littering the shoreline for indications of the kinds that might be located offshore. The duchess was regarded as an expert on dredging: Thomas Pennant, author of *British Zoology*, referred Dr. Charles Cordiner, a clergyman, amateur antiquarian, naturalist, and author, to her so that she could answer his questions about the various ways to dredge.[39] As an experienced field collector, she was an able interpreter of shell matter strewn across the beach, and a capable director of dredging activities since she was the first to discover many offshore species this way.[40]

SHELLS AS OBJECTS OF DISPLAY

Once shells were gathered from the shore, what was the next step in preparing them for arrangement, classification, and display? Mary Delany, a witness to the duchess's collecting practices, mentions that she had begun her shell collection in her youth by killing "a thousand snails."[41] How did she do this? Techniques for removing the living or dead animal from the shell and cleaning it were much-discussed topics in the literature on collecting and preparing shell specimens for display. Even recipe books and advice manuals on the decorative arts contained information on how to clean shells that were to be used in shellwork, a decorative art form that involved arranging and sticking shells to frames, candlesticks, and chandeliers, or to ceilings, window frames, and grotto walls, and using them to make flower sculptures and architectural details such as floral festoons. The recommended way to kill snails and other animals in their shells was to heat them slowly in a large pot of water until they either crawled out or died, and then, after cooling the shells by transferring them to a basin of cold water, to use a "stout pin" or a sharp knife to pull out an animal stubborn enough to have remained in its shell.[42] The techniques that shell workers and collectors used to prepare and polish the shells did not differ all that much, though collectors were more careful to preserve any external casing and were encouraged to use some restraint when polishing them. Peter Dance notes that eighteenth-century collections, even those designated "scientific collections," tended to highlight the aesthetic appeal of the shells, collectors often choosing to remove the epidermis so that the color of the shell could be heightened by polishing and varnishing.[43]

Advice on how to go about polishing shells can be found in books such as Donovan's *Instructions for collecting and preserving various subjects of natural history* (1794) and Hannah Robertson's *The Young Ladies School of Arts* (1767), both of which explain how to use a combination of aquafortis (nitric acid), emery file, and boars' hair brush to remove the epidermis. Donovan suggests that the shells of "Tellens, Muscles, Snails, &c. [which] come out of the sea slimy or even encrusted with filth, coralline matter, moss," should be steeped in hot water for twenty-four hours "to soften the filth or crust, then brush them well, with brushes that are not too hard," but if this does not work, "put them in weak acid, observing to dip them into cold water every minute; strong soap may also be used with a rag of woolen or linen, to rub them, and when cleaned finish them with a soft brush and fine emery."[44] Similarly, Robertson's book, written nearly thirty years earlier, states:

> A shell that is rough, foul, and crusty, or covered with a tartareous coat, must be left a whole day steeping in hot water; when it has imbibed a large quantity of this, it is to be rubbed with rough emery on a stick, or with the blade of a knife, in order to get off the coat; after this it may be dipped in diluted aquafortis, spirit of salt, or any other acid; and after remaining a few moments in it, be again plunged into common water: this will greatly add to the speed of the work; after this, it is to be well rubbed with linen cloths impregnated with common soap; and when by these several means it is made perfectly clean, the polishing is to be finished with fine emery and a hair brush; if after this the shell, when dry, appears not to have so good a polish as was desired, it must be rubbed over with a solution of gum arabic . . .

> When a shell is covered with a thick and greasy epidermis, as is the case with several of the mussels, and tellinae, in this case, aquafortis will do no service, as it will not touch the skin; then a rough brush and coarse emery are to be used; and if this does not succeed, seal-skin, or, as the workmen call it, fish-skin, and pumice-stone are to be taken in to assist.
>
> When a shell has a thick crust, which will not give way to any of these means, the only way left is to plunge it several times into strong aquafortis, till the stubborn crust is fully eroded. The limpet, auris marina, the helmet shells, and several other species are of this kind, and must have this sort of management; but as the design is to shew the hidden beauties under the crust, and not to destroy the natural beauty and polish of the inside of the shell, the method of using the aquafortis must be this, a long piece of wax must be provided, and one end of it made perfectly to cover the whole mouth of the shell . . . and the mouth being stopped by the wax, the liquor cannot get into the inside to spoil it . . .
>
> The shell is to be plunged into the aquafortis, and after remaining a few minutes in it, is to be taken out and plunged into common water. . . . When the repeated dippings into the aquafortis shew that the coat is sufficiently eaten away, then the shell is to be wrought carefully with fine emery and brush . . . In the effort of work, the operator must always have the caution to wear gloves, otherwise the least touch of the aquafortis will burn the fingers and turn them yellow; and often, if it is not regarded, will eat off the skin and nails.[45]

Though Robertson's audience is marked as gendered – the book was written for "young ladies" – the instructions are remarkably similar to those in Donovan's little book, which implies that its readers were men, though it may have attracted some women who, like the duchess, collected shells for the sake of natural history and were not necessarily interested in making them into floral bouquets and festoons.

Whether the duchess polished her own shells is not known, but Mary Delany certainly did, though with the help of several people. At the height of her shell-working days, primarily during her years in Ireland while married to the Rev. Patrick Delany, she told her brother: "I have been very busy in cleaning my new shells, and arranging them in my cabinet, and adding those I brought with me, – and now they make a *dazzling show*."[46] She also wrote to a friend about the dirty business of preparing shells for display:

> Covered with dust and wearied with the toils of cleaning and new arranging my cabinet of shells, throwing out rubbish, adding my new acquisitions, all which has been the work of yesterday and this morning to the present hour of one – (which

could not have been accomplished without the good assistance of my Mrs Hamilton and Bushe, who have toiled like horses) . . .[47]

Delany probably cleaned the shells herself with her friends and servants rather than paying someone else, since this could be quite expensive, and she had a restricted income. "The price for cleaning shells," according to Humphrey, was based on "the time they take from 3° [pence] to 5/ [shillings]. For Hammer Oysters 2.6."[48] Humphrey complained about the time, labor, and "trouble, which I assure you is not a little" that would go into cleaning shells, "one of the Sun shells for Instance will take me two days in cleaning, and do nothing else"; he argued that this extra work justified higher prices for cleaned and polished shells.[49] Even collectors, who were much wealthier than Delany and who could afford the prices, cleaned their own shells. Henry Seymer, for example, did not trust others to administer the potentially corrosive chemicals properly. Pulteney had asked for Seymer's advice on cleaning shells, and he responded:

> Nothing but practice can teach You how, & when, to use ye spt [spirit] of Salt, as it will have a very different effect on shells that have a natural high polish; roughens all Cones irrecoverably, and yet may be used successfully on some Olives, & Cyprea; if You have any subjects You want to clean, put them in yr pocket when You come next, and You shall see the process.[50]

Cleaning shells properly required skill and experience, for, as da Costa warned, "these processes must be judiciously managed according to the attendant circumstances, which it is impossible to regulate in writing";[51] and Swainson cautioned against using acids, "as their application requires much skill, and will prove destructive in the hands of inexperienced persons."[52]

Once the shells had been cleaned, the next step involved sorting them into generic types, putting them into little boxes, and then placing them in drawers, all arranged so that they would be readily available for the time-consuming process of classification. One of the methods for sorting and storing fossils and shells involved turning playing cards into little boxes. Mary Delany describes how one evening's entertainment was spoiled for some of the duchess's guests: they had desired to play quadrille but were disappointed to discover that all her playing cards had been transformed into specimen trays. Apparently, the duchess, who had bought this particular pack of cards for the purpose of entertaining her guests, could not resist using them to sort her specimens: "Fossils were examined, and sorted, a loud cry for patty-pans all exhausted, the *choice cards* were seized upon for the purpose, the *packthread cut*, the paper stript off, and behold appeared fair, unspotted, virgin cards!"⁵³ Delany's purported surprise at the use of playing cards to serve the needs of natural history is somewhat disingenuous because she herself enlisted cards to make trays for her own shell collection. She even sent a pattern for making card boxes designed to hold and divide shells to her young friend Mary Hamilton, while congratulating her on taking up the "science of '*conchyliology*'."⁵⁴

These card trays appear in *A Catalogue of the Portland Museum*; for instance, lot 2073 consists of "Twenty-five cards containing various species of univalves," and lot 2193 has "Seven cards of rare univalves." Guy Wilkins explains in his study of the Cracherode Collection that though it had been assumed that the cards referred to in several eighteenth-century natural history auction catalogues meant shells glued onto cardboard mounts, "it is now known that the cards mentioned so frequently were in fact early plain-backed playing cards utilized by collectors for specimen trays." The use of playing cards was not uncommon; the Muséum National d'Histoire Naturelle in Paris has several such card boxes filled with shells,⁵⁵ and Wilkins notes that there are some playing card trays in Thomas Pennant's shell collection in the British Museum. Instructions, dating from 1756, for making such trays indicate that collectors would purchase cards that were discounted due to misprints: "Cards can be bought . . . at 4½ d. and 6d. a pound . . . They are such as have small blemishes not fit to be put in ye packs." The duchess, at least on this one occasion that Delany describes, seems to have ignored expense in her need for more specimen trays, having exhausted all other materials. According to Wilkins, "trade and invitation cards were also used for this purpose, until at a late date properly made cardboard trays were supplied by natural history dealers."⁵⁶ The making of trays from cards continued into the nineteenth century. William Swainson provides a helpful description of how to make shell specimen boxes out of cards:

> *Card trays* are used by many collectors, who object to boards, as being expensive and cumbrous; hence are substituted square boxes, made of cards in the following manner: – Parallel to the four sides of the card, a straight line is cut by the point of a penknife, sufficiently deep to admit of one half of its substance being cut through, and folded back without difficulty: the space between the edge and the cut line will, of course, constitute the depth of the box, and may be varied according to the fancy of the collector, or the natural of the specimens it is to hold: when these four sides are cut, the corresponding corners are taken out by the scissors, and the sides bent up and united by pasted slips of paper. The bottom of the box may afterwards be covered either with black paper or velvet, and the specimen placed within. It is better to affix small or minute shells with gum water, or all the specimens of a moderate

size may be secured in the same way. A series of generic types should be arranged in the first drawers, and the indigenous species may be distinguished by their names being written upon pink-coloured paper.[57]

Little boxes made of playing cards, cardboard, and even of tin were then placed in drawers in cabinets specially built to hold natural history specimens. Bulstrode and Whitehall were filled with many cabinets designed specifically for this purpose. The Portland auction catalogue lists twenty cabinets, several made of mahogany. For sale were: "a small mahogany shell cabinet, with 7 drawers, and covers," "a cabinet for shells, with drawers, veneered with fine wood, and folding doors," "two very beautiful mahogany cabinets, with drawers, of beautiful wood, with upper parts of plate glass, the back plate silvered," and "an exceedingly handsome large cabinet, with 36 drawers and folding doors."[58] These cabinets sold for approximately £150 in total, quite a substantial sum. Years after the Portland auction, the duchess's cabinets continued to be valued by collectors. One advertised as having a Portland provenance sold at auction in 1804: "a large well finished Mahogany Cabinet with folding doors, containing 48 drawers, lined with cork for Insects, with sliding glazed frames, originally made for the exotic Insects of the Portland Museum."[59]

At Bulstrode and Whitehall drawers holding shells and fossils would be taken out of the cabinets, brought into drawing rooms, and offered to the duchess's interested visitors as a source of amusement and edification. A young visitor to Bulstrode, Mary Hamilton, described one such scene in December 1783, in which she joined the duchess, Mrs. Delany, and Mr. Lightfoot, who were busy discussing the East India tax, in the drawing room where Mr. Lind was "at his table and manuscripts":

> [W]e had a barrel of West India shells to look over. I took Mr. Lightfoot in, making him believe there were *oysters* coming to eat: this occasion'd much mirth. The Dss pronounced the shells to be "*good for nothing*;" afterwards was so good to look [took?] out some fossils and shells for me out of her own drawers; Mr. Agnew [the gardener] came and assisted to sort them out. I begun my *card* almanack wch Mrs. Delany gave me to copy. . . . After dinner ye Duchess had a box of shells brought; we look'd y^m [them] over together, and she gave me ye box and its contents; this employ'd us till Mrs. Delany came up from her room. At 7 tea: we had all our tables, and I finish'd my almanack . . .[60]

This is a convivial scene where young and old, men and women, servants and peers are gathered around talking, reading, and carrying out various collecting and sorting activi-

34   Box of shells: mahogany box with pine base containing three lift-out trays, each holding a collection of small cardboard and glass boxes of varying shapes accommodating different types of shells, probably made in Britain, 1780–1820, h. 16 cm, w. 39 cm, d. 26.5 cm. London, Victoria and Albert Museum. Photo © Victoria and Albert Museum, London

ties involving data management, antiquarian archiving, and the manipulation and evaluation of natural objects. Tables, drawers, and boxes were used as aids. Apparently, it was not only boxes of shells that traveled, but also drawers, as Mary Delany wrote to Miss Hamilton:

> I send for your inspection one of my drawers of shells, and wish most heartily I cou'd have brought them; but I must content myself with the pleasant feeling of sending you a moment's amusement; you may keep them as long as they can be of use to you, knowing them in safe hands.

Referring to the little card boxes, she added: "the mode of the trays I have found answer the purpose better than any," the purpose being here to sort the shells into species (pl. 34).[61]

Cabinets made expressly for holding and displaying shells were quite expensive. Daniel Solander, as keeper of the natural history collections at the British Museum, had to get permission from the trustees to have more made. He wrote in the official museum diary: "the Shell Cabinets have been sent home by the Cabinet Maker, properly completed according to the Estimate."[62] Cabinets, especially those designed for shell collections, often had two tiers, each of twenty drawers of varying depths, ranging from 1.5 to 2.5 inches (3.8–6.4 cm) for most shells, reserving space on top of the cabinet for larger specimens, preferably under glass classes or domes. Mary Delany was delighted with her new cabinet designed to hold her shell collection – "all my beauties":

> We are very busy in settling all my drawers of shells, sorting and cleaning them. I have a new cabinet with whole glass doors and glass on the side and shelves within, of whimsical shapes, to hold all my beauties. One large drawer underneath for the register drawer, and my little chest of drawers I have placed in my closet within my bedchamber, from whence I send you this letter.[63]

The drawers would have been lined with white paper or, better still, "carded cotton (such as used by jewellers), cut to the size of each drawer: this is a very good way of showing them to advantage, but every movement disturbs their position: besides, small specimens get entangled in the cotton, and are frequently lost, unless put into little card boxes (either round or square)."[64]

William Swainson, writing nearly sixty years after Solander, complained about the tendency of London cabinetmakers to place folding mahogany doors over the drawers: "This does not appear to us in good taste, for the whole immediately reminds one of a tented bedstead. Doors, also, made entirely of wood, give a heavy and wardrobe-like appearance." The duchess owned several cabinets with folding doors, and Swainson's criticism may have been rooted in a Victorian's distaste for Georgian aesthetic and architectural preferences rather than any technical consideration; after all, true to Victorian sensibilities he suggested that, instead of wood, the doors should be made out of "rich purple silk defended by brass wire." He was also critical of cabinetmakers who used cedar for the cabinets because this "exudes a resinous gum, which collects, not only on the wood itself, but on the objects contained within, and inevitably spoils them." He cautioned that whatever kind of "wood is used, it must be old and well seasoned. From carelessness in this respect, we have seen cabinets that have cost large sums, completely spoiled." Apparently, Solander did a good job in ordering cabinets since Swainson encouraged his readers to visit the British Museum to see some of the "best made cabinets."[65]

SCIENCE AND AESTHETICS

Cabinets could be used either to house a scientific shell collection, such as the duchess's, or one, such as Mary Delany's, meant to please the collector's aesthetic sensibilities and display a fashionable taste for natural history. Collecting, according to William Swainson, could be guided by principles of selection – "certain rules" and "a regular plan" – or "by the mere direction of fancy." He believed that private collections "may thus be classed under two heads: –1. Those intended to illustrate some scientific object; 2. Those formed upon no plan, intended merely for the gratification of the eye." These latter collections "are not scientific" because they are "formed without reference to any general or connecting plan."[66] He argues that certain underlying principles of selection and arrangement determine whether a shell collection would be considered scientific:

> If the object of the collector is to possess the most beautiful examples of a species, either as regards intensity of colour, perfection in its preservation, or in its size, he will find a princely fortune requisite to pursue his plan, at least to any extent. For the objects of science it is sufficient that the specimen is perfect, and that it represents the usual appearance of the species. Yet no scientific naturalist will reject a specimen because it may be slightly injured, seeing it is better to have some acquaintance with one of the forms of nature than none at all; at the same time he will be cautious in drawing hasty conclusions from such imperfect sources; the single valve of a bivalve should find a place in his cabinet, until a perfect example can be procured . . . Such specimens need not, however, be mixed with the general collection, but kept in drawers by themselves.[67]

Writing in the late 1830s, Swainson could make such sweeping generalizations about collectors' aims and motives with confidence; but such sharp distinctions between collections that were scientific and collections that were meant to dazzle the eye would not have been made so easily in the 1770s and 1780s.[68]

Swainson's characterization of the collector who is concerned primarily with the beauty or perfection of a shell might appear to be applicable to Henry Seymer, who wrote early in his career as a collector that although he wished someday to organize his collection around Linnean principles, he "considerd shells as beautiful objects only, & pleasing on account of their variety, and delicate forms," adding: "I suppose not one collector in 500 considers them otherwise."[69] He was famously concerned about possessing perfect specimens:

> How few shells we see in their perfect state for my own part, so many inconveniences attend me, as to situation, want of disinterested persons to purchase for me, &c: that I sincerely wish I had never seen a shell, for to have any thing only in a midling way never would satisfy me, nor ever will.[70]

However, to characterize Seymer as a collector primarily concerned with aesthetics, interested only in specimens that "will make a more shewy figure in my drawers," is to do him a disservice.[71] Much of his correspondence with Pulteney is concerned with classification, and he can be seen early on in this process, struggling with Linnaeus, angry and frustrated at first but gradually mastering the nomenclature and learning to classify his collection. He began with having "at least one drawer" arranged according to Linnaean classes, and gradually over the course of a decade brought Linnaean order to his collection.

Swainson's sharp distinctions between aesthetics and science apply to the duchess's collecting no more than they do to Seymer's. Part of the problem with the duchess's reputation is that, given the vast numbers and kinds of objects she collected, she can be perceived as having been more of a connoisseur or a virtuoso than a naturalist. To demonstrate how entangled the scientific and the aesthetic were in Enlightenment shell collecting, it is helpful to turn to an eighteenth-century shell expert's thoughts on the subject. Of the texts from that period that dispensed advice on collecting and preparing shells as natural history specimens, da Costa's *Elements of Conchology* is the only one that touches on the conflict of interests between collectors who want beautiful specimens and those who want scientifically significant ones. He offers a compromise to mitigate this tension:

> The scientific collectors, or naturalists, are always desirous of having the shells in their rough state, or just as they were fished. This method, though extremely useful, is not to be absolutely followed; not only because their beauties would be lost, but also on account that the species differing in colours could never be truly defined. However, as a medium, I would advise all collectors to have some Shells of each genus in their rough state, while the others should display their beauties by all the accomplishments of art: and a more easy medium may be kept in bivalves, by one shell or valve being rough, while the other is polished.[72]

Recognizing that many of his readers were not primarily interested in having a scientific collection, he urges them to treat their shells as more than a decorative item, and to consider their contribution to natural knowledge:

Numbers of Shells have an outer skin, or pellicle, different from the Shell itself, called the Epidermis. In regard to Natural History, it is really of use to know the nature, colour of this Epidermis, as it often characterizes some species, as much as any other part. I would therefore recommend a due notice or observation on it: and also to preserve some specimens covered with it; to enrich your cabinet with specimens for Knowledge as well as for beauty.[73]

The duchess's shell collection contained specimens with epidermis or pellicle, these objects being listed in the auction catalogue of the Portland Museum.[74] She also possessed a few examples of a shell's animal or its parts preserved in alcohol. According to Humphrey, Mrs. Le Cocq sent to the duchess a "large shell of that species of Bulla called by Linnaeus *lignaria*, together with a curious internal part".[75] This suggests that her interests extended well beyond aesthetics to what da Costa, and William Swainson fifty years later, would have called the scientific, for not only did she possess the pellicles of shells, but she also displayed an interest in the anatomy of the creatures that resided in the shells.

The fact that Margaret Cavendish Bentinck collected shells complete with their pellicle as well as the mollusks themselves also suggests that she knew the arguments that were circulating about whether it made zoological sense to classify mollusks based on their shells rather than their anatomy. Linnaean taxonomy for shells was based on their surface qualities, a convenient method since the shells could survive being buffeted by the ocean waves and transported from one side of the world to the other, whereas the creature inside rarely survived intact. In opposition to Linnaean practice, Michel Adanson in his book *Histoire Naturelle du Sénégal, Coquillages* (1757) suggested that anatomy should determine classification, and while da Costa agreed with him in principle, it simply proved too difficult (at least in this period) to organize a classificatory system around the organism that produced the shell. Da Costa explained a "great debate among Naturalists":

> whether the methodical system or arrangement of testaceous Animals should be formed from the Animals themselves, or from their habitations, or Shells. The former method seems most scientifical; but the latter, from the Shells, is universally followed, for many reasons.

One reason was that not every person interested in shells would be able to engage in the highly technical procedures that anatomical investigations required:

> Accurate descriptions of Animals, whose parts are not easily seen or obvious, and anatomical researches, are not in the capacity of every one to make; nor are the

particular parts and their respective functions so easily cognizable to any, but expert, assiduous, and philosophical enquiries.

Da Costa here was espousing the Enlightenment's embrace of the "liberal" arts and sciences, an approach to knowledge that preferred forms of inquiry available to the general educated public rather than a narrow specialized knowledge based on expertise. He embraced Linnaean systems (though he quibbled with the names that Linnaeus gave to the shell's physical features) with their emphasis on surface traits and what was easily observable without the assistance of specialized equipment such as microscopes. "Thus all ranks of animals are arranged into systems by obvious and external, not by scientific characters." He was, of course, "well aware of the arguments alleged against it, viz. that, as long as we study only the very Shells, those empty habitations, those spoils or remains only of the animals, the present sole objects of our researches and collections, we consider these beings but partially," and he acknowledged that "the animals that inhabit them should certainly guide us in our methodical arrangement." Systems built upon their anatomy, however, would exclude most amateur shell collectors, even those who were serious about Linnaean classification and were hard at work building scientifically significant collections.[76] Da Costa defended this system because it worked for a range of abilities among collectors. Those such as Henry Seymer could muddle through Linnaean taxonomies, naming his shells and calling upon Pulteney or the duchess to aid him when he got stuck or found Linnaeus confusing. Conchologists, comprising all sorts of people, could label their shells using Linnaeus's terms, even though this terminology did not make much sense in terms of describing the "animals that inhabit them," which, as da Costa reminds us, "are the very fabricators of the inhabitations, and give them their forms, bulk, hardness, colours, and all the other particulars of elegance, we admire."[77]

The subtle distinctions between scientific and connoisseurial collections were eclipsed in the nineteenth century with the professionalization of the study of nature, as demonstrated in William Swainson's hardened distinctions between fancy and system as the ruling principles of collection and arrangement. Swainson, however, was careful not to say that the private collector – in short, the amateur – was incapable of making scientific contributions:

> Private collections . . . it is usual to imagine that in the formation of these every naturalist may follow the bent of his own fancy, and such undoubtedly be true. Yet, if he is in the pursuit of science, he will derive lasting advantage from proceeding upon some one regular plan, adapted to facilitate that line of study he may intend to pursue.

Even collections "formed by amateur collectors for the gratification of the eye, or the decoration of the drawing-room" should not "be despised by the scientific naturalist" since they could arouse the curiosity and interest of "youthful minds" and foster a taste for the study of natural history.[78]

CLASSIFICATION

Although William Swainson's analysis of the kinds of collecting – his breakdown into public and private, amateur and scientific – can be useful in categorizing collecting practices, his nineteenth-century terms must be used with care. In the late eighteenth century the distinctions between a connoisseur's collection and a "scientific" collection could blur because collectors of natural history specimens were mostly amateur naturalists, and, within this group, there was a range of collecting practices from the primarily decorative, as in Mary Delany's case, to Seymer's obsession with perfect specimens to Pulteney's "scientific" collection. Since most twentieth- and twenty-first-century commentators describe the duchess as a connoisseur, meaning someone with a bit of knowledge, a measure of good taste, and an abundance of money, it is important to point out that, in addition to being a specimen hunter, she was also involved in classifying her collection. What did she think about this process and what principles of arrangement did she employ to give her collection coherence and order? The answers to these questions will help us to understand the duchess's relationship to her collection and glimpse the ambitions she had for it.

As early as the 1740s, when she was in her twenties, she already knew the scientific names that naturalists used to identify shells. This was a source of amusement for Elizabeth Montagu, who would often tease Lady Margaret (when they were close friends in their teens and twenties) about her knowledge of this specialized language and her love of natural history. In 1743 she wrote to her about a proposed visit to a shell grotto:

> On Monday we think of going to Lady Fane's grotto, at Basildon, of which I will give your Grace the best account I can. One who bears so true a respect to every individual shell, cannot but truly venerate a number of them happily met together, and therefore I will give your Grace a particular description of them; I mean of their appearance; as to their names, unless I had their godfather Shaw with me, I shall not be able to tell them.[79]

Montagu acknowledges her disadvantage in that she does not know naturalists' names for mollusks; without the assistance of Dr. Thomas Shaw, professor of Greek and an amateur naturalist, she warns the duchess that she can provide only a description of how the shells look. After the visit, she dutifully wrote again, reporting on the Shell Room:

> The first room is fitted up entirely with shells, the sides and ceiling in beautiful mosaic, a rich cornice of flowers in baskets and cornucopias, and the little yellow sea snail is so disposed in shades as to resemble knots of ribbon which seem to tye up some of the bunches of flowers.[80]

Montagu recognized that the shells in Lady Fane's grotto were subject to two distinct but intersecting systems of meaning-making – the aesthetic and the scientific. The shells were beautiful, and when placed together in a design or pattern added luminosity, texture, and color to the grotto; individually, they were representatives of a variety of molluscan species. Although for her shells were primarily associated with decorative artwork and the practices of aesthetic display, she knew that the duchess would have been interested in the grotto both as a crafted element in a designed landscape and as a site in which shells of possible scientific interest might have been displayed.[81]

What were these "names" that Dr. Shaw and the duchess of Portland knew and that Elizabeth Montagu did not? What terminology did the duchess use to identify shells? By the early 1770s she was using Linnaean binomial nomenclature (genus and species) and Linnaean descriptions of the shell's morphological features, its outward appearance, to classify her shell collection. Because it is based on morphology, specifically the external features of biological entities, Linnaean taxonomy is an artificial system and does not take into account function, as in an organism's physiology, or genealogy in terms of the phylogenic relationships between species. It is a tool with which to name species, to differentiate between them, and to organize that information; it does not pretend to provide more than a convenient system of classification nor are there embedded within it theories of causation or evolution. John Lightfoot may have been instrumental in helping the duchess learn the new Linnaean system as it applied to shells, for Mary Delany mentions that in January 1771 "the little Philosopher, Mr. Lightfoot came to town . . . and the science of shells went on prosperously."[82] Her use of the word "science" highlights the fact that Lightfoot and the duchess were working to master Linnaean taxonomy as it applied to shells. Both Delany and the duchess were already expert at using Linnaean names and characteristics to classify plants, for in 1735 Linnaeus had published the first of many editions of *Systema Naturae*, which covered the vegetable, mineral, and animal kingdoms, with the exception of mollusks. At Bulstrode, for instance, Mrs. Delany wrote a manuscript flora "after the sexual system of Linnaeus," meaning the Linnaean botanical taxonomy built on counting the number of pistils and stamens, the generative parts of a flower, as a way of determining its genus and species.[83] Also, Georg Ehret, the great botanical artist who knew Linnaeus and drew the illustrative tables for the *Systema Naturae*, lived at Bulstrode in the late 1760s, drawing hundreds of plant illustrations there. Mary Delany wrote:

> en attendant we have Mr. Ehret, who goes out in search of curiosities in the fungus way, as this is now their season, and reads us a lecture on them an hour before tea, whilst her Grace examines all the celebrated authors to find out their classes. This is productive of much learning and of excellent observations from Mr. Ehret, uttered in *such* a *dialect* as sometimes puzzles me (though he calls it English) to find out what foreign language it is.[84]

Linnaeus published many editions of the *Systema Naturae*, but only in the tenth (1758) and twelfth (1767) did he address what he called *Vermes*, or worms, the class within which mollusks lay. Thomas Pennant, who adopted Linnaean systematics for his multi-volume *British Zoology* (1768–77), explains what *Vermes* are:

> Slow, soft, expanding, tenacious of life, sometimes capable of being new formed from a part: enliveners of wet places; without head or feet; hermaphroditical; to be distinguished by their feelers.
>
> Not improperly called by the ancients, *imperfect animals*; being destitute of head, ears, nose, and feet, and for the most part of eyes; most different from insects; from which LINNAEUS has long since removed these works of Nature.
>
> They may be divided into Intestine, Soft, Testaceous, Lithophytes, and Zoophytes.[85]

Linnaeus divided *Vermes* into eight orders, with *Mollusca* referring to slug-like creatures without shells and *Testacea* for animals with shells. *Testacea* is further divided into three groups: multivalve (e.g., barnacles), bivalve (e.g., mussels), and univalve (e.g., snails). Each of these categories is further divided into genera and species. He lists fourteen different genera among the bivalves or *Conchae*, and also fourteen genera within the univalves or *Cochlea*, with a total of 703 species within *Testacea*. These numbers were considered low even by Linnaeus's contemporaries, in particular Solander, who argued that Linnaeus simply did not have sufficiently large collections of shells to construct a system that would be able to handle newly discovered species from around the globe. Time has proved Solander right, and what Linnaeus regarded as a species would today be considered as wide open a category as a genus. Modern-day malacologists, supporting

Solander, suggest that Linnaeus had a "limited understanding of mollusks" and thus his knowledge of this group was "superficial."[86]

What did it mean to use Linnaean taxonomy to classify shells? Because this was a classificatory system based on nonessential features in an organism's anatomy rather than natural differences, Linnaeus focused on the morphological characters, or the features of the shell, as a way to recognize similarities and to distinguish differences between specimens and thereby determine their genus and species. As Thomas Martyn advises in *The Universal Conchologist*, students of conchology need to pay "scrupulous and minute attention to the figure, mouth, extremities, and convolutions of those shells which he classes in the respective families."[87] Conchologists eager to identify their specimens took note of the shell's size ("twice the size of a grain of wheat"),[88] shape (e.g., long, narrow, striated, heart-shaped), texture (e.g., brittle, smooth, glossy, opaque), and color ("a light greenish colour, with a faint tint of brown or rosy").[89] More specifically, the identifying features of shells to which English-speaking conchologists paid special attention were: "operculum," "epidermis," "umbilicus," "aperture," "columella," "spire," "lip," "mouth," "tooth," "valve," "beak," "ear," "girdle," "wing," "gutter," "slope," and "hinge;" terms still in use today to identify the morphological characters of shells.[90]

How did the duchess identify and classify her shells before the publication of the 1758 and 1767 editions of Linnaeus's *Systema Naturae*? It is unlikely that she used such names as Egyptian Pyramid, Poached Egg, Ear Shell, Fool's Cap, White Tower of Babel, and Midas's Ear, though these were the names that most dealers, including George Humphrey, used, since she distanced herself from such vernacular names, saying politely: "I am not sufficiently acquainted with the names the Dealers give to shells." She used Latin names – *Solen*, *Tellina*, *Mytilus*, *Turbo*, *Helix* – as was the custom in most European seventeenth- and eighteenth-century conchology publications. The pre-Linnaean reference books she drew on were either written in Latin or in French or Italian, though the shell names were usually in Latin. She owned proof pages of Martin Lister's *Historiae conchyliorum* (1685–92) and helped fund the reproduction of this book in 1770. In her correspondence with Pulteney, she cited Filippo Buonanni's *Ricreatione dell'Occhio e della Mente, nell' Osseruation' delle Chiocciole* (1681), Niccolò Gualtieri's *Index testarum conchyliorum* (1742), and Antoine-Joseph Dezallier d'Argenville's *L'Histoire naturelle eclaircie dans . . . la conchyliologie* (1742), which mixed Latin and French names for shells, and she probably also owned Georg Eberhard Rumpf's *D'Amboinsche rariteitkamer* (1705), all considered classics in conchology. Linnaeus based his own work on these texts, writing that the "principal authors for a serious study of this order are Bonannus, Lister, Rumphius, d'Argenville, and Gualtieri. Whoever possesses these may dispense with the rest."[91] The

Systema Naturae (tenth edition, 1758) was published without illustrations, but he referred to illustrations in the aforementioned books by table and figure number so that his readers could compare their specimens against those represented there. A typical citation would read, in this case for *Pinna nobilis*, a species he numbered as 224:

> 224. *P. testa striata: squamis canaliculato tubulosis subimbricatis.*
> *Bonan. recr.* 2. *f.* 24.
> *Argenv. conch. t.* 25. *f.* B.
> *Habitat in M. Mediterraneo.*[92]

Linnaeus's *Systema Naturae* did not replace the older shell books – Buonanni and d'Argenville are referred to in the above passage – but instead made them even more pertinent to the process of classifying shells according to his system. The duchess would have had to open up her copies of these books to make sure that the shell she wanted to identify as a *Pinna nobilis* looked like the images referred to by Linnaeus. Classifying a shell, then, would require the collector to have both old and new conchological books, and some collectors, such as Pulteney, who did not possess all of these rare and costly books, had to rely on the generosity of fellow collectors for loans. It will be seen in Chapter Five that this dependence on access to rare books could be a source of tension between collectors.

British naturalists embraced Linnaean taxonomy, with its notion of species "as fixed and discrete entities distinguished by their formal discontinuity,"[93] finding it easy to use and helpful in organizing information about the plethora of new species that were brought to Britain from all over the globe. Based on the reproductive system of plants, Linnaeus's botanical system was celebrated in a range of publications, in poetry with Erasmus Darwin's long poem, *The Loves of the Plants* (1789), and in lavishly illustrated botanical books such as Robert Thornton's *Temple of Flora: A New Illustration of the Sexual System of Linnaeus* (1799) and Richard Pulteney's *The Progress of Botany in England from its Origin to the Introduction of the Linnaean System* (1790), a historical overview of the history of botany in England.[94] Though very popular in Britain, Linnaean taxonomy was also criticized for being salacious in its terminology and therefore an inappropriate tool of instruction for young people who were interested in natural history. Linnaean botanical nomenclature has received much attention from historians, some of whom argue that Linnaeus's personality, an odd mixture of the prurient and the prudish, seeped into his understanding of the natural world.[95] Linnaeus based his botanical nomenclature on parts of the fruitification of a plant, calling the stamens "husbands" and the pistils "wives" or "concubines," metaphors that anchored his system of differences and similarities to the reproductive anatomy of plants. While its use among women and children caused some consternation, this "sexual system of classification" was at least tied to the anatomy of the flowering plant. When it comes to the parts of shells, however, particularly for bivalves, there is no correspondence between his sexualized terminology for the shell's physical features and the reproductive parts of the animal, the oyster or scallop, for instance, that produced the shell. The terms were based on what Linnaeus perceived to be a similarity between a bivalve's form and the shape of a woman's nether regions (pl. 35). Stephen Jay Gould described Linnaeus's *Fundamenta Testaceologiae* (1771) as containing "one of the most remarkable paragraphs in the history of systematics," noting the bizarre vulgarity of his terminology:

> He [Linnaeus] regards the hinge between the two valves (cardo) as a defining character, and he then writes: *Protuberantiae insigniores extra cardinem vocantur Nates* – or "the notable proturberances above the hinge are called buttocks." He then names all the adjacent parts for every prominent feature of sexual anatomy in human females – *ut metaphora continueter* ("so that the metaphor may be continued"). Clams have a *hymen* (the flexible ligament connecting the two valves at top), *vulva*, *labia*, and *pubes* culminating in a *mons veneris* (various features at the top of the shell behind the umbo – our modern term for Linnaeus's buttocks); and, in front of the umbo, an *anus*.[96]

This terminology, for instance, calling the hinge that joins the two halves of the bivalve the hymen, and the two halves of the shell the buttocks, has no scientific merit; descriptive of neither the animal's structure nor function, it was purely metaphorical, based on a visual and verbal pun – and a smutty one at that.

Despite Linnaeus's stature and weight with British conchologists, some naturalists were repulsed by the terms he used to describe the parts of a shell, da Costa protesting both in private and in print that the terms were vulgar and disgusting. In *Elements of Conchology*, for instance, he wrote: "Science should be chaste and delicate. Ribaldry at times has been passed for wit; but Linnaeus alone passes it for terms of science. His merit in this part of natural history is, in my opinion, much debased thereby." He wanted Linnaeus's "unjustifiable and very indecent terms" to be "exploded with indignation." With the goal of making technical names "chaste" and thus appropriate for women and children, da Costa offered his own terms for use in conchology. These "will render descriptions proper, intelligible, and decent; by which the science may become useful, easy, and adapted to all capacities, and to both sexes."[97] In his private correspondence he also complained about Linnaeus's inaccuracies:

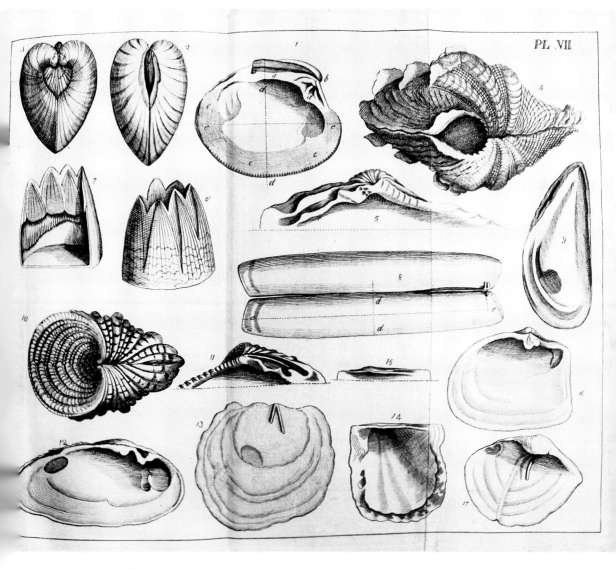

35   Emanuel Mendes da Costa, *Elements of Conchology: or, An Introduction to the Knowledge of Shells* (London, 1776), Plate 7. By permission of University of Glasgow Library, Special Collections

Linné is so extremely erroneous in many quotations, & at the same time, so concise or obscure, and so presumptuous in changing long established names, & intruding new fantastic ones, that the more I look into him or consult him, the less I like & follow him: by too often discovering his little accuracy or knowledge.[98]

This criticism of Linnaeus's shell terminology did not only arise out of prudishness, as some have suggested, but rather, as Gould convincingly argues, because of the inaccuracy of the metaphor and its failure to convey knowledge about the physiology and anatomy of the organism under investigation. As Gould wrote, "The top of a clam is not the bottom of a person – and supposed visual similarities can only be misleading."[99]

Despite da Costa's criticisms, however, British conchologists continued to use and promote the Linnaean system of nomenclature for mollusks well into the nineteenth century. More than once was da Costa scolded by the British conchological establishment for his critical remarks concerning Linnaeus's language and skills as a taxonomist, Pulteney writing to him: "I cannot conclude my letter without observing that I think you have in various instances in your letters to me treated Linnaeus in a manner very unbecoming the conduct of a scientific man."[100] In their "Historical Account of Testaceological Writers," Dr. Maton and the Rev. Rackett register their annoyance at da Costa's attempt in *Elements of Conchology* to revise Linnaeus's terminology while keeping his categories:

> it might be expected that a system, in which he professes to differ materially from that great naturalist, would have contained some important improvements. It is worthy of remark, however, that, after abusive strictures on the Linnean system, Mr. da Costa builds his own chiefly on the general characters which Linnaeus himself has made use of.[101]

Even the duchess faulted da Costa for coming up with such "strange names" in his attempt to avoid Linnaean double entendres, though this did not stop her from supporting his work by becoming a subscriber to *British Conchology*.

Despite the sexism implicit within Linnaean nomenclature and the not-so-latent misogyny ever present within the field of conchology, the duchess embraced Linnaean taxonomy and worked hard at identifying specimens and organizing her collection using his terms. Her letters to Pulteney are full of concern over the proper classification of her shells. Though she acknowledges that she was dependent on Lightfoot, and later Solander, to name new ones, she did the work of sorting and identifying her shells more than competently and was able to recognize a new shell. In addition to referring to Linnaeus's *Systema Naturae*, she also used shell books that employed Linnaean categories written by contemporaries, many of whom she knew personally; she knew more than enough to recognize and correct their mistakes. Though she did not confront Pennant with his errors, she confided in Pulteney that he had made some mistakes in the fourth volume of *British Zoology* (1777), asking: "have you seen Pennants book of shells I think it will be of use tho' there are mistakes, & some might be added to it."[102] She complained to Pulteney on another occasion that Humphrey's and da Costa's series, *Conchology; or, The Natural History of Shells* (1770–72), was full of errors: "Humphreys and da Costa give strange names to the Shells the latter shou'd know better but he is being Inaccurate in his descriptions."[103]

Linnaeus's name is sprinkled throughout the duchess's correspondence with Pulteney, and Linnaean terminology, characters, and binomial nomenclature appear in the lists and queries attached to the little boxes of shells that ebbed and flowed between them. She sent him duplicate shells, along with a list of their names, and he would send her shells he had trouble identifying along with a list of questions.[104] In April 1772 she sent him a box of shells accompanied by a list that she had made of them, doing her best to give them Linnaean names, though she worried that without the Rev. Lightfoot's help she may have made errors:

> I have sent you a Box of Shells according to the inclosed Catalogue by the Blandford Waggon Car. pd. which set out at this day April 9th & I hope will arrive safe. I fear you will find I have made many mistakes for want of assistance but hope you will pardon it. You will observe some Chasms but my *Solens Tellina*'s have not been named; when Mr Lightfoot was in Town before Xmas he intended putting the Linnean names to them but the Asterias took up so much time that it was not in his power. I have stuck on the names to each shell but the small ones are put in separate papers with the name of each Shell, that you will be carefull in unpacking the little Boxes. I hope the tender ones will not be broke & you will have as much amusement in looking them over as I had in sorting them.[105]

In another letter to Pulteney sent in June the same year, she wrote with her usual mix of self-deprecation and polite criticism, lodged this time at Linnaeus in her attempt to clear up some confusion in identifying shells that Pulteney had sent her for classification. She mentions that there is a problem with classifying the *Mytilus frons* as such, and hopes that Solander will have time to deal with finding the proper genus for this shell, something that eventually he did get to and which became a part of his larger project to revise and expand Linnaeus's generic categories.

Linnaeus presented problems as a guide to classifying one's collection because, as the duchess suggested, he based his categories on collections much smaller than hers, and insufficiently large for him to see the range and variety of genera and species. The duchess's letters to Pulteney subtly critiqued Linnaeus's shortfalls while managing to maintain a proper feminine demeanor. She wrote: "I have endeavoured to follow your directions & am sorry I have been able to mark so few in the Linnaeus but there are many hundreds he has not seen & consequently made many mistakes in his annotations."[106] The criticism that Linnaeus had simply not seen enough shells to construct a coherent and useful system of classification punctuates the Pulteney–Seymer–Portland correspondence, erupting in moments of frustration. Though a staunch defender of Linnaeus, Pulteney acknowledged to the duchess that there was a serious weakness in the Linnaean molluscan taxonomic scheme: "how oft have I lamented that Linnaeus, when he formed the conchological part of his system, had not Access to such a Collection as your Grace's!"[107] Seymer went so far as to accuse Linnaeus of being careless and inaccurate in his descriptions and references. In one exchange between Pulteney and Seymer over the proper name for a shell, Seymer blames Linnaeus for their trouble in identifying the specimen: "the figures refered to by Lin: in Arg: & Gualt: for the [Murex] Hippocastaneum are as different as two shells of the same Genus can be"; in other words, Linnaeus's description of this sea-snail shell contained references to illustrations that depicted very different-looking shells.[108] Seymer had a running disagreement with Pulteney over Linnaeus, with Seymer venting his frustration with his *Systema Naturae*: "Linnaeus must surely have collected his Synonyms [type specimens] in a hurry or he never could have made so many palpable mistakes . . . sometimes he refers to the figures in three Authors, each of which differ notoriously."[109]

Although the duchess was quite modest about her mastery of Linnaean taxonomy and willing to admit what she did not know, both Pulteney and Henry Seymer often deferred to her expertise. In the 1770s Seymer wrote to Pulteney about his failed efforts in identifying some of his shells and his hopes that the duchess would help him with the Linnaean system: "If the Ds does not assist me greatly, I shall jumble them together as formerly."[110] Pulteney would often send her boxes of shells to examine and compare with those in her cabinet, hoping that she would pick up any mistakes. He wrote:

> I have packed up in a Box, and sent by the Coach today as single Valve (for I have not a pair) of that Shell which I have always taken for the *Venus pensylvanica*; and I shall be glad to have it determined from your Grace's Cabinet . . . but as your Grace can . . . recollect such, from having several times honoured my Cabinet with your Inspection, you will easily correct such Errors where they occur [see pl. 7].[111]

On another occasion, Pulteney sent the duchess ninety-eight shells and a corresponding numbered list, asking questions about each shell, to which the duchess responded on the right-hand side of the list. For example, Pulteney wrote: "If this shell is named Dr

P will be obliged to the Duchess of Portland, for the general term." Next to this query the duchess replied: "Mya" and "not described," meaning that this shell was "nondescript," in other words, its species belonged to the genus *Mya* but it had not yet been named or described.[112] Though Solander (and Lightfoot on occasion) did the work of naming new species, the duchess was sufficiently involved to know what specimens had been named and how many were left to do. Though she may have passed on Pulteney's lists of unidentified shells to Solander when he came to Whitehall on his weekly visit, she was in charge of this flow of information and was clearly up-to-date with Solander's progress in classifying her collection (pl. 36).[113]

Pulteney would also send her wish lists of the shells he wanted in the hope that she would send him any duplicates in her collection: "I take the liberty to trouble your Grace with a list of those Sea Shells mentioned by Mr. Pennant in his Book & not referred to Linnaeus by him. If Duplicates of any of these should occur to Your Grace may I beg the favour of a Duplicate."[114] In addition to such requests, he loaned the duchess shells to fill out the few "chasms" she had in her collection, so that Solander would have as complete a range of shells as possible. Solander was engaged in a significant recalibration of the Linnaean classification system for mollusks, rearranging species under different genera, a project that was left unfinished by his untimely death in 1782.[115]

The duchess of Portland's commitment to the science of shells can be seen most clearly in her ambition to publish a catalogue of her shell collection, containing the most up-to-date descriptions and names. This would have been the definitive conchological catalogue of her generation, and as part of the process Solander spent one day a week at her Whitehall townhouse naming new shells: "Dr Solander has begun to range my Shells & name the non-descripts but the work will go slowly on as his time is so much taken up at the *Museum*."[116] Her intention was to make her collection available to the world for consultation, hoping that others would benefit from its massive scale and its most complete-to-date series of species within the genera laid out by Linnaeus. Solander's death in 1782 and her own in 1785 put a stop to these ambitions. In her will she stipulated that all her natural history specimens were to be sold at auction, and Solander's names and shell descriptions were what Lightfoot worked from when he wrote the auction catalogue.[117] What would have been an important contribution to the natural history of conchology and a lasting testament to Margaret Cavendish Bentinck's accomplishments as a naturalist and a shell collector became a sales catalogue, confounding even more thoroughly the contradictory yet colluding regimes of value represented by science and the marketplace. Within the sales catalogue, amid

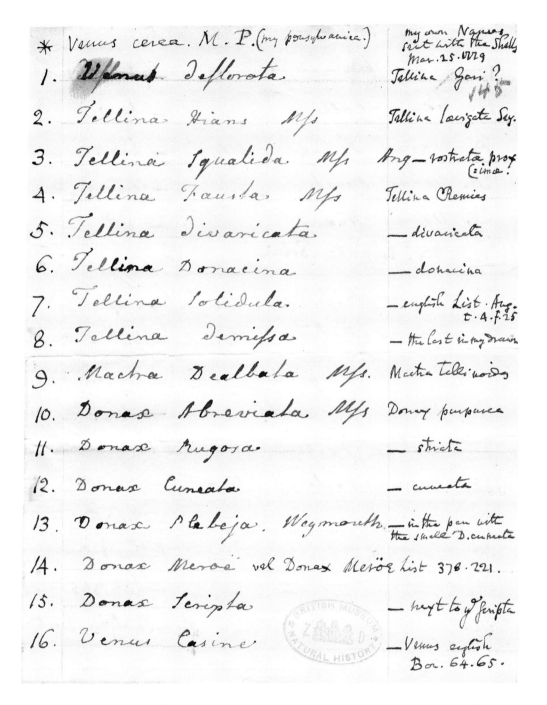

36   List of shells exchanged between Richard Pulteney and the duchess of Portland, 25 March 1779. London, Natural History Museum. © Natural History Museum, London

shells enumerated, named, and described, scientific value competed with the capricious but value-enhancing mechanisms of celebrity so that buyers, with very different motives, vied with each other, eager to purchase, for instance, shells the duchess had owned, shells Solander had named, and shells, as will be seen in the next chapter, that Cook had retrieved from the Pacific.

CHAPTER THREE

Blandford March 23.
1772.

Madam

I return your Grace my most respectfull Thanks for the Honour of your last obliging Letter, and the present of new Zeeland Shells, which I received safe. I do not think myself qualified to speak about conchological matters, having so lately entered into that Field, otherwise I should suppose that some of these Species are such as have before reached England; some from the Patagonian Coast, and others from the East Indies.

Since I received your Grace's Favour, I have seen in the Hands of Mr. Seymer a considerable Number of Shells, that were brought home in the Endeavour, and which were sent down from Mr. Humphry's Shop, for Mr. Seymers Inspection, with the Dealers Names and the prices affixed. Some of these Mr. Seymer has retained, and by this means I have got the Dealers Names to several of the new Zeeland Shells, which your Grace was so obliging as to send me. I wish the English Dealers would adopt more scientifick Names to Shells, and draw up their Catalogues, more in the way of the Dutch, who seem to be much more forward in their knowledge of Shells, than our Dealers if one may judge by the list of some of their sales. What Mr. Humphrys has called the Cut-out Buccinum, & the spotted Whelk are I presume new Species. What he calls the Club is I imagine a Strombus. The Chambered Patella & the Ducks-bill Patella are singular. The latter I should think answers pretty well to the Patella Unguis p.1260. of Linnæus, but I doubt whether it be the shell there described, as it does not exactly answer to Rumphius's figure. Petivers Book I have not. I have amused myself by endeavouring to describe them after the Manner of Linnæus, but have not succeeded to please myself at all. How oft have I lamented, that Linnæus when he formed the conchological part of his System, had not Access to such a Collection as your Grace's!

PATRONAGE, BROKERS, AND NETWORKS OF EXCHANGE

Shells from all over the world were sold at the Portland auction, with specimens from the South Pacific fetching especially high prices. For sale were shells associated with Cook's voyages, shells from Australia, New Zealand, Tahiti, New Caledonia, the New Hebrides, Tonga, the Marquesas, and the Society Islands, as well as from the northwestern coast of North America. Among the most desirable Pacific shells sold was the duchess of Portland's "scarlet Anomia," which, as the catalogue notes, was "from *New Zealand, extremely scarce*," and which was bought for £3 15s. by the agent Dillon acting on behalf of Monsieur de Calonne, the French foreign minister.[1] Also causing a stir was lot 3832, a hammer oyster shell, bought for £4 4s., its value no doubt increased by its association with Cook: "A very large and fine specimen of the white variety of Ostrea Malleus L., brought by *Capt. Cooke* from the *Coral Reef*, off *Endeavour River*, on the Coast of *New Holland – very rare*." The natural history dealer George Humphrey bought many lots throughout the long auction, especially during the last days, when his purchases made up the majority of the shell sales, including the white oyster mentioned above and "an exceedingly fine and large Cypraea Aurora, S. or the Orange Cowry, from the *Friendly Isles* in the *South-Seas, extremely scarce*." He also bought "a large and a small specimen of Voluta incompta, S. from the *South Seas, extremely scarce*," both of which appear in Martyn's illustrated *Universal Conchologist* (see pl. 8).[2]

How did the duchess of Portland acquire her magnificent and very valuable collection of Pacific shells? Did she purchase them from individuals directly involved in Cook's voyages or from intermediary dealers, or is it possible that she received them as gifts from Cook himself? It has been commonly assumed that she resorted exclusively to purchasing her natural history specimens, buying them from dealers and paying agents to collect them on her behalf, and that she did not participate in patronage networks and their complex forms of gift exchange. In fact, as will be seen, she acquired specimens in all these ways.

Shell exchanges within early modern European societies were very complex, involving gift as well as commodity economies, archaic as well as modern forms of exchange. Much has been written in the field of economic anthropology on the complexity of shell exchanges in the South Pacific, beginning with Malinowski's description of the Kula ritual of exchange in the Trobriand Islands.[3] By comparison, little has been written on shell exchanges in eighteenth-century European society, which was undergoing the transition from a quasi-feudal, hierarchical social and economic structure to a modern, capitalized, commercialized economy structured around the contract. Shell collecting in late eighteenth-century Britain was a site of contradiction because these two economies – one based on hierarchy, patronage, and gift exchange, and the other organized around the commercialized exchanges of the marketplace – overlapped and interpenetrated, causing confusion as participants bought, sold, traded, gave, and loaned shells to fellow collectors.

This chapter explores the duchess's participation in natural history networks, some of which were commercial in nature, and some involving reciprocity and return within patronage's vertical ties. The first half of the chapter examines the patronage dynamics within her network of natural history correspondents. The second half examines the duchess's methods of acquisition, particularly of Cook-related specimens, which involved the commercial sector and a network of natural history dealers and their contacts.

SHELL NETWORKS

Although the duchess of Portland purchased shells from dealers and attended auctions, she also relied on gifts, trades, and loans to fill out her shell collection. During this era, such non-commercial forms of exchange as the gift or loan would have moved within what historian Harold Perkin has called the "old society's vertical friendships," a social system structured hierarchically, akin to but not entirely coextensive with the patron–client relation.[4] The duchess participated in these hierarchical exchanges in a variety of ways. As a patron, she would have received gifts from people who would expect something in return, either similar objects or, more likely, favors, for instance, social acceptance, preferential treatment, and the exertion of influence on the giver's behalf.[5] As a patron, she benefited from honorific gifts, receiving shells from a variety of people, ranging from her gardener J. Agnew to the amateur collector J. T. Swainson to the professional naturalist Thomas Pennant and Sir Joseph Banks, President of the Royal Society, all of whom visited her at Bulstrode. She, in turn, gave shells to friends beneath her on the social register, British shells to Swainson and exotic shells to Richard Pulteney, knowing that he would appreciate the South Pacific shells that he could not afford to buy for himself. But gifts were not the only paths by which shells moved between people in non-commercial exchanges. Loans and trades were also prevalent, and these could be quite complex and even more perplexing than gift exchanges, with questions arising about the value of an object, the equivalence between objects, and fairness in exchange. To demonstrate the ways in which the duchess of Portland participated in patronage networks, I begin with an overview of her shell network and then describe a series of encounters between the duchess and Henry Seymer, as mediated by their mutual friend and fellow shell collector Richard Pulteney.

In charting the exchange of shells, one finds flexible, expandable, and inclusive networks of like-minded people who loaned and gave each other shell specimens, as well as expensive illustrated books on conchology and catalogues from natural history auctions, and freely passed on information about how they had acquired items in their own collections. The particular network of exchange that the duchess participated in consisted of physicians, country gentlemen, clergymen, professional taxonomists, amateur naturalists, natural history dealers, professional specimen hunters, members of the Royal Society, university professors, natural history artists, and even her gardener. Paula Findlen's description of the way in which early modern Italian naturalists used "the rhetoric of friendship to facilitate mutual communication and exchange" is equally apt for the way in which conchologists exchanged shells in late eighteenth-century Britain.[6] For instance, the duchess met John Timothy Swainson in the early 1780s when he was a young Customs officer stationed in Margate and she, in her seventies, was visiting the seaside resort. Despite their class, age, and gender differences, they became "friends" in Findlen's sense: they met to examine each other's collections and to discuss classification; they may have gone specimen hunting together, even on dredging expeditions; and they became correspondents and sent each other gifts of shells. The duchess, for example, wrote to Swainson thanking him for sending her a "small box" of minute shells "which have afforded me much pleasure as there are several that seem to be new." She gave him gifts as well and readily acknowledged his

> I am afraid you will think I have been very idle in not returning your shells sooner but I have been a good deal taken up as I propose leaving town soon. I have sent you the box of shells & according to your Obliging request have detain'd a Helix N 41, which I had not in my Collection & for which I return you many thanks.[13]

In the complex process of loaning and exchanging shells, mistakes sometimes occurred. Of such an instance, the duchess wrote to Pulteney: "I have been guilty of such a Blunder that I am quite ashamed of my self which I never discovered till I got home in bringing the Box of Shells I intended for Mr Seymer home & leaving with him that which was for my self, it is inexcusable."[14] Her disarmingly apologetic tone presumably had an ameliorating effect on the tensions and anxieties produced by the differences in the social standing of these collectors.

This network of exchange, created and sustained primarily by gifts and loans of shells, was also the mechanism by which other exchanges occurred. Not only would gifts of shells make the journey from the duchess's residence in London to Dorset, but she would also send venison and other game to Pulteney from Bulstrode. For instance, in August 1771 she sent him half of a buck and announced that she hoped to see him while she was in Weymouth. Pulteney was not a landowner, and had no access to game other than as gifts from the landed classes. Gifts of game were thus reminders of hierarchical social relations and signified a particular kind of paternalism, one that was simultaneously kind and condescending. They signaled that the benefactors' net of largesse had been thrown over the recipients, enclosing them in a dynamic of gratitude and deference, and yet feelings of warmth and genuine concern were also transmitted with these gifts, perhaps mitigating the more condescending aspects of this paternalistic dynamic.

The relationship between the duchess and Pulteney exemplifies that special kind of friendship that existed between naturalists, a friendship based on equality of minds and the free exchange of ideas, specimens, and publications, free in the sense of unencumbered by social restraints arising out of notions of difference based on rank. Such pleasant interactions between shell collectors, however, were not always the norm. Social rank often did play a role in shaping interactions, dividing individuals from each other and causing ill will when the "great" took advantage of their social position to ignore the requests and gifts of those lower down on the social ladder, or when those below took the generosity of their social "betters" for granted. Complicating scientific exchange was the fact that, unlike Findlen's Italian naturalists who were members of the "educated elite," British shell networks included not only aristocratic amateur naturalists and professional scientific experts, but also dealers and professional specimen hunters, as well as those whose work brought them into contact with shells – gardeners, sailors, and fishermen. Even among professed equals, trouble often erupted around the exchange of gifts and the notion of equity in exchange. Maneuvering within the world of shell collectors in late eighteenth-century Britain required an acute awareness of social protocols and a firm grasp of the implicit social hierarchies that dictated who communicated directly with whom and how.

Pulteney was someone to whom everyone could write – those above him on the social scale, like the duchess and Joseph Banks, and those below, like Emanuel Mendes da Costa, of whom more will be said below. Pulteney could take the duchess's requests for specific shells and send them out to people with whom she did not correspond. In addition, he would coach other amateur naturalists in the proper protocol of address, as well as aid those too afraid to approach the duchess directly. Henry Seymer, Pulteney's friend and fellow Dorset resident, asked him for advice on addressing a letter to the duchess, because he was "not used to correspond with such great folks" and was worried about "how to begin & end the letter."[15] Seymer was so nervous about meeting the duchess that he told Pulteney he had second thoughts about the proposed visit. Despite these trepidations, a meeting was arranged and went off well. After the duchess had visited Seymer in Dorset, she wrote to Mary Delany, expressing her delight with the meeting:

> I returned last Saturday from Mr Seymer's at Hanford, he inquired much after you; he and Mrs Seymer expressed such concern that they had not the pleasure of being introduced to you till a day or two before they left Bath. They are exceedingly good sort of people, very obliging and good-humoured, and he is as generous as a prince, has given me fosils and butterflys without end, and I hope to improve by his instructions.[16]

Seymer's gift of butterflies and fossils was the first of many gifts that the duchess received from him; these were eventually reciprocated, as will be seen. Despite this and several further visits to Seymer at his Dorset home, the duchess tended to communicate with him mostly through Pulteney. Although she must have sent specimens accompanied by notes directly to Seymer, she also wrote letters to Pulteney asking him to convey her regards and thanks to Seymer, and whatever information she had about various classification problems.[17]

> I return you Mr Seymer's Letter & beg you will be so good to make my best compts to him & that I shall be very glad of waiting on him at Hanford & that if my Health

> & business will permit I flatter my self that sometime next month I may have the pleasure of seeing him & you. The Cone I have return'd[.] it is seldom to be found in perfection[.] I have not one better than Mr Seymer's I shall be extreamly obliged to him for any duplicates he can spare of the Hampshire fossils & beg my bst compts to him.[18]

She may have borrowed this cone shell because she needed to see a perfect specimen for classification purposes. Seymer was known as a collector who was interested in having only perfect specimens, while Pulteney was satisfied with anything that would enable him to carry out his project of classifying Dorset shells.

Loaning specimens was not as simple a transaction as one might assume. Sometimes a shell could become merged in another collection, the collector forgetting that it had been a loan rather than a gift; for example, Thomas Pennant gave the duchess a shell that had been loaned to him by William Hudson, who then asked for its return (see Chapter Five). Pulteney's experience was even more distressing. He loaned the duchess several rare shells, but ended up losing them because she died while they were in her possession. He had no means of retrieving them, suffering in silence when he saw them advertised in the auction catalogue, and learning later that they had been sold. In 1800, having long outlived the duchess and other members of this shell network, Pulteney described to George Montagu, author of *British Testacea* (1803), what had happened to his shells: "unfortunately there were many English Shells of mine in the hands of da Costa & of the late Duchess of Portland as well as of Mr. Ingham Foster at the Time of their Decease not one of wch could I secure."[19] Montagu had asked for the loan of shells so that they could be drawn for the engraved illustrations in his book, and Pulteney needed to justify his hesitant response. He explained that he had done most of his collecting more than twenty years earlier, and was now too old to replace the missing specimens, needing them for his own publication, on the natural history of Dorset:

> you will not be surprised Sir [when] I confess that I feel great reluctance at the Thought of exhausting my small Cabinet further especially as I must make a point of preserving such shells as are entered in my List of Dorset [illegible] . . . no hope of replacing them at my Time of Life.[20]

Loans were sites of potential misunderstandings, as Pulteney and Pennant knew all too well. More troublesome, however, was the exchange of duplicates, because equivalence, based on some loose notion of parity between specimens, was difficult to determine.

PROTOCOLS OF EXCHANGE

The tensions, complaints, and disappointments that natural history collectors expressed when dealing with each other as they exchanged specimens, books, and information can reveal much about the way in which networks operated. They can also help to ascertain the degree to which collaboration and exchange succeeded in producing natural knowledge.[21] Accusations of misbehavior, particularly in the form of inequity of exchange, negligence as a correspondent, and even the imposition of dubious specimens,[22] point to structural problems within this network of naturalists, natural history collectors, and dealers, exposing the differing regimes of value at work within the community, where specimens were simultaneously objects of disinterested scientific curiosity, passionate possession, and potential profit.[23] What also emerges in complaints about the conduct of collectors are questions about fairness and equity, questions that reveal anxiety and hostility over the disparity between the reality of hierarchical social relations, as in patron–client ties, and the idealized horizontal ties between professed equals, both of which were complicated by the commercialized forms of exchange.

Seymer's correspondence with Pulteney reveals an acute awareness of the problems that could arise when collectors exchanged specimens. Though he was clearly fond of Pulteney and quite generous to him, like the duchess sending him gifts of game – "the Pigeons & Fish were caught this morning, & I hope will be acceptable"[24] – he would often joke about the anxieties and jealousies that were generated when shells circulated as loans to help with classification. Mistakes could easily occur, and loans absorbed into a borrower's collection. Seymer admitted as much: "I found, as you expected, some of the shells You was so kind to give me, among mine."[25] In a note that accompanied some shells that he was returning to Pulteney, he wrote:

> Now Sir give me leave to scold a little at Yr not taking notice of my not returning, with the last, N° 26 which I found next morning under my table, & I have returnd with these; I know you must miss it, but hope You could not think me so mean as to smuggle the specimen; I should be very unhappy if I thought my friend I valued could entertain such an Idea of me. I have sent two or three of the sorts You wanted, & shall be glad to increase Yr collection when convenient opportunities offer.[26]

Seymer here teases his friend, accusing Pulteney in jest of suspecting him of stealing – "smuggling" – one of his loaned specimens.

Seymer was adamant that exchanges involving one's duplicates or unique specimens should be mutually beneficial and based on parity. He sent just such a proposition to Pulteney:

> Before we finish our negotiation for the present; give me leave to offer one more exchange, which will be agreeable to me, if it proves equally so to you; to this end I have sent two fine specimens of the Spondylus, wch cost me a guinea each, either of these, You chuse, I shall be glad to exchange for Y^r [illegible name of a shell] however don't let any particular regard for me influence you to do what is disagreeable to Yrself.

Seymer's emphasis on abstract notions of equivalence, along with his invocation of price as a measure of value, is linked to his refusal to let sentimental ties of friendship affect the exchange, telling Pulteney to look to his own interest and disregard their friendship in this "negotiation." He wrote similarly of another proposed swap:

> I now send You a specimen of the *Trochus solaris* much larger than that you sent, too which I beg Yr permission to exchange and For this reason only because it fits exactly the cells in my upper drawer, which this I now send is too large for & I cant well do without one of them among my smalls shells, the exchange I think can't be to Yr disadvantage tho, the last was if however You do not approve of it, I can easily return Yrs again.²⁷

His language here is derived from the concept of the contract — a mutually agreed-upon exchange based on both parties' self-interest — and he rejected the affect-laden language involving gratitude, deference, and obligation that accompanied patronage's hierarchical exchanges. He even disapproved of Pulteney giving him shells:

> I am much oblige'd to You for Yr kind offer of yr shells sent yesterday, most of which however as I know You have not duplicates, I cannot mortify You so much as to accept, they be of much more consequence, & properer in Yr collection than mine. . . . I know I should not without regret part with a Unique, & as I know You are full as fond of these subjects as myself, judge You would feel what I should on the same occasion.

Here he is refusing to take any of Pulteney's single shell specimens, and insists on taking only duplicates, all of which was consonant with his belief that exchange should be productive of mutual satisfaction:

> I had just such a friendly offer last week from Mr Foster of some curious New Zealanders, but declin[']d it for the same reason, tho I would give any money for the subjects, or be glad to make exchanges, if it could possibly be done to both our satisfactions —.

Seymer concludes this negotiation by announcing that he has "taken yr helix, decollata, & Ampullacea, as I know You have a duplicate remaining, of each," thus maintaining his code of exchange.[28]

Preferring to purchase or trade shells, Henry Seymer was quite vocal in his letters about the troubles generated by gift exchanges, especially those involving the duchess. Gift giving, in general, is complicated by expectations of return, reciprocity, and equivalence, and for Seymer, who thought in terms of parity of exchange, it could prove problematic.[29] At their first meeting, he showered the duchess with gifts of natural history specimens – shells, fossils, and insects – hoping that they would be reciprocated, and she was, indeed, delighted, writing: "he is as generous as a prince, has given me fossils and butterflys without end." These gifts were entirely strategic – a kind of priming of the pump of generosity – and he fully expected that she would reciprocate with some of the rare South Pacific shells in her enormous collections. He wrote to Pulteney while eagerly awaiting the duchess's promised return of a packet of shells:

> I had a letter not long since from ye [the] Ds, who desir'd me to give her compt [compliments] to You, & to say that she had not forgot her promise, relating to some more shells, I hope it wont be long before she settles in town, & begins looking over the Catalogue I troubled her with which she has kindly promis'd to do attentively & to send me whatever duplicates she can possibly spare, she was in raptures with a little fossil & two or three new flys I lately sent her.[30]

Six months later, Seymer told Pulteney that he had received a letter from the duchess: "she has not forgot her promise of adding to Yr collection of shells; I have had about 20 from her lately, but not four worth two pence, not one new: but she promises to send more soon."[31] He again wrote to Pulteney a month later, quite upset by what he saw as her failure to reciprocate:

> I confess I am disappointed, because I could not think it possible, that a person who had so very great a collection, & who had seen mine, could suppose that such trifling things would be worth my acceptance, especially as she knew I had seen, those she sent You, which were infinitely superior, but the scripture says there is no confidence to be plac'd in Princes, & let me add Dss. [S]he took from me ye [the] labour of some years, in the Insects, & I must own, as I knew her ability, I did expect a better return, but there is nothing like experience; & I shall be more cautious how I part with things *to strangers* for the future; I was told indeed, by more than one, that this would be the case but I could not believe it. I find the pocket is the only friend one can depend on

in most cases, but I want opportunitys of purchasing, & am unfortunately situated. [S]oon however I expect a little cargo from ye [the] East Indies, if it scapes ye [the] Custom house Officers, which I have good reason to hope it has; I shall follow Yr maxim, Expect nothing & You shall not be disappointed.[32]

Seymer's lavish gifts of specimens and the duchess's paltry return could be interpreted as a betrayal of trust (certainly this is how he saw her actions initially); but her failure to return in kind could also be read as signaling her willingness to be indebted to him, a position from which Seymer would benefit.[33]

Less tangible gifts also circulated in the network from the duchess's residences in London and Buckinghamshire to Dorset, where Pulteney and Seymer lived – gifts that took the form of favors and the exercise of influence by powerful people. Seymer made use of his relationship with the duchess to request her assistance in acquiring a position in the East India Company for his younger son. Her circle of influence included her son, William Cavendish-Bentinck, third duke of Portland, who had political ties and influence of his own; she wrote to him, asking him to intercede with the directors of the Company on behalf of Seymer, a man "from whom I have received a great many civilities."[34] This network was the means by which the younger Seymer acquired a position in India, but it was sadly also the means by which the bad news of his untimely death was circulated. The duchess received a letter from General Clavering at Fort William in Calcutta announcing Edward Seymer's death from illness and requesting that she convey the news to Mr. Seymer. She then sent Clavering's letter with her own to Pulteney, asking him to go to Seymer's home to break the news face to face and to convey her "great concern": "Knowing the regard you have for Mr Seymer & Mrs Seymer & the whole family made me apply to you as the properest person to acquaint them of the unhappy event (if they are as yet ignorant of it)."[35] Pulteney responded that the Seymers had already received the bad news about the loss of their son, and had asked him to convey to the duchess their "Thanks . . . for your Grace's kind application to the General & all your other kind Endeavours to promote this young gentleman's interest."[36] Although her efforts to promote young Seymer's interests were aborted by his death, she had acknowledged her obligation to the senior Seymer, demonstrating, despite his premature condemnation of "Princes," the strength and complexity of these relationships.

Seymer's habits of accumulation and exchange were based on commercial notions of equivalence, a kind of thinking that could abstract from one object its value in monetary terms and then apply it to another. When Seymer said, for instance, that the insects and fossils he had given to the duchess on their first meeting was "the labour of some years" and that he had expected "a better return," he was using the language of commerce and financial investment, and was clearly more comfortable buying specimens than participating in the more complicated gift exchange system, finding its levels of indebtedness and obligation frustrating and sometimes baffling. Though a landed gentleman, he was comfortable with the commercial model and was content to derive value from the abstracting power of money. The duchess, on the other hand, operated in a different economic register, one that was hierarchical in structure and relied on a dynamic of indebtedness. Early in their relationship, Seymer had been unable to fathom why she did not reciprocate in kind, and read this as a sign of deficiency and negligence. His system of exchange demanded equivalence, that objects of equal value be exchanged within a certain time frame. Moreover, initially he could not see why the duchess was allowing herself to be indebted to him and what that meant. Was she simply being lazy and selfish, as Seymer had suggested? I would like to consider an alternative explanation and to suggest that the duchess's failure to reciprocate in kind signaled her willingness to be indebted to Seymer, and that this indebtedness could be interpreted as an acknowledgment of her friendship, and I mean friendship in the eighteenth-century sense of the word – not a horizontal tie between equals but an enduring hierarchical relation. It could be interpreted as a gift to him: initiating a relationship that was going to be long term, as well as steeped in social and affective ties, which, indeed, turned out to be the case.

At its highest level shell collecting required social skills that could navigate complex practices, whether bidding at auction or gift giving within patronage circles. Because shell exchanges operated within differing regimes of value – as gifts, as commodities, as loans or trades – the methods for acquiring them were multiple, complex, and contradictory. Seymer's initial frustration and confusion over the duchess's method of exchange is a measure of the complexity of this culture of collecting. Early in their relationship, he had mistakenly thought that shell exchanges could be rendered more convenient and transparent in a commercial context, and he voiced a preference for the "pocket" – buying specimens outright – rather than waiting for the duchess to reciprocate. He could not anticipate the complexity of the duchess's engagement in the circuits of exchange that comprehended both the commodity and the gift.

FLEXIBLE HIERARCHIES

The duchess's correspondence conveys the complexity of interactions within this network of exchange, in particular its hierarchical nature and the way in which certain figures acted as nodes. The Pulteney–Seymer–Portland exchange, however, was a very small part of a much larger network that extended to a wide range of people interested in natural history, and shells in particular. As has been seen, luminaries, such as Sir Joseph Banks and Daniel Solander were actively engaged in the exchange of shells and information with the duchess, and these men, along with Seymer, were, in turn, involved with natural history dealers, men such as George Humphrey, Jacob Forster, and Ingham Foster, as well as with professional naturalists who also dealt in specimens, such as Johann Reinhold Forster and Emanuel Mendes da Costa. It was the acquisition and classification of shells that drove these men and women to forge relationships with people with whom they would not normally communicate.

The diversity of social rank involved in the exchange of shells is demonstrated by Emanuel Mendes da Costa (1717–1791), an intriguing figure who, for obvious and less obvious reasons, occupied a liminal position within the shell network. A Fellow of the Royal Society from 1747, he was elected to its clerkship in February 1763, but was dismissed for embezzling funds in December 1767 and imprisoned in the King's Bench until the end of 1772. One of his duties as clerk was to collect members' dues, and apparently he borrowed from this fund of ready cash to buy natural history specimens and costly books. He was the author of *A Natural History of Fossils* (1757), *Elements of Conchology* (1776), and *Historia Naturalis Testaceorum Britanniae; or, The British Conchology* (1778). The duchess of Portland was a subscriber to *A Natural History of Fossils* and *The British Conchology*, and thought well enough of his work to send the latter as a gift to J. T. Swainson.[37] She read his books carefully and wrote a few words about him to Pulteney, critiquing the accuracy of some of his shell descriptions, but on the whole approved of his work on British shells, a topic in which only a few naturalists and collectors took an interest. Because exotic shells, especially those from the South Pacific, took everyone's fancy in the 1770s and 1780s, da Costa was worried that his work on native shells would meet with ridicule. His correspondence with Pulteney is peppered with references to the duchess's British shells, some of which were very rare and of great interest to him.

Da Costa was very eager to maintain ties with Pulteney, who held a central position within the network of shell exchange and was a known correspondent of the duchess and Sir Joseph Banks. One of the ways da Costa did this was to offer to procure shells for him, since he lived in London and had easy access to other natural history dealers. Apparently responding to Pulteney's request for the purchase of shells, da Costa wrote:

> On reviewing that said list [of shells], I find that the greater part are either extremely rare or so very high priced . . . that it is not in my power to get you any. [H]owever I shall yet strive to attain some of the others not so rare, and if I am able, shall send you what you can procure.

To make up for his failure to fulfill the order, da Costa offered Pulteney a gift of South Pacific shells:

> However, to demonstrate to you how earnest I am to enjoy your friendship, I shall next week either in a parcel for Mr Sollers from Mr Wilkies, or by a Gentleman and lady who pass through your town to Exeter, send you a few shells, some whereof will be new species from the South Seas, which [I] hope will give you pleasure; & I pray accept them as a tender of my respect for you.[38]

But apparently these shells failed to satisfy Pulteney, who wrote: "not quite the things desired,"[39] though da Costa tried to defend his gift:

> N° 2 is quite a new species of muscle brought home by the latter navigators from New Holland, & unknown to Banks of the first Navigators – these latter navigators brought about eleven new species, undiscover'd by the former ones. I know scarce any Musea except the duchess of Portland that has any Quantity, & None that have a series of English shells.

He persevered in his attempts to please Pulteney with gifts of interesting shells: "depend upon it . . . whatever shells come into my hands I think worthy of your acceptance, I will send you, & will not be dismay'd by this my first attempt."[40] And apparently he was successful in his endeavors, for Pulteney later wrote: "The rest of the Shells I am thankful to you for as they were in some instances better than my former specimens & therefore illustrate the specific distinctions better."[41]

The above exchange is typical of the Pulteney–da Costa correspondence, dominated by a discussion of how to procure various shells. Most of the requests came from Pulteney, and da Costa seems to have been happy to supply his needs, though on occasion da Costa asked him for Dorset shells, something that Pulteney could provide relatively easily. Da Costa's tone is always grateful, even excessively so: "I would with very great pleasure repay any expence to purchase them [single-valved shells] of the fishermen, or employ one to get some." There is an air of obsequiousness in his prose that reinforces the implicit hierarchy at work within this exchange: the movement of gifts and favors going up to Pulteney and condescending thanks being returned to da Costa. This correspondence cannot be characterized as a free-flowing exchange between equals; one senses that da Costa was tolerated as a correspondent because he was knowledgeable and useful.[42] What he got out of this relationship was to be the correspondent of a well-connected naturalist, with whom he could engage in natural history inquiry and convey his enthusiasm for British conchology, even though he was kept politely at a distance.

An obvious explanation for da Costa's obsequious tone and Pulteney's less than warm manner can be found in da Costa's sentence to the King's Bench Prison for debt; a less apparent explanation might be found in his outsider status as a person of Portuguese Jewish descent. The list of subscribers for his *British Conchology* contains a range of people including Seymer, Pulteney, the dowager duchess of Portland, the dealer George Humphrey, and professional naturalists and doctors, such as Thomas Pennant, George Walker, and John Hunter. Interestingly, also included are names not found among Pulteney's, Solander's, or the duchess's correspondents: Isaac Siqueira, Phineas Serra, David Alvez Rabello, Daniel de Castro, Benjamin d'Aguilar, Moses Isaac Levi, and Joseph Salvador. Like da Costa, these men, with their Hebrew first names and Portuguese surnames, were men of science, amateur naturalists, and physicians who, perhaps because of their heritage or religious and ethnic background, did not circulate with ease among the educated elite in Britain.[43]

Nowhere is da Costa's obsequiousness and marginality more clearly expressed than in his correspondence with Solander, the taxonomist who accompanied Sir Joseph Banks on Cook's first circumnavigation. He wrote to Solander, congratulating him on his return from the voyage:

> As I am confined in the King's Bench I could not pay my respects to you on your safe arrival, with Mr Banks from the great Voyage you undertook for the promotion of Natural History. . . . I do not doubt but that your said Voyage will be the most Scientific and Illustrious Event that has ever happened to that useful and enchanting Study. It may justly be called the Argonautic Expedition for the Study of Nature.

After beginning the letter in this honorific language, he moves onto its real business, which is to ask for Solander's help. He explains that, while imprisoned, he gave six courses on fossils, each course consisting of twenty-seven lectures; this was possible because he occupied "an excellent commodious apartment." He hopes that Solander will recommend these lectures to his friends and reports that he also gave a course of lectures on shells, but did not repeat it since "most people are conversant in that part of Natural History."[44] Cook's voyages, however, would transform the study and collection of shells by introducing new and rare species into the process. In a way, this would alienate da Costa from the shell network even further, at least until the obsession with South Pacific shells ran out of steam in the late 1780s. Though during the 1770s da Costa's role within this network had pivoted on his ability to procure shells from dealers, collectors, and fellow conchological enthusiasts, by the 1780s, with the publication of his books on shell collecting and British shells, he had remade himself into an expert conchologist and was able to recover much of his lost reputation as a naturalist, to the extent that Dr. William Maton and the Rev. Thomas Rackett, authors of "An Historical Account of Testaceological Writers" (1803), could praise him for his contributions to the "natural history of our island."[45]

Disparities in social rank even greater than Pulteney and da Costa's were frequent among those involved in shell exchanges. Examples of fishermen and ordinary sailors contacting and being contacted by "great men" are common. For example, John Marra, a gunner on Cook's *Resolution*, wrote to Sir Joseph Banks:

> Begging Pardon for my Boldness. I take this Opportunity of Acquainting Your Honour of Our Arrival. After a long and Tedious Voyage. Having met with Extraordinary Good Success to the S. & Elsewhere. From many Strange Isles I have Procured Your Honour a few Curiosities As Good as Could be Expected from a Person of My Capacity. Together With a Small Assortment of Shells. Such as Was Esteem'd by pretend'd Judges of Shells. We have many Experimental Men For [*sic*] our Ship that pretend'd to know what was Never known. Not yet Never will be known I have Something Extraordinary to Relate to Your Honour. But – A good Opportunity will Soon Offer I hope. Depend upon it Sir I shall Take Special Care of Sending the Above Mention'd Articles. When in Order and an Opportunity Serves.[46]

Shells are commodities here, and though Marra does not announce that he has shells for sale, this is clearly implied. The roles played by dealers, auction houses, and lone entrepreneurs, such as fishermen and sailors, in the acquisition and circulation of specimens now made the shell network even larger and more diverse, no longer limited to the world of conchologists. Naturalists had to compete with specimen hunters who had no interest in the natural history of the shells, seeing them only as commodities. Johann Reinhold Forster and his son Georg, for instance, the naturalists on board the *Resolution*, often complained that they had to compete with the sailors for shells.

With the exception of Marra's letter to Sir Joseph Banks, the investigation into shell exchanges thus far has focused on those of a disinterested and non-commercial nature, governed by scientific curiosity and a passionate interest in natural history. The way in which the shells were initially acquired, however, complicates this picture: they were, in fact, commodities that were sold in the marketplace for a profit. In Pulteney's correspondence with da Costa, for example, the shells have prices affixed to them, and da Costa mentions the dealers who acted as intermediaries in this economic exchange. Dealers such as George Humphrey had showrooms devoted to shells and other natural history specimens, and smaller shops, such as Mr. Deard's on Pall Mall, sold a combination of toys, jewelry, minerals, fossils, and shells.[47] Auction houses, such as Christie's, Sotheby's, Langford's, and Skinner's, sold the natural history collections of both minor and major collectors. Natural history objects – shells, birds (alive and dead), plants (alive and dead), and artifacts, in particular weapons, from the past and distant lands – were big business.[48]

SHIP CAPTAINS AND AUCTIONS

The duchess's acquisition of Pacific shells, such as the rare white hammer oyster from Australia, needs to be situated within the general enthusiasm sparked by Cook's voyages among natural history collectors, and involves a recovery of the various routes by which Pacific shells moved within natural history networks. The voyages revitalized natural history by introducing botanical and zoological specimens new to Europe. Pulteney wrote to Banks about the impact of the first circumnavigation on his flagging interest in natural history:

> The Passion has been almost extinguished in me, except on particular occasions; among which, I need not say, was your happy and successful return to England, which I confess raised my Curiosity again to a degree that sometimes made me unhappy for want of the power of gratifying it to my wishes.[49]

Cook's voyages also inspired collectors with a kind of mania for what were known as "South Seas" shells. They competed with each other for the privilege of buying specimens from dealers, a few of whom had cornered the market on Cook's curiosities.

While Seymer and Pulteney struggled to acquire shells from the South Pacific, the duchess received them from a variety of sources, some directly from Captain Cook (pl. 38) and his officers, in particular Lieutenant Clerke, who was promoted to captain of the *Discovery* for the third voyage, and some from the naturalists on board, most likely Banks and Solander on the first voyage, and from the Forsters on the second.[50] In 1780, after hearing of Captain Cook's death, which was followed by that of Clerke a few months later, she wrote: "we have greatly regretted Capt Cooks [death] & now Capt Clarke who had made me many promises to collect shells for me poor Man! All those hopes are vanish'd."[51] She had shells from the *Endeavour* voyage (1768–71), particularly from New Zealand, mentioned frequently in letters to Pulteney. In 1771 she even offered to send him duplicates: "I beg to know if you have had any of the Shells out of the *Endeavour* as I hope to be able to procure a few for you if you have not got them

38 John Keyse Sherwin after Nathaniel Dance, *Captain James Cook, FRS*, line engraving, published 1784. © National Portrait Gallery, London

already." He must have replied that he had none, for she responded: "I shall send a little Box by the Blandford Coach of a very few New Zealand Shells & wish it had been in my power to have sent you more of them." A few months later, she apologized for not including more shells: "I return you many thanks for the favour of your Letter & wish it had been in my power to have sent you a greater number of Shells from the S. P. [South Pacific] but my friends did not fulfill their Engagements which is but too often the case."[52] Who exactly these disappointing friends were is not clear, but Banks and Solander are certainly candidates, and she did write later, in 1778, that Clerke had disappointed her: "I hope to be more successful in shells than I have hitherto been but what I hear of Mr Clerke's Collection little is to be expected from it in that way."[53]

She also cultivated relationships with other ship captains. "The Duchess," wrote Mary Delany,

> is at present very happy in the company of Captain Macnamara, captain of an East India man, the Rhoda; he brought her fine corals and is to bring her fine shells; the man seems to have no great judgment about them, and it would divert you to hear the Duchess tutoring him on the subject and coaxing him to bring us the treasures of the deep.[54]

As early as 1751 Horace Walpole had written to a mutual friend, Elizabeth Montagu, joking about the duchess's methods of procurement: "My evening yesterday was employed – how wisely do you think? In what grave occupation! In bawding for the Duchess of Portland, to procure her a scarlet spider from Admiral Boscawen."[55] By the 1760s the duchess had well-established ties to naval officers and commercial captains who sailed to the East and West Indies.[56] Commenting on this, Solander told his mentor Linnaeus that he had asked recently arrived ship captains for natural history specimens, but discovered that the supply of exotic specimens had already been allocated to the duchess:

> I intend to go down the Thames river to meet them [the West India fleet] and to go on board the ships to enquire if they have any natural curiosities; there is no other way of obtaining them here. As soon as they have been in town a couple of days all their curiosities are gone. I have already obtained the names of a couple of captains . . . who have insects. They are destined for the Duchess of Portland, but if I can select the few first, I shall not reproach myself because of that.[57]

He continued this practice – not of pilfering the duchess's specimens, but of going on board ship to inquire after specimens – as a curator of the British Museum, when he went on board the *Resolution* after it had docked at Woolwich. Whether the duchess of Portland would have gone down to the docks to inquire in person is unlikely given her rank, but a note she sent to Pulteney while visiting Margate raises this very possibility: "[H]ave you met with any shells lately? I have had no luck that way . . . several East India ships have gone by this place [Margate] but at too great a distance to inquire if they have any Shells on board."[58] One senses here her frustration and disappointment.

The duchess of Portland acquired shells as gifts from such prominent naturalists as Sir Joseph Banks, who visited her at Bulstrode in 1771 with Solander shortly after their return from Cook's first voyage; and she visited him at his Soho townhouse to see drawings produced by the voyage's professional artist Sydney Parkinson, who died on the way home in 1771.[59] Though she could command the attention of naval officers such as Cook, Clerke, and Boscawen, who gathered specimens for her on their travels, she did not operate on the same level as Sir Joseph Banks, and could not compete with his skilled procurement of specimens through patronage channels. Despite her wealth and her own network of natural history enthusiasts, she was not located at the center of the exchange nodes of institution-based scientific networks. As a woman, she was excluded from membership in such institutions as the Royal Society, and therefore from the informal privileges enjoyed by members. For example, Solander reported to Banks on the arrival of the *Resolution* and its full cargo of natural history specimens and artifacts collected by the Forsters, sending him a list of the intended recipients. This included the British Museum, the Royal Society, Sir Ashton Lever, Marmaduke Tunstall, and Banks, but made no mention of the duchess. As a result of her exclusion from the innermost circle of natural history exchange that had Banks at its center, the duchess of Portland often had to resort to the marketplace for her specimens.

In compensation for not being able to acquire all of her specimens through patronage networks, she made use of dealers and visited showrooms and shops that sold shells.[60] Delany describes an excursion that she made with the duchess to a Mr. Deard's shop in Pall Mall to see

> a curious collection of shells. There were ten small drawers full – the number of shells inconsiderable; not be called a collection, as many shells were wanted but the shells were perfect of their kind, and some rare sorts – and so they had need for the price set on them is three hundred pounds![61]

Although the duchess purchased shells from different dealers, for instance, in 1780 buying from Thomas Martyn a "larger Donax" and an "oriental Cockle" from Cook's third voyage,[62] she seems to have had a long-standing relationship with George Humphrey, who was a collector as well as a dealer. According to William Swainson, whose father was good friends with Humphrey, he was "the chief commercial conchologist in this country" and "had inherited immense collections both in conchology and mineralogy" from his father, who had also been a dealer. Swainson, who knew him from childhood, describes Humphrey as possessing

> a disposition the most amiable, and manners the most gentlemanly, yet unassuming, his company was sought for by all the great collectors and naturalists of his time,

such as the duchess of Portland, the earl of Tankerville, Dr. Fordyce, Mr. Jennings, &c. By these he was esteemed not merely for his scientific acquirements, but as an humble friend, with whom they could hold communion "sweet and large," on their favourite topics.[63]

Humphrey's name appears frequently in the duchess's correspondence, and he wrote frequently to both Seymer and Pulteney, mentioning her purchases. She owned at least one shell collected by the Forsters;[64] how she acquired it is uncertain, though one suspects Humphrey as the intermediary, for he claimed in a letter to Seymer that

> I have laid out with the people of the *Resolution* principally for Shells, near £150. Out of these I select the best of the most rare, containing the principal shells out of 20 different parcels, and charge at the rate of about £50 for them, that is fifteen pounds worth which I have sold the dutchess & those I have sent you.[65]

In 1780 Humphrey noted on a list of shells from the two ships of the third voyage, *Resolution* and *Discovery*, that the duchess had purchased from him a Buccinum for £3 3s. 0d.[66] His name appears in the duchess's note to Pulteney of 1784 from Margate, when she complained about the dearth of new shells: "Humphrey has brought me very few [shells]."[67]

The names of dealers and mentions of auctions dot the duchess's correspondence, underscoring that she could not rely entirely on the patronage system.[68] But even when she had to resort to commercial transactions to obtain specimens, she continued to use whatever influence her privileged social position afforded her to maintain an advantage in the competition. The following example is typical of the way in which she got what she wanted through a combination of unofficial channels and purchasing power. When James West, President of the Royal Society, died intestate, she was informed by a Mr. Hudson, who was an acquaintance of her son's, that West's widow was to administer his estate and was planning to auction off his immense collections of artificial and natural curiosities, classical statuary, coins, books, and fine art. Not wanting to wait, the duchess wrote to Mrs. West asking if she could purchase items before the auction. She wrote to her son, "I must beg my Dearest son will be so good as to inform me in what way I shou'd [satisfy?] Mr Hudson for his trouble," thanking him "for wishing me success in regard to the West Collection." She had not yet received an answer from West's widow and was worried about her ability to gain access to the collections before the auction: "I have some hopes but I fear there will be many competitors. There is nothing surprises me so much as his poverty." Whether she succeeded in purchasing anything prior to auction is not clear, but if not she may have attended the sale.[69]

The duchess did attend art auctions. Mary Delany, for example, mentions that the duchess had been to an auction of European paintings in 1758 and was upset that she had failed to acquire the *Dancing Children*, losing the bid to Sir James Lowther. Despite this setback, "our friend was not quite disappointed, for she got every lot she bid for of the first day's sale, except for the *Children*." She paid £32 for Rembrandt's *Boy's Head*, £76 for a Rubens landscape, and £105 for a Claude Lorrain.[70] These prices are high in comparison to the West auction in 1773, where drawings and paintings by Rembrandt, Poussin, and Rubens sold at much lower prices: a "picture of impure desire" by Rubens sold for £9 9s. 0d., a Rembrandt self-portrait and a portrait of his wife sold respectively for £5 5s. 0d. and £14 14s. 0d., and an Andrea del Sarto painting of the *Holy Family* sold for £10 4s. 0d.[71] West's collection of paintings was sold separately from his curiosities; the duchess was interested in both, since she collected paintings, prints, statuary, and artificial curiosities as well as natural history specimens.

As Robert Altick points out, the number and frequency of auctions featuring collections of artificial and natural curiosities increased as the eighteenth century progressed. "Disposing of private (noncommercial) collections became a thriving business in the course of the century. Virtuosi died, or tired of their toys, or suddenly needed ready cash, or went to the country or abroad, to the profit of both brokers and auctioneers."[72] Although most auctions of the collections of virtuosi consisted of items such as books and manuscripts, fine art, antiquarian and decorative objects, several also sold natural history specimens, and, in the last quarter of the century, the numbers of natural history collections sold at auction increased considerably. They doubled, even tripled, from one or two a year before 1771 to four to five annually by the end of the century, with 1775 and 1789 registering six auctions each, including the auctions in 1789 of Lightfoot's collection and the collection of Ingham Foster, Henry Seymer's agent and George Humphrey's friend and fellow broker.[73] This increase can be explained in part by the impact that Cook's voyages had on collectors' enthusiasm for natural history items, shells in particular. As Peter Dance argues,

> the three circumnavigations of the globe by Captain Cook's vessels were instrumental in shattering the Dutch monopoly in shells and many other commodities. From 1771 onwards European shell collections were enriched by specimens acquired by British voyagers and often the shells passed through a London auction house before ending up in a collector's cabinet.

In fact, not only did shells from Cook's voyages pass through London auction houses on their way into collectors' cabinets, but the same shells would also return to auction houses as collectors died and collections were disassembled, and thus would circulate several times through the auspices of auction houses. As Dance asserts, the period between 1771 and 1822, ending with the auction of the shell collection amassed by Mrs. William Bligh, wife of Captain Bligh of the *Bounty* who had also sailed with Cook, was "dominated by the constant reappearance at auction of material originally collected by men associated with the voyages of Captain Cook."[74]

ACQUIRING COOK'S SHELLS

Many natural history collectors thus acquired the especially valued shells from Cook's voyages through auctions and dealers. The provincial collector Henry Seymer, however, lived too far from London to participate regularly in the auction scene, and the question of how to go about acquiring Pacific shells was a much-discussed topic in his correspondence with Richard Pulteney. Seymer was pessimistic about his chances of acquiring them, complaining of his limited access to the best specimens that had been brought by the *Resolution*, Cook's ship on the second circumnavigation:

> Saturday last I rec[eived] a Catalogue of Shells, brought by the *Resolution* which are to be sold this week at Langfords; it consists of 480 Lots, not all shells, but Arms, ornaments, Utensils &c: of the Natives, intermixt; a Cargo which a dealer, one Jackson, bought at Portsmouth, & of which I dare say he will make 400 pr. Ct. [percent] tho I know most of the capital things have been dispos'd of some time [ago]. Cook and Forster I dare say secur[e]d the best before they came home, & will make a fine penny of them.[75]

His pessimism was justified to some extent, especially for collectors who lived far (or thought that they lived far) from the centers of exchange, whether from London, where dealers put up their precious cargoes for auction, or from port cities, such as Portsmouth and Greenwich, where dealers flocked to inspect ships' cargoes of natural history specimens.

Henry Seymer's desire for these "new" shells from the southern Pacific is palpable in his letters, as is his frustration at not being able to attend the auctions himself, having to rely on intermediaries. He felt that he would never get anything worth having this way, writing to Pulteney:

> I have desir'd Mr Forster [Ingham Foster], to lay out about 10 or 12 Gs [guineas] for me, in what he thinks new, particularizing the Sun shell, & a Cypraea . . . [T]here are

> but three of them in the Sale, & I have limited him to a Guinea, therefore fear shall not get it. There are 11 of the Sun, & I have allowed 3 Gs for a fine one. I don't think however Mr Foster will lay out much for me, as he purchased a Cargo at about 130 £ & promisd me a choice of his duplicates at a moderate price, so that he will buy none for me but such as are not found in his Cargo; he told me he had but one of ye [the] Yellow Cypraea. When they arrive will let You know that we may look them over together; tis impossible however for me ever to procure any capital specimens, this way, as Mr Foster has ingenuously told me that he shall reserve the 2 best of every species, & that I shall certainly have my choice of the rest, which I think is as much as any one ought to expect, or could reasonably desire from a collector.[76]

Another problem was that Seymer feared he would be charged unfairly for Mr. Foster's trouble. He wondered why it was taking so long for his shells to arrive from London, and assumed that it was because Foster was busy trying to figure out how much to charge him:

> I suppose they [shells] will be here some time next week, probably he has delayed sending them till he saw how they went at this Sale, in order to judge how to charge the particular subjects; his however are foul, as bought over, so that twill be difficult to judge of their perfection, therefore, unless they are very reasonable, tis probable I shall purchase but very few of them. They are called S. Sea shells in the Catalogue, but I know there are very many common things among them, wch, were pickd up at different places where they hapned to touch, but ignorant people, such as I am sure half the modern Collectors are, don't consider this at all.[77]

Seymer's insight into the valuation of shells contains an acute analysis of the impact of the marketplace on collecting, and his bitterness, though tempered with irony, is familiar to those, both then and now, who have bought items at auction or through dealers. Well aware of the risks of purchasing shells from people who claimed that their specimens had come from the Pacific, Seymer told Pulteney a story about a mutual friend, an entomologist, who had bought a box of shells as a present for him. Their friend's ignorance of this particular branch of natural history had made him vulnerable to imposition and fraud:

> A worthy friend of mine, was taken in this way lately; He knows nothing at all of shells, but had heard me express a desire of having some of these new ones, consequently he made enquirys, & at last found a Sailor who said he came home in the Resolution, & had some Cargo's of shells to dispose of, my friend pitched upon a small box, containing about 100, for which he payd only half a Guinea, thought

himself quite happy in getting them so very unreasonably. He sent them to me, but such rubbish was never seen, there was not one S. Sea among them, but they were all, the worst specimens of ye [the] most vulgar, worthless, things the West Indies produces, & I dare say had lain by, as unsaleable, in some dealers shop these 20 or more Years, the whole Cargo would not have fetched a shilling at any Auction, as I told my friend who insisted on my sending him my real opinion of them. Had they been Insects he would not have been thus impos'd on, for twas Mr Yeats, this shows how very simple it is to employ persons, in things they don't well understand. . . . You had better not begin this letter 'till Sunday, when, if it should be wet, or You not inclind to go to Church, You may pray at home, & use this by way of Sermon as tis nearly as long as the modern one, & perhaps to You as edifying as many of them.[78]

This cautionary tale about a sailor who may or may not have sailed with Cook, and who was able to dupe a gentleman collector who was an accomplished naturalist, testifies to the complex mingling of desire and ignorance that makes up in varying degrees these kinds of exchanges.

VALUATION

Not only did Seymer assume that Cook and Forster had made "a fine penny" on their natural history specimens, he also suspected that the dealer Jackson, who had gone to Portsmouth to meet the *Resolution*, had made 400 percent on buying and then selling the specimens to collectors. Dealers are mentioned in the Pulteney–Seymer–Portland correspondence frequently and almost always in the context of their high prices; even the duchess complained that she had bought a few shells from Humphrey "at no small price."[79] Seymer emerges as the most critical of Humphrey's prices, perhaps because he was the most dependent on him: "What few valuables I have are only single specimens purchas'd at a very high rate from Humphrey." Of the "new things brought home by Capt Cooke on the *Resolution*," he bought only two or three shells from Humphrey since "the prices were so extravagant & in my opinion so much beyond the value of the subjects." He joked that he was planning to purchase some *Endeavour* shells from another dealer, "& as it is not Humphrey I hope to have them at a reasonable price."[80] Humphrey's clients were also bothered by his lack of attention to Linnaean taxonomy (pl. 39). Seymer and the duchess voiced their irritation with his nomenclature, Seymer describing his sale catalogues as "unintelligible" and the duchess more politely (or ironically) saying: "I am not sufficiently acquainted with the names the Dealers give to Shells."[81]

A Catalogue of Shells chiefly from
the South Sea's sent down to Mr. Seymer
from Mr. Geo: Humphreys.

1775
Sepr. 18

				N°. sent	Price each.
	1.	Small white Buccinum.	new Caledonia	4	3
3.	2.	black brown Periwinkle	S. Seas	4	6
	4.	Common grey Nerites of the East with a yellow mouth	N. Caledonia	2	9
	5.	Small purple Snail	found at Sea	2	9
	6.	new variety of money Cowry	N. Caledonia	4	3
	7.	young Tortoiseshell Cowry	Otaheite		2. 6
	8.	mottled umbilicated Snail	N. Caledonia	2	2. 6
	9.	Small lipp'd Telescope	do	2	2. 6
	10.	Leper or pink mouthed poached Egg Cowry	new Amsterdam	2	8
	11.	White Buccinum	S. Seas		2. 6
	12.	Muricated Periwinkle young	N. Zealand	2	5
	13.	Old do. with Operculum	do		1. 1. 0
	14.	Beauty long Snail	do		10. 6
	15.	do	do		8
	16.	Muscle with greenish Spider	do		1
	17.	do	do		2
	18.	Striped brown Patella	S. Seas		2. 6
	19.	rare triangular Cockle	E. Indies *		8
	20.	White Tower of Babel	do *		6
	21.	Brown notched Snail	N. Zealand	3	2. 6
X	22.	Small banded long Snail	N. Caledonia	2	5. 3
	23.	Olive a new Species	Brasils *	2	5. 3
	24.	Brown & White Murex	S. Seas		15
X	25.	Hairy magellanic Buccinum vide D'avila	Cape Horn		1. 1. 0
	26.	Brown mottled basket Cockle	N. Caledonia		6
	27.	Round pink-edged Cockle young	do		6
	28.	Sharp Edged Egyptian Pyramid	do		10. 6
	29.	Ear Shell without Holes	Dutch E. Indies *		5
	30.	Brown striped spotted mouth 3 spots	N. Caledonia		6
	31.	do	do	4	2. 6

Despite these problems, collectors and naturalists were dependent on dealers and their ability to obtain shells from Cook's voyages.[82] Even Solander, in his role as curator, requested funds from the British Museum's board to purchase shells from Humphrey:

> A Sale of a large Collection of Natural & Artificial Curiosities will soon commence at Humphrey's in St. Martin's Lane, where several specimens which are wanted in the collection of this Museum. Dr Solander begs the Directors of the Committee if he may be permitted to make any purchases there & to what amount.

Two months later, Solander recorded in the museum's diary: "Bought at Humphreys Sale: Carrier Shell – £2: 2: 0; Hammer Oyster – 3: 16: 0."[83] The prices that dealers got in the 1770s for these Pacific shells were high when compared to other auctioned items. Seymer paid Humphrey five guineas for a *Mya truncata*, which seems to have been a lot of money when one considers that a Rembrandt self-portrait, a drawing, went for a similar price in the James West auction.

The purchase price of natural history specimens varied considerably during the two decades after Cook's first circumnavigation. Prices, high and low, were a constant concern because the process of valuation was precarious. It turned out that marketplace transactions were just as unstable and open to constant negotiation as gift giving within the vertical friendships of patronage. Even Seymer, despite his belief in the transparency of commodity exchange when mediated by money, recognized eventually that prices varied with the vagaries of the market. He noted that his agent Foster had delayed assigning prices to the shells he planned to sell him until he had found out how similar shells had sold in a recent auction.[84] Auction catalogues, especially annotated catalogues containing the final sale prices, were thus valuable to collectors trying to evaluate the worth of their own collections and making purchases of new specimens. Pulteney, aware that da Costa frequented natural history auctions in London, asked him if he could send him spare catalogues of recent auctions. Da Costa replied:

> In regards to the Catalogue of Sales of the Shells you are too late in your request for the Prices all those sales I attend, yet I make waste paper of the rest of the Catalogues, except that single one I have prices, which I keep so that I have not any clean catalogues to write the prices in, else could send you scores; but what I can do is, I will lend you all I have, & send them you down in a parcel by Mr Wilkes way. I will also lend you some Dutch catalogues, but we cannot get them priced . . . This I give you only as a temporary remedy, for hence forward, I will always price the catalogues, & present you with the duplicate.[85]

This practice of writing in the catalogue the prices of objects sold at auction was a way to record the fluctuating prices driven by collectors' desires, as well as an attempt to stabilize the valuation process, giving potential buyers the information they needed to compare prices.[86]

Although ostensibly derived from scientific inquiry and the need for taxonomic completeness, the value of natural history specimens was also determined by less "rational" forces. In addition to the love of novelty and rarity, fame also shaped value for collectors. Exotic shells that were associated with Cook's voyages had a certain value, but those that could claim a direct provenance from Cook himself were even more valuable. The impact of Cook's voyages on the imaginations and pocketbooks of natural history collectors is taken up in the next chapter, which describes the networks of exchange made up of Pacific islanders, naval officers, common seamen, naturalists, and dealers, and traces the paths that the shells took to the centers of calculation in London, which included not only the Royal Society and the British Museum, but also the duchess's Whitehall townhouse, where her shell collection was lodged.

CHAPTER FOUR

40 Shells from Sir Joseph Banks's Collection. London, Natural History Museum. © Natural History Museum, London

PACIFIC SHELLS, COOK'S VOYAGES, AND THE QUESTION OF VALUE

Collectors waited eagerly for Captain Cook's ships to return from the South Pacific, hoping to be the first to get the newest, rarest, and most exotic shells, birds, plants, and artifacts yet to be seen. The cargo from the first voyage (1768–71), carried by the *Endeavour*, gave them a taste of what might be available. Only a limited number of natural history specimens circulated beyond the control of Sir Joseph Banks, who had gathered specimens on this voyage for his own collections and those of a few select institutions such as the British Museum, although collecting shells was not high on his list of things to do while in the Pacific. His mollusk collection is still intact, the shells now stored at the Natural History Museum in London in a large cabinet with drawers filled with little tin boxes, a technique that Solander had learned from Linnaeus as a way to organize shells (pl. 40). Collectors, like the duchess, acquired some *Endeavour* shells through gifts from Cook, Lieutenant Clerke, and possibly Banks himself, while also purchasing others from dealers such as Humphrey, who had eagerly bought up what he could of these relatively scarce items. Because shells from the *Endeavour* were in such demand, collectors and dealers, who expected to make huge profits on the sale of the second voyage's natural history objects, looked forward eagerly to the return of the *Resolution*.

When the *Resolution*, Captain Cook's ship on the second voyage, arrived at Woolwich in 1775 after a three-year circumnavigation, the scene was one of festivity, celebration, and anticipation, as important personages flocked to the ship to see what exotic cargoes it held. Though it docked nearly a year after its companion ship, the *Adventure*, expecta-

41   John Hamilton Mortimer, *Captain James Cook, Sir Joseph Banks, Lord Sandwich, Dr. Daniel Solander, and Dr. John Hawksworth*, 1771(?), oil on canvas. Canberra, National Library of Australia PIC R10630 LOC Scr 105

tions ran very high as to the quality and quantity of its cargo of natural history specimens, largely because the *Resolution* carried the naturalists Johann Reinhold Forster and his son Georg on board. Daniel Solander went down to the docks to see for himself and to report to his patron, Sir Joseph Banks, on the ship's bounty of curiosities. Solander accompanied John Montagu, earl of Sandwich and First Lord of the Admiralty, and his large party as they were ferried down the Thames to the shipyards near Greenwich (pl. 41). Solander wrote to Banks:

> Our Expedition down to the Resolution, made yesterday quite a feast to all who were concerned – We set out early from the Tower . . . visited Deptford yard . . . and afterwards to Woolwich, where we took on board Miss Ray & Co. [Lord Sandwich's mistress] and then proceeded to the Galleons where we were welcomed on board of the *Resolution* – and Lord Sandwich made many of them quite happy.

Between the shipboard ceremonies and the dinner provided for the officers by Lord Sandwich in Woolwich, Solander had a short time to peek at "their curious collections":

Mr Clarke [Lieutenant Charles Clerke] shew'd me some drawings of Birds, made by a Midshipman, not bad, which I believe he intends for You [Banks]. I was told that Mr Anderson, one of the Surgeon's Mates, has made a good Botanical Collection, but I did not see him. There were on board 3 live Otaheite [Tahiti] Dogs; the ugliest & most stupid of all the Canine tribe. Forster had on board the following Livestock: a Springe Bock [antelope] from the Cape, a Surikate [meercat], two Eagles, & several small Birds all from the Cape. I believe he intends them for the Queen. . . . Pickersgill made the Ladies sick, by shewing them the New Zealand head, of which 2 or 3 slices were broiled and eat on board the Ship – It is preserved in spirits; and I propose to get it for Hunter [the anatomist], who goes down with me tomorrow on purpose, when we expect the Ship will be in Deptford.[1]

This brief glimpse of how the ship was stuffed with natural history specimens, both alive and dead, natural and artificial, captures the crew's excitement over the impact that they hoped their cargoes of curiosities would have – and did have – on London society in general and collectors in particular.

Ships' crews, from officers down to common seamen, all knew about the natural history business and all hoped to profit from collecting whatever they could lay their hands on, trade for, and transport back home. The collecting craze took hold of Cook's men, who all seem to have imbibed John Woodward's dictum: "neglect not anything; tho' the most ordinary and trivial: the commonest peble or Flint, Cockle or Oystershell, Grass, Moss, Fern, or Thistle will be as useful . . . as any the rarest production of the country."[2] Once back in England, they were eager to sell the items they had collected;[3] Solander reported to Banks that several seamen of the *Resolution* had let him know that their curiosities were for sale. In addition to gathering shells as natural curiosities and scientific specimens, the voyagers also collected artifacts that were fashioned from shells, such as fishhooks and armbands, as well as ones that contained shell matter, mostly in the form of mother of pearl, an important decorative element in the construction of such objects as statues, ceremonial dress, bracelets, and necklaces (pl. 42).[4] Shells that had been worked by Pacific islanders into objects of use and beauty showed up in dealers' showrooms and private museums, and were sold alongside shells as natural history specimens.

The practice of collecting natural history specimens with the aim of selling them or using them to secure emoluments was not limited to common seamen. Once back in Britain, Captain Cook, Lieutenants Clerke and Pickersgill, and the naturalists Johann and Georg Forster gave specimens and artifacts to important people and influential

42   Necklace made of shells, Hawai'i. From the Leverian Museum, possibly collected during Cook's third voyage. London, British Museum, OC1977,Q.4. © The Trustees of the British Museum

institutions. Cook, for instance, presented a Tahitian mourning costume to the British Museum, and the Forsters gave another such costume to the Ashmolean Museum of Oxford University.[5] The Forsters collected natural history specimens to sell and to give away as strategically placed gifts. In his report to Banks about the cargo of curiosities on the *Resolution*, Solander describes an encounter with Forster just after the ship arrived at Woolwich:

Mr Forster overwhelms me with civilities upon your account. . . . he then desir'd me to pick out of his insects two of each species, one for you and one for the Museum, which I did not think proper to refuse. He has very few indeed from the South Seas, but some very fine ones from the Cape. I believe I told you before, that in the rest you are to be the third sharer – 1st Br. Mus. [British Museum]; 2nd Roy Soc [the Royal Society]; 3rd Banks; 4th Tunstal; 5th Lever.[6]

(Tunstal and Lever were well-known collectors; Lever's museum was open to the public for a fee.) The Ashmolean Museum was a recipient of Pacific islanders' artifacts, and Michael Hoare, the elder Forster's biographer, speculates that they were given shortly after Forster received an honorary degree of Doctor of Civil Laws in November 1775.[7] With these well-placed gifts, Forster hoped to secure the interest and favor of great collectors, as well as to set himself up as an authority and expert, to whom they would defer in their need to obtain and classify specimens.[8]

As was seen in the last chapter, shells from Captain James Cook's voyages were especially popular with serious collectors, who as connoisseurs had large collections of beautiful and rare shells; with naturalists eager to see specimens unknown in European scientific circles; and with brokers who hoped to profit from buying and selling shells associated with Cook's voyages. This chapter takes Cook's voyages as its central concern, exploring their role in shaping the interests and desires of natural history collectors, and examining the ways in which Cook's status as a celebrity added value to the objects collected. It also describes in detail the methods by which Pacific shells were gathered, with the goal of reconstructing the movement of shells from the Pacific to the London showrooms of natural history dealers and the cabinets of collectors.[9] The official published accounts, combined with ships' logs and the journals of individuals, can tell us much about the places where the shells were found and the uses to which they were put both by the voyagers themselves and by Pacific islanders. The written record for the natural history collecting activities of common sailors is slight, so we have to rely on the surviving journals, letters, and logs of officers, naturalists, and other "Experimental Gentlemen."[10] Helpful in filling out these narratives are the journals from French voyages of discovery – Bougainville's (1766–69) and D'Entrecasteaux's (1791–93) – and by those who belonged to the First Fleet (1787–88), the ships that brought the first convicts to the newly formed penal colonies in Australia. These narratives, combined with the instructions written by dealers on how to gather and transport natural history specimens, provide insights into the material practices by which Pacific shells were turned into scientific specimens and potentially lucrative commodities.

INSTRUCTIONS FROM DEALERS

Dealers, driven by market demands for exotic shells and eager to obtain specimens from the southern oceans and tropical seas, wrote instructions on how to collect and preserve natural history objects for those who voyaged to these regions. Humphrey wrote a set of instructions entitled "Collecting and Preserving all Kinds of Natural Curiosities" (1776), which provided guidelines on how to clean, store, and transport specimens. He suggested that shells with the animal in them "may be preserved in spirits," but if the animal was dead or did not merit preservation, the shell could be submerged in scalding water, the creature removed, and then the shell dried carefully. Humphrey's pamphlet urges that "shells must never be placed in the heat of the sun . . . otherwise the colours will fade & the skin with which many of them are covered (and which must on no acct. be taken off) will peel and fly off." He also recommended dipping the sharp tips and points of shells in melted beeswax to prevent damage during transport, and provided details on how to pack small, delicate shells:

> A good way to preserve the small shells is to pack them inside the very large ones, which will also save room. . . . Such shells as are tender should be packed in cotton in small boxes. They should be afterwards packed together . . . in strong boxes which should be close filled up to prevent their shaking & closely nailed down, so to remain till they arrive in England.[11]

J. R. Forster wrote "Short Directions for Lovers and Promoters of Natural History" (1771), attaching it to his published catalogue of North American animals to advise readers how to collect, preserve and transport animal and plant specimens. Of shells, he counseled:

> The shells, both those found in fresh water-lakes, ponds, and rivers and those that live only in the ocean, must not be chosen among those that lie on the shores of the sea and fresh waters, and have been broken and injured, or rolled by the waves and exposed to the air and sun and thus calcined; but rather as fresh as possible, and with the animal in it: one or two specimens of which may be preserved in Spirits: from the rest extract the animal, and keep the shell, when perfectly dry and sweet, packed up in cotton, tow, or moss. The same is to be done with the echini or sea-eggs, and other crustaceous animals; especially be careful to preserve their curious spines.[12]

In the early part of the nineteenth century John Mawe, a specimen hunter turned dealer, published a little pamphlet, "A Short Treatise" (1804), on collecting techniques addressed to "Gentlemen Visiting the South Seas, and all Foreign Countries; more particularly Commanders, &c. of Ships, and Gentlemen residing on Shore, with a view to encourage the collecting of Natural History." Even though the treatise was published more than twenty years after Cook's voyages, the description of how to gather shells is more detailed than in the pamphlets of Mawe's predecessors and can shed light on how this was actually done. Like Humphrey and da Costa, Mawe urged the use of a dredge because the "best shells" are those that contain the living animal, and these can be found "in shallow water, before the Birds have seen them" and "out in deep Water," where, in addition to using a dredge, he recommended hiring "People diving for them."[13] He published a much more elaborate guide in 1821, *The Voyager's Companion; or, Shell Collector's Pilot*, which dispensed advice on where to look for specimens, who to deal with locally, and how to store shells properly for the journey home. A world traveler, with "fifteen years . . . at sea," and an expert specimen hunter, Mawe drew upon his own experiences, particularly his time spent in Brazil, to suggest that travelers employ local people with regional expertise. He recommends to his reader that

> whenever opportunity occurs, as ships loading, refreshing, &c. to employ the fishermen on the coast to collect for him; these men are well acquainted with the places where shells may be found, and for a trifling remuneration would gather a supply, which, on his return home, might gratify his friends, or otherwise be turned to advantage.

He encouraged readers to find "an expert negro,"[14] whose knowledge of the interior would be crucial to the collecting of land snails, as the following narrative testifies:

> At the Royal Farm, Santa Cruz, about forty miles from Rio, where I held an official situation, (first administrator), I directed some expert negroes to pick up what snail shells and curious animals they might meet with: these they left at my house as they passed, and, by allowing them a small compensation, I obtained many fine shells, insects, birds, reptiles, and small animals of the monkey, ape, and hedge-hog species.[15]

Mawe also acknowledged another overlooked but talented group of specimen hunters, ships' cooks and cabin boys: it is "only the boys or the cook, who notice these *rarities*, and who made a few pounds by them every voyage." The "finest lot" of Australian shells he ever sold as a dealer had been

> gathered by two boys in Western Port. – A whaler off the coast sent a boat on shore to search for fresh provisions, as birds, animals, &c. whilst the crew were shooting,

the boat grounded among stones and weeds, and during the time before she floated, the boys left in charge of her, employed themselves in gathering the shells entangled in the weeds and about the stones.

Mawe tells how he paid the price the boys asked and gave them each a guinea on top, "to stimulate them to look out for shells on another voyage." He urged travelers to continue to collect shells from the Sandwich Islands (Hawai'i), the Marquesas, New Zealand, and Australia, all locations "from whence we have *many* beautiful and rare shells, chiefly collected by Circumnavigators," promising to "amply remunerate the trouble of collecting."[16]

HEAPS OF SHELLS

European voyagers encountered shells in a variety of ways, primarily as useful and decorative objects, which the South Pacific islanders crafted with great skill and aesthetic flair, but also as a form of food, as in oysters and mussels, which the voyagers observed native peoples eating and ate themselves.[17] The phrase "Heaps of shells," much repeated in voyagers' accounts, refers to the piles of empty shells left over from the meals of mussels, clams, and oysters that islanders prepared on the beach, roasting them over a fire or cooking them on hot stones. Thomas Edgar, master of the *Discovery* on Cook's third voyage (1776–80), noted in his journal that the people of Adventure Bay (Tasmania): "sometimes make their abode by the Sea Side and feed on Muscles, Scollops and other Shellfish which may be very easily got on the Rocks at low Water – for we found many remains of Fires with heeps of Shells above high Water mark."[18] Lieutenant Clerke mentioned in his log of the second voyage that shellfish were eaten in New Caledonia: "I believe their chief subsistence is Fish . . . for they pay great attention to the ebbing of the Tides – are very busy at low water, and I observe, always get abundance of Shell Fish, with which they frequently regale themselves upon the Beach, as soon as they've caught them."[19] J. R. Forster confirmed this observation, though with less detail and stylistic flair, when he remarked in his journal: "They eat a great many Shelfish which they constantly collect in their reefs."[20]

On one occasion Cook interpreted these piles of shells as a sign of a "primitive" society without the tools, means, and know-how to transform raw nature into complex cultural products. In keeping with Enlightenment theories about civilization and its stadial progress from hunter-gatherer societies to more complex social structures involving commerce and manufacture, he read a narrative of underdevelopment into the piles of shells he saw on the shores of southern Patagonia, declaring in his *Endeavour* journal:

> . . . these People must be a very hardy race; they live chiefly on shell fish such as Muscles, which they gather from off the rock, along the sea-shore and this seems to be the work of the Women; their Arms and Bows are neatly made. . . . [F]ew either men or Women are without a necklace or string of Beeds made of small Shells or bones about their necks . . . we could not discover that they had any head or chief, or form of Government, neither have they any usefull or necessary Utentials except it be a Bagg or Basket to gather their Muscles into: in a Word they are perhaps as miserable a set of People as are this day upon Earth.[21]

Cook was not alone in making the connection between eating shellfish and savagery. Jacques Labillardière, author of *An Account of the Voyage in Search of La Pérouse* (1800), the published narrative of D'Entrecasteaux's voyage, interpreted the heaps of shells he saw along the coast of Tasmania as a sign of a less-than-civilized population: "These scattered huts indicated a very scanty population; and the heaps of shells which we found near the sea-shore, shewed that these savages derive their principal means of subsistence from the shell-fish which they find there."[22] Such interpretations suggest that oysters, mussels, and clams were not valued highly by the visiting Europeans as a form of food, not because they were not nutritious, but because they could be gathered and eaten raw immediately, requiring little human intervention to turn them into food.[23] Thus the "heaps of shells" indicated to European travelers that the native populations who consumed shellfish in this manner were less civilized than themselves.

Cook's attempt at anthropological writing has been criticized by the anthropologist Nicholas Thomas for the way in which he extrapolated from what he saw to generate a theory about the Haush people of Patagonia:

> Is it too obvious to say that this is the result of observation, not communication? There is a fundamental difference between seeing people and hearing from them . . . So far as appearances went, perhaps he could not reasonably have been expected to grasp much more than he did. He was right that they ate mussels, but wrong to extrapolate from what they were eating at this particular time to the proposition that they subsisted "chiefly" on shellfish. He did not consider that they might move seasonally, that they might in fact eat a great variety of plants, birds and animals at different times. He saw them as wretched scavengers and imagined that their menu would be limited.[24]

Cook also assumed that their lack of utensils – tools and other useful objects that would enable them to master their environment – made the people "miserable," for without tools they would have been without comfort, or a means to secure their comfort, which for the Enlightenment was an index of civil society. Thomas points out that, despite Cook's declaration, one of the artists on this voyage, Alexander Buchan, made two drawings of the necklaces worn by the Haush people, entitling them "Ornaments used by the People of Terra del Fuego," and writing on the page beneath one figured necklace that it was made "of small Shells beautifully polished" and beneath the other: "Necklace made of pieces of Shells, neatly polished."[25] Cook's negative portrait of the Haush's technological abilities is contradicted by Buchan's illustrations and captions, which praise the delicacy of workmanship. Commenting on their differing assessments, Thomas notes rather dryly: "It would not be the last time that a native people would be denigrated, while their arts were appreciated."[26]

Despite Cook's disparagement of the Haush's cultural attainment based on their predilection for shellfish, his men routinely ate oysters and mussels, and recorded their own efforts at gathering shellfish as food. For instance, while the *Endeavour* was anchored in Botany Bay, Second Lieutenant Gore was "sent out in the morning with a boat to dredge for oysters at the head of the bay"; he successfully performed "this service,"[27] no doubt because two days earlier Cook had seen the "oysters themselves as I rowed over the shoals but being highwater I could not get any having nothing with me to take them up."[28] The voyagers celebrated New Zealand's plentiful coastal waters as well, especially Dusky Sound, for their wealth of seabirds, fish, and shellfish. On his second voyage, Cook wrote enthusiastically about the biodiversity of Dusky Sound, suggesting that its fish and fowl would be of use for subsequent naval expeditions.

> If the Inhabitants of Dusky Bay feel at any time the effects of cold they never can that of hunger, as every corner of the Bay abounds with fish, the Coal fish (as we call it) is here in vast plenty, is larger and better flavoured than I have any where tasted, nor are there any want of Craw and other shell fish.[29]

This enthusiasm for the bounty of the seas surrounding New Zealand is somewhat mitigated by Captain Tobias Furneaux's less sanguine statement about the waters off Queen Charlotte Sound, which, at the northernmost tip of New Zealand, possessed a different ecosystem from the southern Dusky Sound: "In the afternoon we put a dredge overboard in sixty-five fathoms: but caught nothing except a few small scallops, two or three oysters, and broken shells."[30] Cook was more successful than Furneaux in locating edible shellfish in these northern waters, describing the landscape surrounding Mercury Bay as "very convenient both for wooding and watering," and reporting: "in the river there is an immense quantity of oysters and other shell-fish: I have for this reason given it the name of Oyster River."[31] In fact, Cook, or more accurately his interlocutor Hawkesworth, states that "those who visit this coast" of the northern island of Eaheinomauwe "will not fail to find . . . shell-fish in great variety, particularly clams, cockles, and oysters."[32]

On his first voyage, Cook organized shellfish-gathering expeditions along the New Zealand coast. These were closely watched by the inhabitants, who, on at least one occasion, aided his men in their efforts. According to Anne Salmond, Cook put his crew to work, "collecting large supplies of greens, fish and shellfish during their visit to Mercury Bay, including boatloads of succulent oysters." Once friendly relations were established with the local people, "the men assisted the sailors with their work, and women and children helped Banks and Solander to gather plants and rocks for their collections."[33] One of "these children was a young boy called Horeta Te Taniwha, who in later years often reminisced about Cook's visit to Whitianga." Salmond includes his narrative, which was itself collected in the nineteenth century by John White. According to this, Horeta's people were astonished by the Europeans, concluding that they were *tupua* or "goblins" rather than human beings, perhaps because they seemed other than human in their desires:

> In the days long past . . . we lived at Whitianga, and a vessel came there, and when our old men saw the ship they said it was an *atua*, a god, and the people on board were *tupua*, strange beings or "goblins." The ship came to anchor, and the boats pulled on shore. . . . These goblins began to gather oysters, and we gave some kumara, fish, and fern-root to them. These they accepted, and we (the women and children) began to roast cockles for them; and as we saw that these goblins were eating kumara, fish and cockles, we were startled, and said, "Perhaps they are not goblins like the Maori goblins." These goblins went into the forest, and also climbed up the hill to our pa (fort) at Whitianga (Mercury Bay). They collected grasses from the cliffs, and kept knocking at the stones on the beach, and we said, "Why are these acts done by these goblins?" We and the women gathered stones and grass of all sorts, and gave to these goblins. Some of the stones they liked, and put them into their bags, the rest they threw away; and when we gave them the grass and branches of trees they stood and talked to us, or they uttered the words of their language. Perhaps they were asking questions, and, as we did not know their language, we laughed, and these goblins also laughed, so we were pleased.[34]

Once the strangers ate oysters and cockles, they became somewhat less alien and slightly more recognizable as human beings.

At times the boundary between food and scientific specimen blurs in voyagers' narratives. An account written between 1787 and 1789 by Arthur Bowes-Smyth, the surgeon on board *Lady Penrhyn*, the only convict ship of seven in the First Fleet that carried women, displays an interest in natural history collecting, an activity that gave him some intellectual stimulation as the fleet made its way along the coast of New South Wales. "During our stay in this place I frequently made excursions up the country some Miles in company with some others, I generally collected some Natural curiosities – sometimes shot Birds, at other times collected a large quantity of Balsam."[35] After the *Lady Penrhyn* delivered its cargo of convicts, Bowes-Smyth traveled through the South Pacific, where "I made a point of procuring as many of the different Articles that t/s [this] Country afforded as I could, in order to Oblige my friends with them at my return."[36] What began as a specimen hunting expedition would often turn into an opportunity to gather food. While at Tinian, he noted in his journal that he "went in the Jolly Bt. [boat] wt Mr. Anstis on the reef to collect Shells & Coral, got many large Clamps [clams] which were cook'd for Supper & were very good."[37]

At Port Jackson in Australia, the doctor noted the abundance of oysters: "All the Rock near the Water is thickly cover'd with Oysters, which are very small but fine flavour'd, they also adhere to the Branches of ye Mangrove Trees, I have Frequently brot. [brought] the Branch of a Tree thus loaded with Oysters on Board."[38] He appears to have carried on this voyage a copy of Hawkesworth's account of Cook's first voyage, perhaps to guide his collecting of specimens. Impressed by a passage in one of the volumes that mentioned hammer oysters, and well aware that these shells had fetched high prices in London, Bowes-Smyth set out to gather some, with the intention of selling them as specimens. Although he thought that he had located the exact spot in Botany Bay where Cook had found hammer oysters, his search for them ended in frustration. Clearly annoyed that his efforts at dredging proved futile, he cast doubt on Cook's veracity:

> During my Stay in Botany Bay I one day went in the long Boat accompanied by Mr Alltree & Mr Smith (a passenger on board our Ship) to the extream Southern part of the Bay, & I took my Oyster Dredge with me, hoping to get some hamer Oyster, (which Capn Cook mentions, & are sd [said] to be so very valuable for the singular form of the Shell), however of them Nobody saw any during the whole of our stay at Botany Bay, altho' the very spot in which Capn Cook says he found them. – We caught one & only one, very large Oyster with the Dredge, exactly like what we in England call the Kentish Oyster, I opened it, it was good tasted.[39]

Although irritated that he was unable to locate any valuable specimens, Bowes-Smyth did find some solace in eating delicious oysters. As for the doubt that he cast on Cook's veracity, it seems that Bowes-Smyth must have confused Botany Bay with Bustard Bay. Hawkesworth does describe oysters in Botany Bay, but the passage referring to hammer oysters is in the Bustard Bay section of the book: "upon the mud-banks, under mangroves, we found innumerable oysters of various kinds; among others the hammer-oyster, and a large proportion of small pearl-oysters."[40] The journal of Sydney Parkinson, the natural history illustrator on the first voyage, also confirms Cook's assertion: "We found a nautilus pompilius, and some of a curious kind of hammer oysters; as also a number of porpoises. . . . The hills seen in this bay, which was called Bustard Bay, appeared very barren, having nothing upon them but a few diminutive shrubs."[41]

COLLECTING FOR SCIENCE

In his official narratives Cook positioned himself as more interested in shellfish as a ready source of protein for his men than as specimens, while the naturalists on board – Banks to some degree and the Forsters much more so – focused on shells as objects of scientific curiosity. In his journal of the *Endeavour* voyage Banks recounts his attempts to gather specimens along the Patagonian coast under very difficult conditions. Between snow squalls, he managed to go ashore:

> Last night the weather began to moderate[.] And this morn was very fine, so much so that we landed without any difficulty in the bottom of the bay and spent our time very much to our satisfaction in collecting shells and plants. Of the former we found some very scarce and fine particularly limpits of several species: of these we observd as well as the shortness of our time would permit that the limpit with a longish hole at the top of his shell is inhabited by an animal very different from those which have no such holes. Here were also some fine whelks, one particularly with a long tooth, and infinite variety of *Lepades*, *Sertularias*, *Onisci* &c &c &c much greater variety than I have any where seen, but the shortness of our time would not allow us to examine them so we were obligd to content ourselves with taking specimens of as many of them as we could in so short a time scrape together.[42]

Such passages are unfortunately few and far between. After his arrival in Tahiti Banks's narrative was taken up with social interactions, and natural history description, particularly of shells, took a back seat. When he did turn his attention to natural history, it was the natural history of plants rather than of mollusks that was noted, for Banks had always been fond of botany; but apparently, in the Society Isles, he grew even fonder of court intrigue and beautiful women.

The Forsters, on the other hand, naturalists on board the second voyage, were very earnest about fulfilling the scientific duties prescribed to them by the Admiralty: to collect, describe, and illustrate natural history subjects, including the cultural productions and social conduct of the peoples of the Pacific region and Southern oceans. As a result, J. R. Forster's journal is replete with references to the shells he collected. His son's published narrative, based on the father's journal, fills out its staccato style with a more complete picture, recording not only how the pair collected specimens, but also their feelings as professional naturalists when competing for scarce and rare objects, both natural and artificial, with the rest of the crew, including the officers, who had little knowledge of what they were collecting, except that they hoped to sell whatever they found for spectacularly high prices.

The Forsters would go down to the beach to look for shells, wading up to their knees at low tide or taking out a boat to examine the reefs and rocky coastline for shellfish. On several occasions, they accompanied Cook in the pinnace while he took soundings as part of his survey of the New Zealand coastline: "during the time Capt Cook took more bearings, I collected with the Pinnace-men some fine large Turbines, on the rocks," and "we went out with the Capt in the afternoon towards *Point Jackson* . . . On the Rockweeds, which grow close in shore were several fine Shells, of which we collected a good many."[43] At Dusky Sound, J. R. Forster seems to have been pleased with the range of fish and shellfish that he and his son managed to collect at low tide, writing in his journal on 8 April 1773:

> We got some other new fish, viz. a small Lumpfish, a small Blenny, both caught under the rocks at low water mark; some Starfish, Coats of mail water spouts, Centipus, Muscles & Turbines, besides in the Evening a Lamprey with four beards of *Mystaces* was caught. We described all this & drew a new Sole.[44]

The next day, after having gone out in a boat to fish in "the very place where I & Capt Cook had been the first day of our arrival here" and where he "saw vast Schools of small fish, playing on the water," he found upon landing a "vast number of Ear-Shells of a very large Size on the Shore between the Stones. I shot a Waterhen [Western Weka], & collected besides Starfish, small Shells, Coats of Mail, & some Corals."[45] Six weeks later, after a storm that brought strong offshore winds, Forster and Cook went to Long Island, where "I collected a good many Shells on the Seaweeds, which were thrown ashore by the last wind."[46]

In general, the Forsters seemed to have collected their own specimens during their stay in New Zealand, but once they arrived in the Society Islands (Tahiti), they began to rely on local people, in particular women and children, to help them. Since "both sexes were expert swimmers," these "amphibious creatures" excelled at gathering shells on the reefs and in coastal waters.[47] When the Forsters went off to explore the coastline of Huahine, they were accompanied by islanders, who found a good number of shells: "corals, shells, and echini, which the natives had gathered for us on the sea-shore."[48] A brief statement by Georg Forster reveals how much help he and his father received from the Tahitians who collected natural history specimens on their behalf:

> Seeing that I enquired for plants, and other natural curiosities, they brought off several, though sometimes only the leaves without the flowers, and vice versa; however, among them we saw the common species of black night-shade, and a beautiful erythrina, or coral-flower; I also collected by these means many shells, coralines, birds, &c.[49]

The statement contains a veiled acknowledgment of his dependence on locals to collect for him, but this idea is undercut by the phrase "I collected," which puts him in the sentence's subject position to suggest that he was the active agent in the process of collecting. Furthermore, the phrase "by these means" buries the islanders' activities in a prepositional phrase and reduces humans to a process ("means"), thus obscuring the islanders' role in collecting specimens for him.

PACIFIC "MARKETS"

The process of gathering natural history specimens and cultural artifacts involved the Forsters in a complex system of barter and exchange. In Tongatabu in the Friendly Isles they gathered shell specimens themselves and then bought some shell ornaments:

> We picked up a quantity of shells at the foot of the steep rock, where we sometimes waded in water to the knees upon a reef, on account of the flood tide which was advancing. We likewise met with several natives returning from the trading-place, who sold us a number of fish-hooks and ornaments, a fish-net made like our casting-nets, knot of very firm though slender threads, some mats, and pieces of cloth. We likewise purchased of them an apron, consisting of many wheels or stars of plaited coco-nut fibres, about three or four inches in diameter, cohering together by the projecting points, and ornamented with small red feathers and beads cut out of shells.[50]

Like Cook and Banks in their journals, in this passage Georg Forster uses the language of the marketplace – "purchase," "bought," and "sold" – to describe the transactions, writing: "Preparations were made for sailing from this island the next morning, whilst the natives crouded about us with fish, shells, fruit, and cloth, of which we purchased all that was to be had."[51] Money, however, was never involved, because nails or other manufactured goods were exchanged for shells and objects made from shells. Thomas Edgar, for example, who as master was in charge of procuring food through exchanges, wrote that in Tiarraboo "we bought Hogs, Fowls, Sweet Potatoes, Yams, Bread Fruit, Plantains, Cocoa Nuts, Sugar Canes, Apples and a few Fish, for Hatchets, Nails, Beads etc."[52]

The words "purchase" and "bought" belie the complexity of the system, barely grasped by the European voyagers as they stood on the beach exchanging, as they thought, nails for food. They were initially convinced of the value of their own goods, a mix of nails, glass beads, and cloth among other cheaply produced manufactured items. Although phrases such as " so desirous were they of our things" appear frequently in journals and logs,[53] what also creeps in is the recognition that European "things" were often less desirable to the islanders than objects from other Pacific regions. More popular than British-made ironware or textiles were items gathered from other islands, whether natural objects, such as red feathers and mother of pearl, or manufactured ones, such as woven mats and tapa cloth made from mulberry trees. In Tanna, the crew were unable to interest the islanders in trading food or artifacts for British ironware:

> They set no value on our iron-ware, but preferred Taheitee cloth, small pieces of green nephritic stone from New Zeeland, mother of pearl shells, and, above all, pieces of tortoise-shell. For these last they sold their arms; at first, parting only with darts and arrows, but afterwards disposing also of their bows and clubs.[54]

Red feathers, especially in Tahiti, became "a very valuable article of trade."[55] According to Clerke, they were "of the most essential service to us, in purchasing Hogs, Roots, & Curiosities,"[56] so much so that "Our people giving any piece for them – on a supposition that they would turn to good account in Otaheitie."[57]

This instability of value and the vagaries of exchange are also revealed in the following passage, when the Tongans substituted artifacts, much desired among the crew, for the food the captains were hoping to receive.

> Notwithstanding the engaging manners of the natives, we foresaw that we should not make but a very short stay among them, because our captains could not obtain refreshments in any considerable quantity; which might be owning not so much to their

scarcity upon the island, as to the difficulty of making our goods current for such valuable articles, when they could obtain them in exchange for arms and utensils.[58]

The competing desires among those on board the ships complicated the exchanges between Europeans and Pacific islanders, some sailors willing to go without fresh provisions – hogs, coconuts, plantains, and yams – so that they could "purchase" an artifact or specimen that might be worth money in London.

> They had brought indeed a few yams, bananas, coco-nuts, and shaddocks, for sale, but they soon dropt that branch of trade. Our people purchased an incredible number of fish hooks made of mother of pearl, barbed with tortoise-shell, but in shape exactly resembling the Taheitee fish-hooks, called witte-witte; some of which were near seven inches long. They likewise bought their shells, which hung on the breast, their necklaces, bracelets of mother of pearl, and cylindrical sticks for the ear.[59]

Such competition erupted several times on Cook's voyages, and sometimes resulted in an order prohibiting the crew from trading with the islanders. On the third voyage, for instance, Thomas Edgar wrote that in New Zealand "our folks were all so eager after curiosities and with all so much better provided than in any other voyage – that our traffic with the Indians was quite spoilt,"[60] meaning that he was unable to trade for food since he could not compete with his own men, who had brought with them desirable items such as hatchets to trade with the Maori. The situation got out of hand, forcing Cook to issue a temporary restraining order: "About 11 this Forenoon an order from Captn. Cooke was read to the Ships' Company prohibiting all people from purchasing curiosities until further orders."[61]

If tensions between officers and seamen were this high over who had the right to trade with islanders, imagine the intensity of competition between crew members who collected for money and the onboard naturalists – the Forsters – who collected, as they professed, for the sake of science, considering that this was authorized by their commission. They positioned themselves against what they described as the craze for curiosities among the besotted common seamen, and were irritated by the crew's unscientific and totally acquisitive stance towards natural and artificial objects.[62] The elder Forster lashed out at their mercenary motives, "their mean grovelling Passions," with a high-minded disdain, but in reality their activities were uncomfortably close to his own:[63]

> Today a Saylor offered me 6 Shells to sale, all of which were not quite compleat, & he asked half a Gallon brandy for them, which is now worth more than half a Guinea. This shews however what these people think to get for their Curiosities when they

come home, & how difficult it must be for a Man like me, sent out on purpose by Government to collect natural Curiosities, to get these things from the Natives in the Isles, as every Sailor whatsoever buys vast Quantities of Shells, birds, fish, etc. so that the things get dearer & scarcer than one would believe, & often they go to such people, who have made vast Collections, especially of Shells, viz. the Gunner & Carpenter, who have several 1000 Shells; some of these Curiosities are neglected, broke, thrown over board, or lost.[64]

Palpable in this journal entry are the tensions between the educated men of science, Marra's "experimental men," and the common seamen in their competition for curiosities. Even though they were engaged in the same activity, Forster constructed a rhetorical divide between himself and the crew to mark his collecting as legitimate and scientific. As Nicholas Thomas argues, Forster based this on a distinction between his "philosophical interest in natural history" and their "desire for financial advantage."[65]

Georg Forster, perhaps channeling his father's irritation, also condemned the ignorance and rapacity of the crew members. *A Voyage Round the World* contains several passages complaining about the lack of respect shown by their fellow voyagers towards scientific men such as him and his father. One occasion annoyed them particularly. While in New Caledonia, the Forsters had been unable to visit the isle of Ballabeea (Balabio) because they were unwell, having ingested, along with Cook, a virulent poison from a fish they had eaten. They were disappointed at missing the opportunity to explore the island and collect specimens, but hoped to learn about the place from those who had accompanied Mr. Pickersgill. Their queries, however, ended in frustration:

> One of the surgeon's mates, who went on this excursion, collected a prodigious variety of new and curious shells upon the island of Ballabeea, and likewise met with many new species of plants, of which we did not see a single specimen in the districts we had visited; but the meanest and most unreasonable envy taught him to conceal these discoveries from us, though he was utterly incapable of making use of them for the benefit of science. We had therefore more reason than ever to regret that our illness disabled us from sharing the perils of this little excursion.[66]

The footnote that glosses this passage launches into a general complaint against the crew and its officers:

> It will not be improper to acquaint the reader, that we were so situated on board the *Resolution*, as to meet with obstacles in all our researches, from those who might have been expected to give us all manner of assistance. It has always been the fate of science

and philosophy to incur the contempt of ignorance, and this we might have suffered without repining; but as we could not purchase the good will of every petty tyrant with gold, we were studiously debarred [from] the means of drawing the least advantage to science from the observation of others, who of themselves did not know how to make the proper use of a discovery when they had made it. . . . It may seem extraordinary, that men of science, sent out in a ship belonging to the most enlightened nation in the world, should be cramped and deprived of the means of pursuing knowledge, in a manner which would only become a set of barbarians . . . If there had not been a few individuals of a more liberal way of thinking, whose disinterested love for the sciences comforted us from time to time, we should in all probability have fallen victims to that malevolence, which even the positive commands of captain Cook were sometimes insufficient to keep within bounds.[67]

In its self-pitying tone, this passage mirrors one that the senior Forster had written in his journal the year before while confined to the ship in Dusky Sound because of bad weather. He complains that he and his son have the worst cabin on the ship, one that made his duties as a naturalist difficult to perform, all of which he interpreted as indicative of the low esteem in which his activities were held by the *Resolution*'s officers:

> It rained still & was vastly bad foggy weather: We finished some Descriptions, cleaned Shells, brought aside plants etc: but few things would keep, on account of the moist weather: my Cabin was a Magazine of all the various kinds of plants, fish, birds, Shells, Seed etc. hitherto collected: which made it vastly damp, dirty, crammed, & caused very noxious vapours, & an offensive smell, & being just under the Chain-plates & the Ship lying close in Shore under high trees; it was so dark, that I was obliged to light a candle during day, when I wanted to write something: None of the other Cabins on the same deck, was subject to this inconvenience . . . sometimes I was pent in for hours together . . . all the dirt and noise in the whole Ship, was accumulated about it: so that my & my Sons accomodations were the worst in the whole Ship, under all circumstances, in hot & cold climates, in dry& moist weather, at Anchor & at Sea.[68]

J. R. Forster complained not just on paper but face to face about the conditions under which he and his son were consigned to carry out their investigations. In the process, he managed to arouse the crews' dislike to the degree that, according to the astronomer William Wales, "there was scarce a man in the ship with whom he had not quarrelled."[69]

Although the Forsters complained about the ignorance and greed of their fellow voyagers, they too expected to profit from their artificial and natural history specimens, intending some for strategic gift giving and others for sale. Using the rubric of collecting for science, they tried to raise their material practices above mere commoditization and to separate their acquisitive activities from those of the crew. Not all Pacific islanders, however, discriminated between the two forms of collecting they observed. Cook's journal contains an incident in which a native boy mocked the crew's eagerness to trade for Pacific artifacts:

> It was astonishing to see with what eagerness everyone catched at every thing they saw, it even went so far as to become the ridicule of the Natives by offering pieces of sticks, stones and what not to exchange, one waggish Boy took a piece of human excrement on a stick and held it out to every one of our people he met with.[70]

Interestingly, in this passage Cook positions himself outside this collecting frenzy, obscuring his own participation in it and his expectation of profiting from the activity. He alternated between condemnation and bemusement at his seamen's desperate attempts to acquire curiosities. Despite the Forsters' bitter complaints, some sailors and officers clearly succeeded in acquiring items of value.

CELEBRITY, PROVENANCE, AND VALUE

After Cook's death on the third voyage, and popular opinion had confirmed his celebrity status, natural history objects associated with him took on an increased value distinct from their scientific and material qualities. Nicholas Thomas makes the point that, immediately after the first voyage, "Cook was no celebrity," and that it was Banks who was feted by his fellow naturalists and was the object of gossip, newspaper coverage, lampoons, and satirical prints. Cook was noticed by the Admiralty, however, which was impressed with his maps and navigational contributions. His reputation grew with the publication of John Hawkesworth's *An Account of the Voyages . . . for Making Discoveries in the Southern Hemisphere* (1773), a compilation of the *Endeavour* journals and narratives by other sea captains, so that by the time the second voyage was concluded in 1775 he was well known to the public. It was the third voyage that catapulted Cook into the realm of national heroes, his renown growing even greater with the story of his dramatic death in Hawai'i in 1779 (pl. 43).[71]

Cook's posthumous popularity was fueled by the publication of numerous narratives, some of dubious quality, written by participants of the third voyage, as well as the Admiralty's authorized narrative, *A Voyage to the Pacific Ocean* (1784), cobbled together by the

43   Francesco Bartolozzi and William Byrne after John Webber, *The Death of Captain James Cook*, 1784, engraving. London, British Museum. © The Trustees of the British Museum

Rev. John Douglas from Cook's official journals. The preface, written anonymously by a fellow naval officer, praises Cook for his ingenuity, bravery, and humanity:

> The death of this eminent and valuable man was a loss to mankind in general; and particularly to be deplored by every nation that respects useful accomplishments, that honours science, and loves the benevolent and amiable affections of the heart. . . . For, actuated always by the most attentive care and tender compassion for the savages in general, this excellent man was ever assiduously endeavouring, by kind treatment, to dissipate their fears, and court their friendship; overlooking their thefts and treacheries . . .[72]

Cook's reputation as an enlightened explorer reigned supreme in the British imagination, despite eyewitness accounts that his treatment of Pacific islanders was not always as humane as represented in print culture.[73]

His status as a national hero was further consolidated by Philippe Jacques Loutherbourg's pantomime *Omai; or, A Trip round the World*, which opened at the Theatre Royal in Covent Garden in December 1785 and was performed seventy times during the following year.[74] It tapped into the audience's curiosity about Cook and its eagerness to embrace him as a hero. With its dramatic display of *The Apotheosis of Captain Cook*, a painting of grand proportions that descended onto the stage during the final scene, *Omai* used performance, music, and art to confirm the heroic image that had been constructed in the popular press and in the poetry of the likes of Anna Seward, William Cowper, and Hannah More, all of whom celebrated Cook's voyages as emblems of humanitarian and enlightened scientific exploration. The Evangelicals Cowper and More were critical of the potential for violence implicit in such voyages of discovery, and yet they portrayed Cook as an exception to the rule of the cruel explorer, More stressing his "gentle mind": "Thy love of arts, thy love of humankind . . . thy mild and lib'ral plan." Visual artists, responding in part to Loutherbourg's *Apotheosis*, took up the theme of Cook's death at the hands of those very "savages" he had attempted, in the words of the play, "not to conquer but save," and produced paintings of his last moments in Hawai'i, images that circulated widely as engravings, even as decorative motifs on wallpaper and fabric.[75] With the popularity of the pantomime *Omai*, the proliferation of illustrations of Cook's encounters in the Pacific, and official and unofficial publications describing the final voyage, Cook was on everyone's minds when the duchess of Portland's auction was announced in the press and advertised daily in newspapers for more than a month in April 1786. It is clear from the high prices at which Cook-related items were sold that his celebrity status raised the value of all Pacific items, including the natural history specimens.

To illuminate the way in which celebrity creates value, I mention here two examples that capture the intensity of the desire to acquire Cook memorabilia. The first comes from a narrative written in 1792 by George Tobin, Third Lieutenant on the *Providence*, under the command of William Bligh, who was attempting for the second time after the disastrous *Bounty* voyage to transport breadfruit seedlings from Tahiti to the British West Indies. Tobin recounts how he encountered in Tahiti a portrait of Cook that had been painted by Webber on the third voyage and had been left as part of a gift exchange with the chiefly Pomare (pl. 44):

> I cannot refrain from remarking with what friendly care and reverence a picture of Captain Cook by Webber (painted while at Tahiti in his last voyage) was preserved by Pomare. Nothing I suppose could tempt this amiable chief to part with it. Much did I covet the polygraphic secret, to steal the portrait of this immortal navigator, which

44   John Webber, *James Cook*, 1776, oil on canvas. London, National Portrait Gallery.
© National Portrait Gallery, London

was said by those who knew him to be a most striking resemblance. It has been customary for the different commanders of vessels visiting the island to note on the back of this picture the time of their arrival and departure. Some other tablet must now be found, as visits have been so frequent, no more space is left.[76]

Clearly, Pomare valued the portrait as an emblem of his relationship with Cook, with whom he had many complex encounters; in cherishing the portrait, he had become a

45   James Caldwell after William Hodges, *Omai*, circa 1777, engraving. London, British Museum. © The Trustees of the British Museum

collector of Cook memorabilia.[77] Behind Tobin's somewhat ironic revelation of his own impulses to steal the portrait from Pomare lies the recognition of the value that it would have in London, while a contrapuntal note is voiced to the oft-repeated complaints about the alleged thefts perpetrated by Pacific islanders.

Perhaps even more bizarre was the eagerness with which the officers and crew of the *Lady Penrhyn* snatched the opportunity to buy Omai's things when they arrived on Tahiti, learning that he had died (pl. 45). Hearing that his "European articles of Domestic &

Culinary Use which he had been liberally supplied with by the Government upon his return [to Tahiti from Britain] with Captain Cook, were so little esteem[ed], or at least [their] use so little known or attended to by his Countrymen," crew members eagerly offered to buy them. Mr. Watts, for instance, purchased several items, including a "Cast Iron Pot." The circulation and exchange of objects here were dizzying: Cook had brought Omai to England on the second voyage, returning him to Tahiti on the third voyage with British goods, including a horse, armor, weapons, fireworks, toys, and domestic metalware. Omai had valued these objects because they represented Britain and the (fire) power he thought he had acquired there, but they lost their value at his death, and his fellow islanders were eager to sell them back to those they figured would prize them – and so the pot made the journey back to Britain, loaded with significance. The value of the pot for Watts was not to be found in its use but in its provenance, having been owned by Omai, who had journeyed with Cook.[78]

To gauge the effect of Cook's celebrity on the value of objects associated with his voyages, I turn briefly to sales catalogues and museum guides that promoted the association in their descriptions of material culture and natural history specimens. When the dealer George Humphrey went bankrupt in the late 1770s, he was forced to sell most of the natural history specimens, along with several artifacts, drawings, and reference books, that made up his vast collections. The auction catalogue, *Museum Humfredianum; a Catalogue of the Large and Valuable Museum of Mr. George Humphrey* (1779), contains descriptions of shells and "Curiosities from the new discovered Islands in the South Seas." There were shells from places Cook had journeyed to: New Amsterdam, New Zealand, and Otaheite. "A fine partridge tun, South Seas" – and a "clouded Ethiopian crown, New Holland" were listed alongside "a curious basket made of cocoa nut fibres and beaded with shells, Anamokka" and "a grass apron worn by the dancing girls of Otaheite, and a curious ornament made of cocoa nut fibres, formed into a variety of singular shapes, and decorated with shells, &c. Anamokka."[79] Cook-related material was also promoted in the *Museum Calonnianum* (1797), the catalogue for the sale of Monsieur de Calonne's collection of natural history specimens. Humphrey, who handled the sale and wrote the catalogue, explained in the preface that this wealthy man, an ex-minister in the French government, had special access to various "circumnavigators," in addition to having purchased many shells – more than forty of them – from the late duchess of Portland's auction, one of which, a white hammer oyster, is described as having been "brought home by Capt. Cook on his first voyage round the world, and was got on the coral reef off Endeavour River. M. P. 3832," the abbreviation M. P. 3832 referring to lot 3832 of *A Catalogue of the Portland Museum*.[80]

Captain Cook's popularity was used by owners of private collections to draw visitors to their fee-charging museums.[81] The Leverian, the largest of these commercial museums, had several rooms devoted to the display of Pacific artifacts and natural history specimens associated with Cook's voyages. According to the museum guide, the "Sandwich Room" contained "the admirable and curious articles collected by Captain Cook, in his third and unhappily last voyage; for at *Owhyhee*, one of the Sandwich Islands, he was cut off by those persons who had before treated him with a respect approaching adoration." Written over the entrance was: "To the Immortal Memory of Captain Cook." Most of the objects in this room were collected on the third voyage, and the guide repeatedly refers to Cook's death as a tragic event, describing these objects in reverent tones, turning them into relics that memorialize him as the great explorer. The guide notes: "the viewing of such objects cannot fail to excite a melancholy pleasure, while we reflect on his eminent abilities, and his unhappy fate" (see pl. 13).[82]

The auction catalogue of 1806 for the Leverian lists dozens of shells from the South Pacific: lots "1118 – a scarce sun, South Seas"; "1123 – the great cockscomb oyster, Friendly Islands"; "1125 – the great high admiral cone, two inches and a half in length, Friendly Islands, very rare"; "1250 – an iris ear, *New Zealand*"; "1256 – a fine cardium arundincea"; "1389 – A large imperial sun, from *Cloudy Bay*, *New Zealand*, uncoated and polished, very rare."[83] Several of the museum's Native American artifacts from the Pacific Northwest were brought back from the third voyage's exploration of the coastline of what is now British Columbia and Alaska as Cook searched for the Northwest passage. A newspaper advertisement for the Leverian Museum announced: "The public are respectfully informed, that a great variety of very curious articles from Nootka, or King George's Sound, are to be seen at the Museum, the whole of which were collected by Captain Cook."[84]

Less well known to scholars is Daniel Boulter's "museum" in Yarmouth, which mingled Cook-related objects with goods he sold, including cutlery, toys, and jewelry. The bookplate affixed to the *Museum Boulterianum* (1793), a catalogue of the "Curious and Valuable Collection of Natural and Artificial Curiosities in the Extensive Museum of Daniel Boulter," describes Boulter as a "Dealer in curious Books, Antiquities, and Natural Productions." The title page announces that his museum has been "upwards twenty years in collecting" and contains: "Elegant Shells, Corals, Corallines, and other Marine Productions. . . . Dresses, Ornaments, Weapons of War, Fishing-Tackle, and other singular Inventions of the Natives of the lately discovered Islands in the South-Seas." Lot 62, for instance, consisted of a "*Hammer Oyster*, South Seas, Ostrea Malleus, L. [Linnaeus] 1147, n. 207, 7s. 6d"; lot 113, "*Netted Cowry*, Friendly Isles, scarce, Cypraea reticulata, Mart.

[Martyn's *The Universal Conchologist*] v.i. n. 15, 5s."; lot 163, "Spur *Buccinum*, New Zealand, Buccinum Calcar, Mart. v.i. n.10, 5s."; and lot 243, "*Iris Ear*, New Zealand, Haliotis Iris. Mart. v.i. n. 36, 2s."[85] In addition to pages of natural history items, shells included, the catalogue lists what must have been stunning "Indian Dresses," beautifully crafted clothing, basketry, tapa cloth, and ritual items, a "Collection from the new discovered Islands in the South-Seas, by Capt. Cook and others," including a "Beautiful Feathered Cloak, worn by Chiefs of Owhyee, 1£ 1s," "A Curious Helmet of Scarlet and Yellow Feathers, from Sandwich-Islands, 1£ 1s," and "Another [necklace], formed of Cocoa-Nut Beads, Pieces of Shell, and Fishes Teeth, with a flat part of a Volute Shell for a Pendant, from Sandwich-Islands, 5s. Another, made of a Species of Murex, 1s. 6d."; "Another composed of a great Number of Strings of human Hair twisted, with a Pendant in the middle, in form of a Handle of a Cup, carved out of Shell, Sandwich-Islands, 5s"; and "Bracelet made of large Boar's Tusks, worn by Women of superior rank, which Capt. Cook thought elegant, 15s."[86] Cook's presence is repeatedly invoked in Boulter's catalogue to enhance the value of the objects.

William Bullock's museum, established at the turn of the century initially in Liverpool and then moved to London, contained artifacts "brought from the South Seas by Captain Cook," as announced in his guidebook of 1801, a pamphlet entitled *A Companion to Mr. Bullock's Museum, Containing a Description of Upwards of Three Hundred Curiosities*. The *Companion to Mr. Bullock's Extensive Museum* of 1811 describes the contents of the Sandwich Islands Case, which carries the footnote: "Several of the articles in this Case were once the property of the celebrated Captain Cook," among them: "an extremely curious pair of Bracelets, made of boar's teeth, presented by Mr. G. Humphrey" and "A beautiful Fly-Flap, purchased at the sale of the late Leverian Museum," apparently purchased by Lever from David Samwell, an officer on Cook's third voyage. Bullock seems to have purchased the Cook voyage items second or third hand; the descriptions of many of the objects in the museum match those in Boulter's, suggesting that he may have purchased Pacific items from Boulter as well as from the dealer Humphrey. *A Companion to the London Museum . . . by William Bullock* (1814) describes a cloak much like the one Boulter listed in his catalogue. Bullock's description, though, goes well beyond Boulter's brief statement – "Beautiful Feathered Cloak, worn by Chiefs of Owhyee" – recognizing the value-enhancing qualities of narrative and context:

> A superb cloak, made of the black feathers of the Powhee bird, ornamented with a broad checquered border of red and yellow. . . . It is worn by none except the chiefs, and by them only on particular occasions; as they never appeared in them but three

times during Captain Cook's stay in Owyhee; *viz*. at the procession of the king and his people to the ships, on their first arrival; in the tumult when the unfortunate commander fell a victim to their fury and mistaken resentment; and when two of the chiefs brought *his bones* to Captain Clerke.[87]

The linking of specimens to Cook's voyages continued to be a strategy employed decades after his death to interest visitors in a museum's collections. In 1813 Captain James Laskey published a guide to the Hunterian Museum in Glasgow, focusing on those shells with provenances that included not only Cook but also the duchess of Portland. Of the "Genus Ostrea . . . [the] white hammer oyster," he wrote: "This specimen was brought home by Captain Cook from the Coral Reef at Endeavour River, and is very rare. Twelve Guineas have been given for a fine specimen." Of the "Genus Cypraea" and the much-sought-after orange cowry, Laskey's description includes a Cook narrative to heighten its value (see pl. 8):

> They [orange cowries] were met with by Captain Cooke and other Voyagers at the Friendly Isles, where they are much esteemed by the natives, who wear them as ornaments, round their necks. One of these with a hole on the side has been used as such, the other is perfect and rich in color, and is valued from 10 to 15 guineas.

Laskey's text calls upon the duchess of Portland as well as Cook to pique his audience's appetite with celebrity-laced narratives: "Only two other specimens are known, one of which is in the Museum of the late John Hunter, brother to the Doctor, and was originally in the Cabinet of the late Duchess of Portland, and now in the possession of the Royal College of Surgeons, London." Another very rare shell in the Hunterian Museum is an Arca, its "country unknown," and "only one other specimen is in this Country, which originally belonged to the late Duchess of Portland, at whose sale it was purchased by the late Dr. Fordyce, for £ 4:15:0, and at the sale of the Doctor's Cabinet a few years since, it sold above £ 5."[88] As can be seen, shells collected on Cook's voyages, as well as those owned by the duchess, continued to be a topic discussed well into the nineteenth century.

A connection with the duchess further elevated the value of shells associated with Cook's voyages, but if Daniel Solander had named them their value to naturalists and serious shell collectors was increased still more. For example, lot 4039 in the Portland sales catalogue, which consisted of one shell, sold for nearly £9: "A very fine specimen of Voluta pacifica, *S.* brought by *Capt. Cook*, from the *Reef off Endeavour River*, on the *Coast of New Holland*," the *S.* here referring to Solander, who had examined and named it.[89] Though provenance may seem to be a category best left to the connoisseurial practice of authenticating art objects, it did also increase the value of scientific objects, despite notions of scientific value being somehow free from the influence of desire and the vagaries of the marketplace. As Charles Smith notes in his sociological study of auctions,

> Provenance and ownership of items – past, present, and future – are often equally important defining characteristics as price. . . . The importance of provenance and ownership underscores the fact that objects are seldom if ever confronted in isolation. They are rather perceived as embedded in a complex social reality. Central to this context is not only the history of ownership and origins of the items in question but the relationships between the people with an interest in the items.[90]

That natural history specimens could take on qualities associated with and manipulated by the marketplace may seem strange, given commonly held notions of science as "disinterested," not motivated by economic interest and operating in a distinct terrain. Equally strange is the fact that the market value of a zoological specimen could be shaped by its proximity to fame and its possession by a celebrity. Clearly, then and now, science is embedded in complex social realities. It is this social embeddedness, in particular "the relationships between people," that Smith sees as central to the construction of value. It can explain to some extent the divergent meanings attached to shells as they moved across the globe and in and out of different social economies, where they could be valued for their aesthetic and utilitarian qualities, their potential as objects of exchange, or their status as natural history specimens.

EXCESS AND DECLINING VALUE

Although museums continued to use Cook's name to capture visitors' interest in their collections into the nineteenth century, the actual market value of Cook-related objects fell in price with the increase of voyages to the South Pacific. When the first ship, the *Endeavour*, returned to England in 1771, the items collected on board were extremely rare, but with the intensification of collecting on board the two ships of the second voyage, the *Resolution* and the *Adventure*, items that had been unique and novel were more easily acquired. For example, on the first voyage Banks had been eager to collect and bring home the elaborate and dramatic *Heiva*, a Tahitian mourning costume decorated with mother of pearl and worn by the elite during funerals (pl. 46). He was unsuccess-

46   Joseph Collyer after William Ellis, "A Man of Otaheitee in a Mourning Dress," published in *Lady's Magazine*, circa 1785. London, British Museum. © The Trustees of the British Museum

ful, but on the second voyage the Forsters and Cook managed to collect ten costumes. It was this kind of aggressive collecting by participants of the second and third voyages that damaged the market in Pacific artifacts and natural objects. Forster, who had hoped to make a pretty penny (to quote Seymer) from the thousands of natural history specimens he and his son collected on the second voyage, had to compete with seamen such as the often drunken gunner's mate John Marra. Proffering his shell specimens to Sir

Joseph Banks in 1778, Forster was unable to control his petulant tone and irritation at Banks's indifference: "I am sorry that my Shells are of no service to you. There are several of them of which neither Capt. Cook nor any body else got specimens."[91] Banks's lack of interest could have stemmed from their proliferation. Even as early as 1775 the dealer George Humphrey had complained in a letter to Seymer that "plenty makes cheapness," giving the example of the falling prices for shells:

> Jackson [a dealer] gave himself 2/6 each for the Beads [common name for a kind of shell] on board the *Adventure*. You will remember Sir that when you gave me a guinea for a Weed & 1 ½ for a bead those shells were at that time very rare. As the last mentioned ship brought home many of them[,] the price of course fell considerably. Tis the same with all merchantable affairs, plenty makes cheapness. I have Shells by me which cost me 2 & 3 guineas each which I would part with at 1/3 the cost. I can now sell you the weeds at 2/6 each and the beads at 5/ each five of the kind & if any one will take all the weeds I have[,] I will sell them at 3 each, notwithstanding many of them cost me 1/6 and 2/. each.[92]

The *Adventure*, commanded by Tobias Furneaux, had returned home a year earlier than the *Resolution*, which had the effect of diminishing the value of the latter's natural history cargoes. This also contributed to the Forsters' frustration and lessened their ability to interest collectors in their specimens.

With the successive voyages of Bligh and Vancouver, those of the First Fleet in 1787, and French and Russian expeditions in the 1780s and 1790s, the rarity and novelty of Pacific natural and artificial curiosities declined, and the earlier prices they had fetched seemed like a dream. Even during the third voyage, which seemed at times to have been less about discovery and more about revisiting known destinations, for those on board excitement had waned and a sense of exhaustion had set in. The dealer Thomas Martyn, who had been a very aggressive purchaser of second-voyage shells, was disappointed by the number and quality of shells brought back from the third voyage: "It is a little extraordinary so few new species should have been collected, considering the many different Islands in the S. Seas the two ships visited & so many persons employed to gather them," and the specimens collected "do not abound either in variety of the New or many duplicates of the known ones that are valuable." Without a professional naturalist on board and with Clerke ill and dying, natural history collecting may have suffered from a lack of interest or expertise. Thomas Edgar, master of the *Discovery*, commanded by Clerke, wrote in the official journal he was required to keep that to describe the natural history of the Tahitian islands was for him a distasteful task: "of Insects – I can say nothing, & indeed all these things more particularly come under the notice of a Naturalist, a study to me the most insipid – besides the Qualities of these Islands have been minutely described."[93] His narrative continues with an attempt at describing the customs and manners of the region, a usual component of natural history travel narratives, and after a short passage on tapa cloth, he gives up, saying: "the methods of making Cloth has already been described by Captn. Cooke (and indeed most of their Customs). Therefore of them I shall say nothing farther, & indeed it is a very unnecessary piece of business to relate anything more than what happened during our stay amongst them."[94] Also exhausted was the novelty of these voyages for Pacific islanders. The Tahitians in particular, who had been visited by the British repeatedly and for extended periods of time over the course of more than a decade, were tired of them, their needs, and their stuff. The market for European goods there and in New Zealand was saturated. On the last voyage to Charlotte Sound, Edgar noted that a hatchet or an axe was needed to barter for curiosities previously bought by a nail.[95] Ultimately, in spite of Cook himself becoming even more popular after his dramatic death in 1780, the material culture and natural history specimens from voyages began to lose their economic value; as these objects, once special and exotic, proliferated in the centers of Europe, so their prices fell. By the 1790s exotic shells no longer held collectors' attention as they once had. In Britain, the study of native shells, which earlier had been the focus of only a few collectors, became quite popular after the outbreak of the Napoleonic Wars, with the consequent curtailment of foreign travel and a heightened sense of patriotism.

In the 1770s and 1780s shells were bartered for on the beaches of the south seas and traded on board European ships, undergoing various kinds of exchange before assuming their full commodity potential in the London natural history marketplace. Their prices fluctuated from voyage to voyage, declining as more Pacific and southern hemispheric shells entered the market and the demand for exotic shells decreased. The prices were difficult to gauge, belying Seymer's faith in the pocket book as the most transparent and rational mechanism of valuation and exchange. Once collectors had made their purchases, the shells were withdrawn from commodity exchange ("enclaved" in Appadurai's terms), but could continue to circulate. They entered into other forms of exchange, involving the trading of duplicates and loans for classification purposes. The duchess of Portland also lent very scarce specimens to naturalists and dealers who were involved in publishing conchology books and serial publications. Those she lent to Humphrey, Martyn, and Pennant, for instance, entered the realm of representation, taking on textualized lives as objects that illustrated the standard characteristics by which all other mollusks of the same species were measured. Her shells appeared in several conchologiical publications during and even after her lifetime. In fact, apart from the British Museum's lone specimen, the *Cymbiola aulica*, these images are the best surviving representatives of the duchess's shells in terms of solid provenance. There is a way in which these textualized shells live on after their material selves have disappeared from sight, and it is to this process of textualization that I now turn.

CHAPTER FIVE

47　Thomas Martyn, "Checkered Mitre" (detail of pl. 52)

PATRONAGE, PUBLICATION, AND THE ILLUSTRATED CONCHOLOGY BOOK

Shells moved across the globe to make their way to centers of accumulation in London, and they circulated within British natural history circles as gifts, trades, and loans, as ways to fill out the "chasms" in collections and to assist with classification. Another way shells circulated was as loans to authors of natural history books, who would use them as models for the illustrations and verbal descriptions in their publications. The duchess of Portland was quite active in these kinds of exchanges, loaning specimens to Thomas Pennant, for instance, for the fourth volume of *British Zoology* (1777). She supported the production of several important natural history books, as well as publications devoted to conchology, her support taking various forms, including loaning specimens, vetting illustrations for their accuracy, and offering financial support.[1] Her generosity was crucial to the production of a number of illustrated natural history books; the authors recognized that, without her contribution, their publications would have been far less accurate, interesting, and comprehensive.[2]

This chapter explores how the duchess of Portland aided those who produced books that named, described, and pictured shells, activities that were consonant with the Enlightenment's scientific agenda of bringing order to bear on the diversity of Nature. She was thanked by the authors of several important shell books published in the 1770s and 1780s, but these statements of gratitude are often flowery and vague, and shed little light on what she actually did to aid in the publication of the volume. My goal here is to tease out from these volumes and from the correspondence of several naturalists, dealers, and booksellers what the duchess may have done to merit their gratitude (as well as their occasional frustration and hostility). I also explore the processes by which illustrated natural history books were published, focusing on behind-the-scene tensions over the production process, including the necessary access to specimens and the books needed to classify the shells to be depicted and described, and the anxieties over the expense involved in the production of attractive and accurate illustrations.[3]

LOANING SHELLS, PATRONAGE, AND PUBLICATION

Three important conchological books were dedicated to the duchess of Portland: the William Huddesford edition of Martin Lister's *Historiae sive synopsis methodicae conchyliorum* (1770), the fourth volume of Thomas Pennant's *British Zoology* (1777) and George Walker's *Minute and Rare Shells . . . discovered by William Boys* (1784). Walker also dedicated *Minute and Rare Shells* to Sir Joseph Banks, thanking both him and the duchess for their "Very Obliging Attention and Generosity They Have Been Pleased to Bestow Upon its Publication." In the introduction, he is a bit more specific as to what this attention and generosity may have entailed. Despite the difficulty of working with tiny shells, which required the use of a microscope, Walker persevered and was able

> to represent the objects, not only to the satisfaction of my two worthy friends, Mr. Jacob and Mr. Boys, but also of that most noble Lady her Grace the Dutchess Dowager of Portland, who hath been pleased to accept of several of the minute Shells here engraved, and whose generous approbation and assistance in this publication can never be forgotten without the highest gratitude.[4]

Apparently, the duchess examined the drawings and plates to make sure that the shells were properly represented and engraved.

She also seems to have assisted Walker in other ways, which may have included help with the expense of publishing. But having no evidence of this, I turn to something more concrete in the form of a comment that George Montagu made in a footnote to *Testacea Britannica* (1803). Referring to his friend William Boys, whose collection Walker drew and delineated and which Montagu inherited, he wrote:

> From this patron of science [Boys], we have also received a nomenclature by Doctor Solander, of many minute shells he had sent to Bullstrode for the Duchess of Portland; and others received from the Doctor at the same time, from her Grace's cabinet in return. This has been of considerable service to us with respect to synonyms.[5]

Montagu suggests here that Boys, and possibly Walker, exchanged shells with the duchess and that Boys sent his shells to Solander, and vice versa, to check them against those held in the Portland Cabinet and to get the proper names for them. Because shells were fragile and could break easily if not packaged properly, sending them back and forth was always a risky business. Exchanging tiny, microscopic shells would present even more difficulties, as George Walker notes:

> It is also adviseable to place the objects for inspection in a situation where no sudden blast of air can come, otherwise, being very light, they may be unexpectedly blown away, as I have too frequently experienced thereby the loss of several rare specimens; indeed a careless breathing or cough, while they were under examination, hath been attended with the same disagreeable accident.[6]

The duchess loaned and gave shells to several other writers as well, including Thomas Pennant and Thomas Martyn, who wanted to include them in their books – her shells were often so rare that they were the only available specimens. While the exchange of shells between the duchess and Boys seems to have gone on fairly smoothly, much could go wrong in this very complex process.

An awkward exchange between Pennant and the duchess over a Trochus, a snail's shell, reveals the real potential for error.[7] The fourth volume of *British Zoology* includes descriptions of eleven shells belonging to the duchess. The dedication reads: "To the Dutchess Dowager of Portland this work is dedicated, as an grateful acknowledgement of the many favors conferred by her Grace, on her most obliged, and most obedient, humble servant, Thomas Pennant." What exactly were these favors? Loans for this volume, as well as discussion of the illustrations, dominate their correspondence. Apparently, Pennant arranged for one of his illustrators, Peter (or sometimes Pierre) Brown, to draw some of the duchess's shells. Brown was from Denmark and, struggling to find employment as a natural history illustrator, had sought out Pennant for work and advice about his career. The duchess good-naturedly responded to Pennant's request to assist this young artist. She welcomed Brown to her Whitehall residence, gave him shells to draw, and when he sought her approval, she liked his illustrations so much that she wrote to Pennant that she would keep the drawings if he did not want them for his book.

> I am always glad of an opportunity of Obeying your Commands as my stay in Town will be very short[.] I sent to Mr Brown & gave him the two Shells you desired to have drawn & a new *Solen* or *mytilus* which I found last year at Weymouth which they

tell me has not been described but if you don't chuse to take the drawing of that shell I told Mr Brown I wou'd have it.[8]

Brown reported to Pennant that he had sent "three drawings of shells[;] the Duchess examined them with the shells and found them very correct. I have done one very fine Drawing upon vellum of a very scarce shell which Her Grace admired much." He was nervous about whether his illustrations were accurate enough in reproducing what naturalists needed to see in order to identify their specimens. He had had an unnerving encounter with Sir Joseph Banks, who had criticized some of his bird drawings, which Banks had said "consisted of little else than copies from [illegible]. Drawings, which are horid bad, and drawing by a fellow /as he terms it:/ that knew nothing of Natural History." He must have been discouraged by this accusation of making poor copies of badly executed illustrations, but he knew that he must "behave with the Greatest Civility" to Banks because it was his "duty and Business."[9] Brown must have taken solace in the duchess's approval of his shell drawings, but he may have had greater aptitude or experience in this subject, since he had drawn several shell images for the elaborately and beautifully illustrated series called *Conchology*, which was published in the early 1770s.[10]

Pennant sent the duchess a rare snail shell, a land trochus, as a gift, to thank her for her loans of specimens and advice on illustrations. It turns out, however, that this may not have been his to give away. The shell appeared in *British Zoology* as "*T. Terrestris*. Minute, conic, livid. A new species, discovered in the mountains of *Cumberland*, by Mr. *Hudson*. Tab. lxxx. *fig*. 108."[11] After seeing it, Mr. Hudson, the owner – or at least he believed he was the owner – wrote to Pennant demanding the return of his shell, conveying the request in the formal tones of the third person:

> Mr. Hudson wanted to have spoke to Mr. P. in regard to the shells Mr. Hudson lent Mr. P. and which Mr. H. has been informed Mr. P. had presented to the Dss. D. of Portland, particularly the land Trochus, which Mr. H. is loth to believe. Mr. H. therefore intreats the favor of Mr. P, to return them to Mr. Hudson, the first opportunity, that Mr. H. may have it in his power to convince his Friends that Mr. Pennant was incapable of such an Act – Hr. H. has sent to Mr. Whites's a copy of the *Flora Anglica* for Mr. P. which Mr. H. begs his acceptance of.[12]

In this letter, William Hudson, author of *Flora Anglica* (1762), manages to insult Pennant, accusing him of underhand dealing in giving the duchess the trochus that Hudson had "lent" him, while at the same time soliciting Pennant's approval of his recently published book. Hudson's combination of fawning and rudeness disgusted Pennant. Scribbled at the bottom of Hudson's letter is his comment: "Mr Hudson gave me the shell about seven years ago. & now demands it[.] [Wishing?] I had left him as a Apothecary for misrepresenting & abusing Mr Lightfoots Flora Scotia, to promote the sale of his own book." In another letter, Hudson thanks Pennant for responding to his letter about the shell, "which Mr H looked upon as only lent to Mr. P. for the purpose of his having it drawn etc," and Pennant scrawled at the bottom: "I accept his book; forgive him; but cannot employ him again." Eight days after the date on Hudson's first letter, the duchess wrote to Pennant telling him that she had sent the trochus to Hudson, trying to console him by saying that Hudson was not worth the bother: "he is below Mr Pennants notice, to resent his impertinence is doing him too much honour." Hudson continued to employ the legalistic tone of his previous letters, writing to Pennant to acknowledge the return of the land trochus: "Rec'd June 26 1778 from Thos. Pennant Esqr by Mr Hudson of her grace the Dss of Portland a land Trochus. W. Hudson." Clearly, Pennant had been mortified by Hudson's accusation, and it must have been equally awkward to ask the duchess to return his gift, even though she dealt with the situation by taking his part and telling him to ignore Hudson's behavior.[13] Despite this awkward episode, Pennant and the duchess seem to have had a generally congenial collaborative arrangement, with the possible exception of the following incident.

Although Pennant's dedication in the fourth volume of *British Zoology* is full of kind words about the duchess, he managed to slip into six copies a nasty little joke at her expense, playing up the sexual innuendos that were never too far from the surface of conchology. He asked his publisher to include a passage from Virgil about "the fondness of widowed ladies for this shell"; this passage, he claimed, is "only a bit of fun[.] if without much trouble it cd be put p. 75 to the note * tab. III p. 86[.] It may but I wd only have half a dozen leaves printed with it. It not being for the public."[14] Added were two phrases, one in Greek and the other in Latin, the latter reading:

> *Oblongae conchae solenes, et carne jucundâ*
> *Conchylium, viduarum mulierum cupediae.*

Translated this passage reads: "The oblong shellfish (known as) cockles, and mussels with their pleasant meat, are a delicacy for widowed women."[15] Preceding these inserts, which can be found in Sir Joseph Banks's copy of *British Zoology*, is the statement: "They were used as a food by the ancients, *Athenaeus* (from *Sophron*) speaks of them as great delicacies, and particularly grateful to widows" (presumably either because of the purported qualities as an aphrodisiac and/or their phallic shape). Pennant's little jab at the duchess may have been merely a prank, or it could have arisen out of feelings of frustration at his dealings with her, especially in his many requests for specimens, for advice on illustrations, and for the loan of rare and expensive conchology books to help with describing and classifying specimens. His request for the last was denied, even though borrowing books among naturalists was usual practice. Few, including Pulteney and da Costa, could afford these expensive and rare volumes, and they would have had to rely on borrowing them from wealthier fellow collectors and naturalists. Seymer loaned books to Pulteney while he was working on his publication of the natural history of Dorset. Though Pennant's request to borrow some of the duchess's books was not out of the ordinary, her refusal appears in the cold, formal language of the third person (apparently bad news often traveled in this disembodied voice):

> The Dss Dow^{er} of Portland presents her Compts to Mr Pennant . . . she is truly concern'd that it is not in her power to Oblige Mr Pennant in lending him the books he wants out of the Library at Bullstrode as she makes it a rule never to remove any books from thence[.] if it had been her own property she wd very readily have sent them to him.[16]

In this letter she thus positioned herself as the steward of her family's estate and possessions, drawing attention to her high social rank. This refusal to comply with a request that would have furthered the cause of natural history must have annoyed Pennant, and may have led to his private joke about widows and their desires in *British Zoology*. It is not surprising that Sir Joseph Banks possessed one of the six copies containing the joke; such high jinks were typical of the kind of misogyny that the duchess must have encountered often, surfacing frequently as a reaction to her power and influence in this male-dominated sphere.

THE BUSINESS OF PUBLISHING ILLUSTRATED NATURAL HISTORY BOOKS

Another publication for which the duchess lent shell specimens was *Conchology; or, The Natural History of Shells* (1770–72), a serial containing plates and descriptions published on a monthly basis as specimens became available to be described and drawn. The title page of the first volume reads: "to be published in Monthly Numbers, each Number containing Two Copper Plates, & Four Pages of Letter-Press; with their description in English and French." An issue with uncolored engravings sold for 3 shillings, and "Some Copies will be coloured after Nature, for the Curious who desire it, at the Price of Five

Shillings." They were to be sold by Benjamin White, Peter Elmsly, and "Mr Humphrey, Dealer in Shells."[17] In the preface, the "Editor" explains the way in which the numbers would be published, basically according to how many specimens of new species collectors were willing to lend for inclusion:

> The Editor begs leave to acquaint the Curious, that it is impossible to fix the Extent of his Work, as it will depend on the Quantity of new Species that occur; but he assures them, that he shall neither spare Expence, or be wanting in unwearied Application to render it complete, and hopes that on the Publication of the Numbers, they will judge of its Merits, and of its being more perfect than any other Work of Conchology hitherto offered to the Learned.

In the third issue, however, it became clear that this ambitious undertaking was suffering from a lack of coordination, the descriptions taking longer to produce than the illustrations. Apologizing for the lack of descriptions, the preface states: "it will be impossible at all Times for the Descriptions to keep pace with the Engravings." It indicates that the "Editor" will soldier on with the task and solicits the aid of collectors, asking them to deliver the shells to be described and drawn to Mr. Humphrey:

> There now only remains to solicit of the Collectors an Access to their Cabinets, to acquire the proper Opportunities of perfecting his intended Plan; and should any Ladies or Gentlemen possess any non-descript Shells in their Collections, and chuse to have them engraved and described, if they will honour the Editor to send them either to the Booksellers Messrs. White and Elmsley, or to Mr. Humphrey, to be conveyed to him, he will return them safe, and gratefully acknowledge the Favour, by adding to the Description the Collector's Name (if permitted) to whom he is obliged.[18]

Though the British Museum is the only collection cited here – mentioned four times in this third issue – the duchess was involved in lending shells for the production of *Conchology*: Emanuel Mendes da Costa's voluminous correspondence, held in the British Library, contains a series of letters between him and Humphrey about the making of the publication and the duchess's role in it. The letters shed light on the processes by which such shell publications were made, as well as offering some insight into the collaborative nature of authorship of the illustrated natural history publication.[19]

The authorship of *Conchology* has now been disputed for more than two hundred years. Some have argued that it was George Humphrey's project, others that da Costa was the author, while a few have suggested that it was a collaborative effort on the part of both Humphrey and da Costa.[20] This last position, espoused by Tom Iredale in 1915 and seconded by J. Wilfrid Jackson in 1937, makes sense since neither of these authors could have produced *Conchology* on their own, given the peculiar circumstances surrounding its publication. As for writing the shell names and descriptions, some of Humphrey's contemporaries doubted that he had the scientific expertise to classify and describe non-descript shells; one need only to look at the language he used in *Museum Humfredianum* (1779), the auction catalogue of his own collections, to see the dominant use of colloquial or local names of shells, such as "Persian robe," "clouded Ethiopian crown," and " crowned melon," with little recourse to Latin terminology.[21] As P. J. P. Whitehead argues, da Costa is the more likely candidate as the author of the preface and the shell descriptions, given his expertise as a taxonomist and his track record of book publishing (in the 1750s he had issued a book on fossils). But da Costa could not have been solely responsible for *Conchology* since he was in prison at the time it was written, illustrated, and published, and he could not have managed the stages of production from within the confines of the King's Bench. Whitehead describes these activities as proper to an editor, and reserves the title of author for da Costa because he probably wrote the shell descriptions.[22] That the authorship of this series has been a source of confusion and disagreement testifies to the complexity of the process of publishing an illustrated natural history text and the instability of the category "author," a construct that locates creativity within the individual and, in doing so, belies the collaborative nature of research, writing, and publication.

Humphrey's and da Costa's own writings provide contradictory evidence about the authorship of *Conchology*. Humphrey claimed authorship by attaching his name to these volumes in his catalogue, *Museum Humfredianum*, as in "Humphrey's *Conchology*"; and since he also helped to prepare the auction catalogue for the duchess of Portland's sale, it is not surprising that *Conchology* is also listed there with Humphrey as its author.[23] Da Costa, on the other hand, attested to his role as author in a letter written to an old friend, the Rev. William Borlase, in which he describes his activities while incarcerated:

> I have also made a full or descriptive catalogue of all the Elegant & various Collection of Native fossils of my esteemed friend Mr Ingham Foster & am now describing his extraneous fossils. I have besides publish'd an English Edition of Cronstedt's Mineralogy[,] Made a couple of (seven) Lectures on Shells and had begun a folio Natural History of Shells which as it did not meet with Encouragement suitable to the Expense I discontinued[.] These with other minute matters have enabled me to pass

near four years in confinement with more tranquility than in the great circle of the world . . . Thus my Dear Sir I have given you an Account of what I have done since my misfortune . . .[24]

One may suspect da Costa of being self-serving in this missive, and therefore an unreliable narrator, but contemporaries in the know assumed that Humphrey and da Costa were working together on this publication. Referring to *Conchology*, the duchess complained to Pulteney in April 1772: "Humphreys and da Costa give strange names to the Shells, the latter shou'd know better but he is being inaccurate in his descriptions."[25] Internal evidence supports da Costa's assertion that he wrote the descriptions for *Conchology*. A comparison of the prose style of the descriptions in *Conchology* with those he wrote for his later shell book, *The British Conchology* (1778), reveals an unmistakable similarity in style and content. The description of the shell with the name "large Fool's Cap" in *Conchology* reads:

> Shell – rather thin and semipellucid
> Size – when full grown is that of a Walnut, but small ones less than Filberds are most commonly met with;
> Shape – conic, with a broad Basis, but it is however very raised or cropt.
> . . .
> Inside – finely glazed, and is of a Milk-white, and a beautiful rosy Blush or Carnation Colour.[26]

The same species, *P. pileus morionis major*, is described in *The British Conchology* as:

> The *shell* is rather thin and semipellucid, when full grown, of the size of a Walnut; but small ones, of the size of filberds, are most commonly met with. The *shape* is conic, very elevated or copped. . . . The inside is smooth, finely glazed or glossy white, with a beautiful rosy blush or carnation colour.[27]

Of course, da Costa could have plagiarized *Conchology*, since copying passages wholesale from other books was commonplace in eighteenth-century book production, especially in reference books. Even George Montagu, one of the most scrupulous and scientific of shell book authors, borrowed much from his sources, including Pulteney's catalogue of Dorset shells. It seems unlikely, however, that da Costa plagiarized *Conchology* since it contained very few British shells and would not have been a logical choice as a source for *The British Conchology*. It is more likely that he wrote one description that was used twice. Additionally, some of the illustrations in *The British Conchology* bear da Costa's inscription "taken from the anonymous new Conchology," suggesting that he possessed, or had access to, the plates and perhaps the drawings of these shells.[28]

Because of his incarceration, da Costa could not have overseen the complicated process of drawing, engraving, and printing illustrations that formed the cornerstone of *Conchology*. This was George Humphrey's task, as several letters between him and da Costa testify. A constant theme that runs through the letters is the difficulty of procuring access to specimens located in prominent collections, such as the British Museum, and the frustration of negotiating with notable collectors for loans of rare, beautiful, and valuable specimens to draw and describe. The process of gaining access to an aristocrat to ask a favor was notoriously time-consuming, involving what was known in common parlance as "dancing in attendance" on His or Her Grace. Waiting in the anteroom of an aristocrat's palatial townhouse was a scene that appeared with frequency in satiric portraits of patrons and of those who depended on them for support and resources.[29] Waiting for the duchess to respond to his request to borrow specimens, Humphrey wrote to da Costa in February 1770: "I have not been sent for to the Duchess since her coming to Town, but expect a Message from her Grace every Day."[30]

Gaining access to shell specimens and reference books in the British Museum was even more complicated. Humphrey asked da Costa to draft a letter petitioning for admittance: "You will therefore please to find me a form of an Application, that I may transcribe it & send it to Dr Maty before Friday." He consulted da Costa about tipping the museum official who would accompany him: "As there will be an Officer to attend me what must I give him for attendance, or is there a Customary fee for it?"[31] Finally, he asked him to include a request to permit artists, one of whom was his own brother, to draw some of the museum's specimens (pl. 48).[32] Here is da Costa's draft, which Humphrey had to copy, sign, and submit:

> Your petitioner being now about publishing a natural history of Shells is extremely desirous of rendering it a perfect work he, therefore, with humble submission prays for your honours leave to inspect & consult the collections of shells in this noble museum at such hours & times your Honours shall pleasure to direct and that are not inconvenient for the Gentlemen who are keepers thereof to attend[.]
>
> His petition therefore is that your Honours permission be granted him to see & describe such shells as he cannot find in his & other Collections in this Metropolis & to bring with him a draughtsman to design or draw such shells in order to their being engraved and also to grant a permission for the Reading Room to consult, collate,

48 Emanuel Mendes da Costa, Plate 4 from *Conchology; or, The Natural History of Shells* (London, 1770–71). By permission of University of Glasgow Library, Special Collections

& quote such authors on this branch of natural History as are difficult to be obtained elsewhere.

The desire of your Petitioner has to cultivate Natural History and his attachment for this branch of it is the sole Cause of his troubling the Honourable Trustees of this noble foundation, & further he begs leave, to ensure them that his Gratitude shall attend his Thanks by humbly presenting A copy of his intended work on its publication.

Your Honours Most Devoted & Most Obliged Humble Servant GH[33]

Humphrey revised this letter a little, standardizing the punctuation and spelling, and removing some of the more abject quality of tone, before sending it to the British Museum staff.[34]

Publishing illustrated natural history books was a complicated and expensive enterprise. Procuring specimens was a difficult matter, and financing the project was equally daunting. Since neither da Costa nor Humphrey had funds to pay for the publication process, they opted for a subscription scheme, hoping to get enough subscribers to pay up front for the production of the series. Having seen their advertisement, Henry Seymer, the duchess's correspondent and fellow collector, offered them advice on how to handle subscribers. In a letter to Humphrey, he cautioned him to focus only on the "beautiful" and "rare" shells that would appeal to connoisseurs and provide a framework that was more manageable than the proposed depiction of every species:

> I would therefore omit the best common subjects, that have no beauty; and figure only the very rare, and well known ones, that will [illegible word] please the Eye, and Add to the beauty of the plates; such as the tygers, Cloth of Gold, Brunette; – Spectres, Night officers, flambeaux, &c &c &c of every Genus; if you could publish a Number every fortnight as was mentiond in your proposal Febr. 1st 1769 it would greatly encourage persons to subscribe as then upon the plan I have proposed, it would be finishd within three years & persons almost of Any Age might hope to see the Completion of it and this would not hinder your going on further if subjects occurrd, as Edwards did with his birds for when persons have purchased any work of this kind, if the Author goes on ever so long with new Specimens, there will be no doubt of their continuing to subscribe, till the whole is finish'd. I should therefore set out Another Advertisement setting forth your intention to publish for the future only the most rare, or beautiful subjects with an hint of the time, in which you think it will be finish'd . . .

Seymer worried that the expense and time "your Hist. of Shells will take up" would put off potential subscribers:

> if you proceed on the plan of figuring every variety you meet with[,] I am very certain the Work must be endless (this I know by my own Collection which is nothing compared with Dr Fothergills or the Duchess of Portland) and that most people will be deterrd for this only reason from Subscribing to it.[35]

In addition to drumming up subscribers with advertisements, da Costa tried to tap into a wider market by making arrangements with European booksellers and natural history dealers to sell *Conchology* at a rate that would also allow him to make a slight profit. To J. C. Sepp, a bookseller in Amsterdam, he wrote:

> I am extremely rejoiced that you accept of the proposal to become my correspondent in regard to the sale of History of Shells I am now publishing here in numbers[.] I find your conditions are 20 pc [percent] commission on the Coloured copies & 25 pc on the black or Uncoloured to which I agree and do not [doubt] you will act with all assiduity in the sale of them. . . . The prices we sell each coloured copy here in London is five shillings sterling and each uncoloured or black one we sell at three Shillings. I therefore think you should sell each coloured copy at 6 s – & each uncoloured one at 4 s else I should be a great Loser by the Expences of sending them to Amsterdam[.] I hope you will receive these 18 copies very soon and safe . . . I have sent these as a meer trial. . . . I was glad you are a Collector & Love Natural History . . .[36]

Though da Costa had an extensive network of European correspondents with whom he exchanged ideas, information, advice, contacts, and even specimens, he was only marginally successful in promoting the sale of *Conchology*.

Managing the artists and engravers while coordinating their efforts with printers was a complex task that was often too much for Humphrey and da Costa to handle. The need to speed up the process was a source of constant concern; panic surfaces throughout the letters as deadlines loom. At one point Humphrey wrote to da Costa trying to calm him down about their failure to meet deadlines, saying: "We are not in so bad a pickle as you imagine (tho' bad enough)." Da Costa was anxious about the slow engraving process, and Humphrey attempted to cheer him up, writing that there was still time to complete the various tasks of illustrating and describing the shells for the first volume (pl. 49):

> Plate 6 is engraved, and 5 will be done in about a week or 8 Days; the *Masks* which are for the 7th plate (which I suppose you have forgot) have been drawn twice, tho' some of them must be redrawn; It [is] his wish [for] you to settle for the 8th plate which is to contain all the Ears; and as soon as you have look'd them out [over?] shou'd be glad you will let me know & I will find Bearer for them. You shall have Gualtieri in a day or two. Herewith you will receive Valentini which I have got at last; it cost me 1.2 – You will find however that we are time enough for quoting it.[37]

Da Costa not only had to write the descriptions, which involved examining those in other conchology texts, such as Niccolò Gualtieri's *Index Testarum Conchyliorum* (1742),

49 Emanuel Mendes da Costa, Plate 5 from *Conchology; or, The Natural History of Shells* (London, 1770–71). By permission of University of Glasgow Library, Special Collections

Michael Bernhard Valentini's *Museum Museorum* (1704–14), and his *Amphitheatrum Zootomicum* (1720), but he also had to examine the drawings and engravings, checking them against the shells for accuracy, and ensure that the illustrations corresponded correctly to the text. Drawings, page proofs, and proof plates traveled back and forth between Humphrey and da Costa: "Mr H has also sent 2 proof prints of plates to which Mr DC will be pleased to number & return per bearer."³⁸ Trouble with timing and coordinating of all these tasks plagued the project. Humphrey wrote:

> Martini you shall have in a day or two. I also herewith find you the Shells of Plate 5 that I have. I think I gave you the Descriptions of the rest, if not, send me a word & I will send them per Letter. I have about 10 or 12 new Species of Shells by the Endeavour, 2 or 3 of them are Ears & Limpets I will endeavour to see you soon. & shall put the varieties in my pocket. I also want to consult you abt a plate No XII with a * to consist of these Limpets & Ears by way of appendix to that Class. Tis not too late as it will be some time before you reach so far with the Description.³⁹

A month later, Humphrey was less confident about juggling all these demands in a timely manner (pl. 50):

> Marzelle [draftsman and engraver] has promis'd me the 7ᵗʰ Plate next Monday . . . I find it will be impossible to recover the work to the regular [illegible word] time of publication till we are a little more settled, in spite of all my Endeavours some cursed Delay of the Drawer, Engraver or Colourer keeps it back. Perhaps you will add "Or attendance on Duchesses." . . . I had an add' [additional] Subscriber (Mrs Delany an acquaintance of the Duchess').⁴⁰

Delays caused by artists, engravers, and printers, compounded by trouble getting hold of specimens, prevented Humphrey and da Costa from attaining their goal of publishing issues at regular intervals.⁴¹ By the time the third volume was published, all the plates were finished, but da Costa had fallen behind in his descriptions, and the plates were published with only half of their accompanying captions. The expense and effort to carry on with the series proved too much for them. Da Costa explained the reasons in his *Elements of Conchology* (1776):

> A new anonymous conchology began to be published in this Metropolis in 1770, in folio, illustrated with copper plates. It was to be published in monthly numbers, and each number to contain two plates of Shells, with their descriptions in English and French. It was also intended to be a General Natural History of Shells, and to include the figures of all the known species, common as well as rare, beautiful, or otherwise; and some copies were designed to be accurately coloured for the use of the curious. Six numbers of it were published, comprehending the families of the Limpets, Sea-Ears, and Worms; but not meeting with suitable encouragement, the authors have laid it aside, at least for the present.⁴²

50 Emanuel Mendes da Costa, Plate 7 from *Conchology; or, The Natural History of Shells* (London, 1770–71). By permission of University of Glasgow Library, Special Collections

The failure of *Conchology* can be traced to several problems. Obviously, one was its expense, and another was its conception. As Maton and Rackett suggested in their "Historical Account of Testaceological Writers" (1804), this serial publication "was on too extensive scale to admit of being completed; we mean the '*Conchology or Natural History of Shells*,' which was published, anonymously, in folio numbers, but never proceeded beyond twenty-six pages of letterpress and twelve plates."[43] Designed primarily for connoisseurs, it emphasized the illustrations, which were large, elaborate, and beautifully colored. *Conchology* appealed to the eye, as Seymer counseled, and had much in common with mid-eighteenth-century Continental conchology books and auction catalogues.[44] Its appeal to connoisseurs shows Humphrey's influence over the project; he was, after all, a dealer in natural history specimens, and such a book would help his business, stimulating collectors' desires to obtain the figured specimens, and teaching them what to aim for. Da Costa's descriptions may have been too scientific for this audience, and yet the shells did not adhere strictly to Linnaean nomenclature, making the volumes less attractive to naturalists. Lacking focus, *Conchology* reflects the goals of its two creators, which were sometimes at odds with each other. For these reasons, perhaps, the series failed to capture a sufficiently large readership to make it a financial success, and so the project was discontinued, though da Costa did go on to write two more shell reference books, *Elements of Conchology* (1776) and *The British Conchology* (1778). Less ambitious in scope, they were far more successful.

Elements of Conchology was probably based on the series of lectures that da Costa gave while incarcerated in the King's Bench, where he had the space to give a few series of lectures on fossils and shells.[45] He continued to rely on specimens from friends, Pulteney contributing some for the illustrations in *The British Conchology*, and also continued his working relationship with Pierre Brown, who did illustrations for both *Elements of Conchology* and *The British Conchology*. Whether the duchess continued to supply him with specimens for these two books is uncertain, but she did subscribe to *The British Conchology*, finding the latter so useful despite its odd nomenclature that she gave a copy of it to the budding conchologist John Timothy Swainson.[46]

Da Costa referred to his years in jail as resulting from "my errors" and "my misfortune," and was initially much concerned that he had lost the good opinion of his fellow naturalists.[47] Commenting on his behavior, Dru Drury wrote to a fellow naturalist, Dr. Pallas, with whom da Costa also communicated, and described in immoderate tones da Costa's disgrace:

> Da Costa's temper and principle was sufficient to overturn a Kingdom. I imagine you have heard his *Fate*. If not, I will tell you. He is no longer librarian to ye Royal Society. His is dismissed from thence with ignominy and disgrace. . . . Hence ye periodical work he intended to publish . . . is entirely stopt: the circumstance I must own I am very sorry for on account of Natural History in general. But if it can not be promoted by men of better principles than him it is better perhaps for it to lye dormant.[48]

This condemnation provides another clue to the authorship of *Conchology*, and would explain why it was published anonymously. While an air of ignominy hung over da Costa's reputation, his name would not have helped the search for subscribers or the sale of the volumes. Rather than give up the project, it seems that he worked out with Humphrey a way to publish without calling attention to his contribution.

Judging from da Costa's continuous correspondence with notable naturalists after and even during his incarceration, it is clear that Humphrey, Dr. John Fothergill, and Ingham Foster remained his friends; the Rev. William Borlase and Pulteney seem to have forgiven him, and the duchess contributed shell specimens for *Conchology*, knowing that Humphrey and da Costa were partners in this publishing venture. Da Costa may have forfeited his position and fellowship of the Royal Society, and his place within polite society – already marginal given his Jewish heritage and religion – but he did not, as some historians suggest, lose the respect of his fellow conchologists.[49] His *British Conchology* remained the standard text for classification purposes into the first decades of the nineteenth century, and the plates were slated for reissue by the Rev. Thomas Rackett in 1813 as *Triton Britannicus*, a second edition of Pulteney's "Catalogues."[50] As the popularity of studying British shells increased in the last years of the eighteenth century, so did da Costa's reputation as a naturalist. A measure of this is the way in which Edward Donovan's advertisement for *The Natural History of British Shells* (1799–1803) proudly proclaims his acquisition of da Costa's collection of shells and notes, which formed the basis of this illustrated edition:

> [W]e have fortunately added Selections from several cabinets of great Celebrity, and in particular the entire Collection of the late EMANUEL MENDES DA COSTA, Author of the *Historia Naturalis Testaceorum Britanniae*. This came into our possession at the death of that eminent Naturalist, and is of the utmost importance to the scientific World. It contains not only the individual Specimens he has figured and described, but many extraordinary and interesting *Varieties, new and rare Species, MS. notes, &c.* designed for an improved Edition of that Author's Work, but never published. We readily acknowledge the assistance derived from this source of information, and shall state specifically every instance in which we avail ourselves of it. Many faults in his Works are corrected; many obscure passages are explained; but in general we are indebted to his Notes for detecting them. The new Genera of DA COSTA, as well as the specific names and Characters, are throughout reduced to the more approved System of Linnaeus; yet as many Admirers of his ingenious Arrangement may still remain, a separate Index will be appropriated to elucidate it, with all the Shells he has described in his BRITISH CONCHOLOGY.[51]

For Donovan, Pulteney, and George Montagu, da Costa's *British Conchology* was the foundation of their work. Certainly, some naturalists quarreled with his attempts to revise Linnaean nomenclature and to remove what he felt were vulgar and inaccurate terms. All in all, though, his descriptions were accurate, and clearly and vividly written; they were ultimately helpful to naturalists and conchologists eager to identify their collections of British shells. In an overview of the conchological writers of the eighteenth century, Maton and Rackett sum up da Costa's career: "*British Conchology* was the work that conferred most reputation on this writer; and it formed a valuable addition to the natural history of our island."[52] Judging by its availability today in science libraries, it must have been published in large numbers, and despite da Costa's tinkering with Linnaean nomenclature, was a much-used reference book throughout the nineteenth century and into the twentieth.

THE UNIVERSAL CONCHOLOGIST AND OTHER SHELL BOOKS

The troubles that Humphrey and da Costa faced in producing *Conchology* were typical of illustrated natural history books. A decade later, Thomas Martyn, natural history dealer, illustrator, printer, and bookseller, took up the challenge of producing an illustrated shell book. Drawing on his own collection of shells and those of other famous collectors, including the duchess of Portland, Martyn actually succeeded in producing a magnificent publication, *The Universal Conchologist*, originally published in two volumes in 1784 with a second, deluxe edition of four volumes issued in 1787 and 1789, which in the words of Peter Dance was his "magnum opus and lasting memorial."[53] Unlike the ill-conceived *Conchology*, Martyn's publication had the well-defined goal of picturing southern hemispheric shells: "The work will commence with the figures of the shells (most of them rare and non-descript) that have been collected by the several officers of the ships under the command of Captains Byron, Wallace, Cook, and others, in the different voyages made to the South Seas." He did not bother with description or textual references or taxonomic quandaries, since these concerns would take care of themselves over time:

> The long descriptions and details of the generation and properties of Shells, given by most writers on Conchology, are wholly omitted here; and the utmost care has been taken that each figure, by being an exact and faithful transcript from nature, shall be sufficiently explanatory of the subject which it represents.

Martyn focused exclusively on the visual aspect of the shells; what was important to him was making beautiful – and, more importantly, accurate – illustrations, so that collectors could compare their shells with those pictured; in this way they would learn what was available and on what to direct their collecting endeavors.[54]

Martyn, like Humphrey, saw connoisseurs as the market for his book. As an active buyer and broker of shells from Cook's expeditions, he knew who possessed the voyages' best shells. To augment his own purchases from newly docked ships, he asked collectors such as the duchess of Portland and Dr. John Fothergill to loan him shells to portray in the volume.[55] Unlike *Conchology*, which rarely cited the collection that a pictured shell came from, keeping this information confidential, Martyn's *Universal Conchologist* begins with an impressive table that indicates both a shell's provenance (New Caledonia, Otahiti, etc.) and the British collection in which it was housed (pl. 51). Martyn also included in the preface brief biographies of the shell collectors who had supplied specimens for many of the images, as well as a brief assessment of the quality and range of their collections. Of one collector, he wrote:

> If acuteness of judgment in the selection of specimens, joined to a critical knowledge in the arrangement of Shells, suffice to constitute a Conchologist, John Smith Budgen, Esq. has every pretension; consequently his collection is not only very extensive, but admirable in its suite, and shews what genius accompanied with industry may accomplish.

These descriptions are an excellent source of information on women collectors. For instance, he describes Miss Fordyce, daughter of Dr. George Fordyce, both of whom purchased quite a few of the duchess's shells at the Portland auction:

> Few ladies have studied more attentively, or succeeded more happily in this branch than Miss Fordyce, as every subject in her truly capital collection is a proof of that peculiar care and discernment requisite to form so rich a suite of Shells, including some unique, and many others of the scarcest species. The whole are classed and arranged with the most critical judgment and elegant taste, affording at the same time instruction and delight.[56]

Part of the pleasure of this sumptuous publication was this connoisseurial knowledge concerning provenance: in short, who possessed what and where it currently resided. Martyn had plans for several volumes, publishing one volume of forty plates in 1784, and an expanded two-volume version in 1789. He was, however, worried that he could not sustain this project of publishing an image of every known shell from the southern

51 "Explanatory Table," showing shells and their collectors, from Thomas Martyn, *The Universal Conchologist* (London, 1784). London, British Library, 37.g.8 (2). © The British Library Board

hemisphere, given the great decline in the number of new and rare shells on the third voyage. As early as 1780 he complained to Henry Seymer about the dearth of new molluscan specimens:

> It is a little extraordinary so few new species should have been collected, considering the many different Islands in the S. Seas the two ships visited & so many persons employed to gather them, & I may venture to affirm that I have purchased, amounting to 400 guineas, more than 2 thirds of the whole brought home. Nevertheless I do not abound either in variety of the new or many duplicates of the known ones that are valuable.[57]

Martyn's plan to produce a multi-volume publication depicting Pacific and southern hemispheric shells was abandoned, but, according to Dance, "in the attempt he produced a work which, for beauty, has seldom been surpassed in the history of conchological iconography."[58]

Well aware of the enormous expense and other difficulties that could plague the production of shell books, Martyn explained in the preface to the expanded version of *The Universal Conchologist* (1789) why there were no conchology books of merit:

> This in part may be ascribed to the employment of draughtsmen, painters, engravers, and colourists, ill qualified for this business; or who, however adequate to such an attempt, have nevertheless neglected to give that minute attention to the execution of their respective departments which the exigence of the subject required.[59]

To achieve his goal of printing a shell book filled with beautifully drawn, engraved, and colored pictures, he established what he called a "seminary" for boys, who would be responsible for producing illustrations for this and other natural history books. The seminary mingled workshop conditions with aspects of an art academy and the social codes of a charity school for the poor. Eager to "acquaint the public with the nature and principles of a private establishment which he has formed for the instruction of youth, in the art of illustrating and painting subjects of natural history," Martyn explained in his preface that it functioned much like a seminary, training boys who came from humble homes to be art workers.[60] He described the rules of conduct expected of his "scholars":

> To yield an implicit obedience to the Master, and to apply themselves to their several occupations with assiduity, and in silence; To maintain a strict cordiality among themselves; To be more ready to amend their own failings than to expose those of their fellows; To avoid with scrupulous attention all indecent words and dishonourable actions; To cultivate a love of truth, and entertain a modest opinion of their own merits; and to respect those of others. In short, this little seminary is governed by the dictates of religion and virtue; and the several duties both towards God and man are here strongly enforced; since the founder and conductor of it would feel it a nobler boast to have educated one good citizen, than any number of artists however ingenious.[61]

Martyn's little "academy" of poor boys enabled him to avoid the expense and trouble of dealing with artists, whom he considered notoriously difficult to work with on natural history illustration. Some were too proud to demean themselves by drawing natural history subjects, and others would not or could not copy a specimen faithfully, leading Martyn to conclude that there were "few artists (we may indeed say none) who particularly devote the application of their talents to this particular branch of the art." To employ artists capable of "executing the work according to the author's ideas, would eventually have been attended with an expense so great, as in its necessary consequences would more than have trebled the present price of each volume." Surprisingly frank about the economics of his drawing academy/workshop, he wrote: "the labor of boys . . . is always cheaper than that of men." Its establishment depended on being able to spot talented boys who would be grateful for the training and work: "he had to find for the execution of his purpose such hands as, possessing abilities adequate to the end, could not, from their situations in life, be more profitably employed in other occupations," meaning that the boys would not walk away from their job as his illustrators to search for more lucrative employment.[62]

Martyn's unusual solution to the problems associated with illustrating natural history books reveals the complexity of the process and the numbers of people involved in production, including draftsmen, etchers, printers, and colorists, many of whom were children and women, all no doubt poorly paid for their labor.[63] These art workers produced lovely illustrations, such as the one depicting the duchess's shell, a "Checkered Mitre" (pl. 52). But though clever, Martyn was not immune to the problems that plagued the production of illustrated books. Even with his little academy of industrious boys, he was unable to cover the costs of production. In their overview of "Testaceological Writers," Maton and Rackett described *The Universal Conchologist* as "one of the most beautiful and costly conchological works this country has ever seen," but:

> [B]efore this ingenious artist had completed his two volumes of South Sea shells, he discovered the impossibility of procuring purchasers sufficient to compensate him for his labour and expense, – a misfortune generally experienced by private individuals who embark in such extensive and sumptuous undertakings. He, therefore, did not proceed beyond 160 plates; which, however, as they include all the species then known to the southern navigators, may be considered as constituting a complete work, so far as it goes, and it was all that Mr. Martyn had absolutely engaged himself to execute.[64]

The duchess's shells are represented minimally in the large-format 1789 edition of *The Universal Conchologist*, since they were sold at auction after her death in 1785, before Martyn completed it.

52 Thomas Martyn, "Checkered Mitre," Plate 19 of *The Universal Conchologist* (London, 1784). London, British Library, 37.g.8 (2). © The British Library Board

The duchess's contribution to the shell publications discussed above consisted primarily of loaning specimens for depiction and subscribing to the volumes. It is thanks to Martyn's efforts and those of such shell-book authors as Pennant, Walker, da Costa, Humphrey, and Donovan that we have illustrations of the duchess's shells, which is some compensation for the fact that the whereabouts of her shells is quite uncertain, a topic examined in the next chapter. She may also have been actively involved in the publication of William Huddesford's reprint (1770) of Martin Lister's classic *Historiae sive synopsis methodicae conchyliorum*, initially published in four volumes between 1685 and 1688, which is dedicated to her. A clue to what she did to deserve this dedication can be found in *A Catalogue of the Portland Museum*. Lot 2668 is described as "A quarto volume, containing a variety of original drawings of Shells, by Lister and his daughters." Apparently, the duchess owned not only these drawings but also Lister's own annotated copy of his four volumes, now housed at the Linnean Society. It was this copy that Huddesford used to produce a modern edition of Lister's text.[65]

As the study of British shells grew in popularity, so the importance of Lister's works increased. He was one of the very first naturalists to describe English shells, and his books with their focus on the island's animal life became foundational texts for classification. A measure of Lister's importance is shown by a request from Pulteney asking da Costa to purchase the *Historiae animalium Angliae* (1678), an important book for taxonomists; the duchess owned a copy, which was used by Solander and Lightfoot in their descriptions of her specimens. In 1776 da Costa wrote to Pulteney with the bad news that the book was impossible to find for sale:

> In regards to Lister's *Historiae Animalium Angliae*, that you desire me to buy for you, It is now become a very scarce book; there is not a bookseller of note in this Metropolis, but I have been to, tho as yet without Effect. During the Winter, as Sales of books come on, I may likely get it, & will watch narrowly for it. However, . . . til then (for I know you earnestly desire to peruse & consult it,) I have borrowed said book of my friend Mr. Foster, with permission to lend it to you in a very few days, by way of Mr Wilkie. Your perusal of it will not only be a great satisfaction to you, but I am also convinced I shall greatly gain by it. The Nerit you lately found on your Coast is a pretty shell, & is not very rare on our shores, I call it the Chain snail, you will find it in Lister.[66]

The comment that the nerite was less rare than Pulteney thought is slightly barbed, a way of asking how Pulteney could possibly classify specimens without a proper library. It is an odd comment, given that Pulteney had asked da Costa to purchase Lister's book for exactly that reason.

Lister's shell book, *Historiae sive synopsis methodicae conchyliorum*, was even more sought after as a reference book and was quite scarce. It is unclear whose idea it was to reissue the book; the project was most likely a collaborative one, involving the natural history dealer Ingham Foster (da Costa's friend and Humphrey's brother-in-law), Huddesford, curator of the Ashmolean Museum at Oxford, and the duchess of Portland. Huddesford was the editor, but, as da Costa rightly points out, he did not do much editing.[67] It is, in short, a reproduction or what might be called today a facsimile. Da Costa, who had been eagerly awaiting the publication of this valuable reference book, was extremely annoyed by the final product since it contained no new commentary on Lister's original text – no attempt to update it or to critique Lister's descriptions. He was frustrated by Huddesford's failure to take advantage of the opportunity to revise a book that was nearly a hundred years old. He complained to his friend Ingham Foster that Huddesford's edition failed to honor the "Original Author" since it did not provide "Natural, Critical, and Historical Notes." Da Costa wished that Huddesford had

> given some Historical Account of the Original Work either from proofs or probabilities by Collating different Copies or Manuscripts of it and by the variations (as print collectors express it) . . . Mr. Huddesford . . . might have informed us of the State of the Copies in the public Libraries of England as those in the Royal Society[,] in the British Museum[,] the Royal College of Physicians . . . and those he could find in private Libraries . . .

He went on to chastise Huddesford for not giving "some Account of the very plates themselves in the Ashmolean Museum under his custody and to have informed the Curious whether he has reprinted all he found or whether he rejected any[,] what rule or authority he governed himself by in numbering plates." This lengthy critique, including also Huddesford's failure to describe the book's reception by naturalists since publication, cannot be ascribed merely to crankiness or, as da Costa says, to "being an Invidious Critick."[68] His standards for a scholarly edition are surprisingly modern, since they mirror nearly exactly the methods and tenets espoused by modern textual scholarship.

Huddesford was ill equipped to do the kind of textual analysis and reception studies that da Costa thought the volume deserved: "It is a very great pity the Illustrious University of Oxford when they took the republication under their patronage did not appoint a person or persons more conversant in this branch of Natural History to revise[,] correct and enrich it with proper Illustrations." Not well versed in natural history, Huddesford had merely inherited his father's position as curator of the natural history collections of the Ashmolean, which had been neglected under his father's tenure and had deteriorated significantly. Unlike his father, who regarded the post as a sinecure, Huddesford took his job seriously and worked hard to restore the museum to order and to bring it up to current scientific standards, dusting off its reputation as merely a hodge-podge of dusty, old curiosities.[69] Perhaps his editorship of the Lister reprint was an attempt to call attention to the work being done at the Ashmolean and to bring it and its curator to the notice of the scientific networks centered in London: the British Museum, Royal Society, Greenwich Observatory, and Kew Gardens.

CATALOGUING THE PORTLAND MUSEUM

The conchology book with which the duchess was most involved, of course, was the catalogue of her own shell collection, which due to Solander's sudden death in 1782 and her own in 1785 never came into being. It appears only as a ghostly presence in the sales catalogue that accompanied the auction of her collections of natural history specimens, fine and decorative arts, gems and jewels: *A Catalogue of the Portland Museum, lately the property of the Duchess Dowager of Portland*. This auction catalogue contains vestiges of the years of labor that went into amassing the shell collection, the knowledge that governed the choices that shaped it, and the work of classification that the duchess and Solander engaged in as they worked toward the goal of producing a scientific publication describing and naming her molluscan specimens. Their goal was the dissemination of accumulated knowledge, for, as the preface notes, "It was indeed in the Intention of the enlightened Possessor to have had *every unknown* Species described and published to the World."[70] Since the duchess's shell collection was one of the largest, if not the largest, collection of its kind in the world, the catalogue would have been a landmark in the print culture of conchology, an invaluable guide for taxonomists, and an important source of information for collectors and naturalists alike. Solander worked on and off for four years on the collection, depending on when the duchess was in town. He spent one day a week, usually a Tuesday, staying from 11 in the morning to 6 in the evening, pausing only for a bowl of soup. According to Pulteney, it took Solander fifty days to classify all the "Bivalves except the *Mytili* & *Pinna*," and "the Volutes took 20 days six Hours every Day. . . . The Buccina were not finished when the Ds. left town in 1781 for the summer[.] 12 Days had then been spent on them."[71] At the same time Solander was also at work revising Linnaeus's molluscan taxonomy, renaming general as well as specific categories; his goal was to publish his own taxonomy based on annotations, refinements, and redefinitions of Linnaeus's. This larger project permeated his cataloguing of the duchess's shells, so that even in its diminished and bastardized version the sales catalogue contains Solander's ideas. *A Catalogue of the Portland Museum* is therefore a valuable tool in

53 Daniel Solander, "Ostrea peregrina," manuscript descriptions of animals written on slips and systematically arranged. London, Natural History Museum. © Natural History Museum, London

reconstructing his classificatory schemes and remains an important source for recovering his thinking about natural categories.

Another source can be found in Solander's own manuscripts and annotations, and on the little slips of paper on which he wrote down names, all of which were kept at the British Museum, and which naturalists consulted as they worked on their own catalogues of shells and other zoological matter (pl. 53). Edward Donovan, compiler of *The Natural History of British Shells* (1799–1803), clearly had access to Solander's manuscripts since his descriptions contain several references to Solander's work. Despite his respect for Solander's expertise as a taxonomist, Donovan registers on one occasion his critique of the choice of the name "edentula" for a species of volute, a sea snail, that has an "outer lip" which is "slightly denticulated," as the teeth-like pattern on the edge of the shell is described:[72]

> Dr. Solander, who, it is well known to the scientific conchologist, intended to have published a catalogue of that Museum [Portland], it appears, on a reference to his posthumous papers, called this species edentula; a name which, without detracting from the merit of that able naturalist, it must be allowed is by no means applicable.

Donovan seems to have consulted Banks's shell reference books to determine the proper name of another shell, and in doing so found a note by Solander: "from a ms. note in one of the copies of Lister's work, in the library of Sir Joseph Banks, we find the late Dr. Solander intended to have named it [*Arca noae*] specifically *fusca*, had he lived to publish his new arrangement of Conchology."[73] Several collectors, dealers, and naturalists worked from Solander's manuscripts, and also made copies of his molluscan slips. George Humphrey copied the information he found there into a set of notebooks, which the Linnean Society now possesses: "three small notebooks containing Solander's notes on genera of molluscs, copied out by George Humphrey, the shell dealer."[74]

Despite the loss of Solander and his expertise, the duchess forged ahead with plans to publish a catalogue of her collection. Fearing that she would publish Solander's half-completed descriptions, Sir Joseph Banks wrote to her in 1782 to express his concern about her plans. His outrage at what he thought would be a disservice to his dear friend was conveyed in politely muted and deferential terms:

> As I understand from Mr Lightfoot that your Grace is absolutely determined to publish the very unfinish'd descriptions of Shells which my deceas'd friend, Dr. Solander, made from your Collection, it would be impertinent in me to offer any advice on the subject. As, however, the memory of a Man with whom I have spent so many years in close connexion cannot but be an object of importance to my Feelings, & as I trust that your grace feels the remembrance of his friendship almost as warmly as myself, I shall venture to make one request relative to the mode of doing it.

His request was for her to employ someone "who is capable and will undertake" to finish the desiderata and that these additions be "printed in a Character [font] different from that in which the original writing of our friend is printed." This way "every reader will gain a true knowledge of what really is the work of Solander; & what ever merit the Editor may have will like wise be justly attributed to himself."[75] He closed the letter with a suggestion that he would be willing to "correct the press, for the purpose of ascertaining this to be always done," referring to the typeface that would distinguish Solander's names and descriptions from those of the person who would take over the completion of this manuscript.

The duchess responded, trying to quiet Banks's fears by assuring him that she too wished to honor and respect Solander's work:

> my motive was to shew my real regard for him, & to throw a light upon a branch of natural history that no one but himself was capable of, & therefore I most intirely

agree with your Idea that not even the smallest particle shou'd be added to Dr Solander's work but in a different Character.

She explained how she had planned to have Lightfoot "correct the Press," but she readily accepted "your Obliging offer (if it is not given you too much trouble)." She concluded her letter most graciously by stating: "I shall be better satisfied to have it published in the way most agreable to you," which ultimately is an acknowledgment of Banks's very close relationship to Solander and his right to protect Solander's reputation and legacy as Britain's leading taxonomist.[76]

The duchess did indeed turn to Lightfoot to finish Solander's descriptions of her shells. Lightfoot was at work on preparing the manuscript for publication in August 1784, when she wrote to Pulteney, explaining that she still needed the shells she had borrowed from him: "I have not return'd your Shells[.] they are all safe in the Box as Dr Solander left them[.] I am in some hopes Mr Lightfoot if he can afford time will go over them for the press."[77] In November 1784 Banks wrote to a fellow naturalist of Solander's death and accomplishments, noting that "the descriptions he made of them [the duchess's 'Gastropods'] he delivered to the Duchess and we daily expect to see them made public in the same state he left them, by Mr Lightfoot."[78] Solander's descriptions and new names, however, as well as Lightfoot's additions and emendations, did not find their way into a state-of-the-art book on conchology as the duchess had hoped; instead, they became part of the auction catalogue. With the duchess's death, the likelihood that Solander's descriptions would be published in a scientific document had ended; despite its importance, no one else seemed prepared to bring this work to publication. Banks, for all his bluster about honoring his friend's reputation and guarding the integrity of his work, was unable or unwilling to publish it. Solander's work, however, continued to circulate after his death in manuscript form, appearing, along with references to his new molluscan names published in *A Catalogue of the Portland Museum*, in the citations of subsequent shell publications.[79]

If the duchess had lived long enough to bring the catalogue of her own collection to publication, she, in collaboration with Solander and Lightfoot, would have produced a landmark in the print culture of conchology. Even without Solander's finished descriptions, the publication would have had scientific significance as a natural history reference book and would have had an enormous impact on shell collecting. Only the *Catalogue of the Portland Museum* remains of this ambitious project, which would have secured Solander's reputation as one of Europe's greatest taxonomists and the duchess's reputation as a serious naturalist who had amassed a scientifically significant collection of mollusks. But this was a catalogue for an auction that dismantled, dispersed, and, in short, destroyed the duchess's shell collection.

As it turned out, her contribution to the print culture of conchology was that of a patron, not an author: she lent specimens to be drawn and described; she offered advice on the naming of specimens and gave them to Solander to work on; she compared the illustrations against the specimens, judging the scientific accuracy of the drawings; and she could always be counted on as a subscriber. These activities, important to the making of these illustrated natural history books, reflect the collaborative nature of book publishing and the way in which specimens and information flowed within naturalists' networks. Without publication, Margaret Cavendish Bentinck's work as a naturalist was invisible except to those who knew her; and this recognition of her abilities as a specimen hunter and as an expert in Linnaean systematics was lost over time.

CHAPTER SIX

54   View of the house and museum of the duchess of Portland in the Privy Garden, Whitehall (detail), 1796, watercolor. Inscription at bottom reads: "The House of the late Dowager Duchess of Portland as it appeared May 1796, by John Bromley." London, British Museum. © The Trustees of the British Museum

THE DISPERSAL OF THE COLLECTION

Margaret Cavendish Bentinck, dowager duchess of Portland, died suddenly on 17 July 1785 aged seventy years. By the following April her natural history specimens, all of which had taken her decades to accumulate – shells, insects, birds, fossils, minerals, gems, and corals – had been divided up into lots and were ready to be sold at auction, along with her collections of china, prints, furniture, snuffboxes, sculpture, and other fine and decorative art objects. The auction took thirty-eight days, and thousands of shells were sold, packaged as small lots designed to attract potential buyers eager to own something rare and beautiful from the duchess's collections. Curiosity seekers, serious naturalists, antiquarians and connoisseurs, dealers acting as agents, and brokers speculating on shell futures all sought out the auction as an event not to be missed. Promoted in newspapers weeks before it took place, the auction and the events leading up to it proved quite newsworthy, with much speculation about the duchess's finances and the reasons for the auction, as well as daily reports and advertisements describing the ongoing sale. The excitement over this event, the almost giddy carnivalesque atmosphere evoked within the gossip columns of newspapers, belies the real grief that those closest to the duchess, particularly Mary Delany and John Lightfoot, experienced with the loss of their patron and friend. The impact of her death on the study of natural history was perhaps most acutely felt by Lightfoot, who was eloquent in describing how heartrending it was to watch the dissolution of her natural history collections, in particular, the breaking up of her shell collection. This chapter recounts the events leading up to the auction, describes the auction itself, and speculates about the paths that the duchess's shells took as they exited the collection, dispersed in bits and pieces, to hundreds of buyers, some well known, like George Humphrey and Emanuel Mendes da Costa, and others about whom very little is known.

THE DUCHESS'S DEATH AND ITS AFTER EFFECTS

In July 1785 London newspapers were full of reports of Margaret Cavendish Bentinck's death and its impact on her children, her will, and the fate of her collections. Typical of the reportage is the following news item from the *London Chronicle*:

> Last Sunday died at Bulstrode in Bucks., the Duchess Dowager of Portland. By her death, above 14,000 L. per annum descends to the present Duke of Portland. Her Grace's jointure was 10,000 per annum. The Duchess Dowager of Portland has left to her servants considerable legacies. The lowest has not got less than 200 L. They were all brought up to her chamber after her death, and the will was read to them by the Duke. Her Grace never, but on death, changed a domestic for many years past. On account of her Grace's death, the Duke did not yesterday appear in the House of Lords.[1]

A week later the *Morning Chronicle and London Advertiser* carried speculations about the fate of her collections and her residence, Bulstrode:

> The Duchess of Portland's collections of different curiosities will certainly be sold. In botanical science and in gems she had a few things of greater value than in almost any other private cabinet. There is one pink pearl, which is said to have cost her above 1200 guineas.
>
> Bulstrode house and park come most opportunely to the Duke of Portland as he had no place in the Country at a less distance than Welbeck. The grounds at Bulstrode are very beautiful and well kept up, but the house, originally very ill laid out, has been so much neglected, that to repair and modernize it would cost as much as a new villa.[2]

Countering these narratives about the duchess's will being read over her body as the servants wept is the Rev. John Lightfoot's report to his friend Sir John Cullum, who was eager for insider knowledge of the duchess's demise and its after-effects. As a member of the duchess's household at Bulstrode, her chaplain no less, as well as a fellow botanizer, Lightfoot was in a position to know exactly what had happened during the weeks before and after her death at Bulstrode. Lightfoot wrote to Sir John on 17 July:

> The Report you read in the Papers of her Grace's Liberality to her Servants is without Foundation. Not a Tittle as yet has transpired respecting even a Will much less its Contents. Every Door & every Cabinet at Bullstrode is seal'd up till after the Funeral, which is expected will be at the end of this week at Westminster Abbey.

To another correspondent he wrote how much she would be missed and how people, such as himself, Mary Delany, her staff, and even Bulstrode's neighboring communities, would be forever changed by this event:

> With great Concern & Grief I inform you that last Night I lost my noble & invaluable Friend, the Duchess Dow^{er} of Portland. . . . Three parishes, who were constantly fed at her Gate, are overwhelm'd with Tears; all her Domesticks are sobbing privately in Corners; but poor Mrs. Delany's Affliction is beyond Expression, & such as will in all Probability soon put a Period to her Life.[3]

A month later Lightfoot responded to Pennant's inquiry about what was to become of the duchess's natural history collections; he explained that her will, written in 1771, prescribed the sale of her specimens:

> Her other three children, Lady Weymouth, Lady Stamford & Ld Edward Bentinck, she constitutes her Executors, & to those (after the following Legacies), she leaves all her Personals of every Kind, her rich Museum, with all the Subjects in natural History, to be sold, & divided equally between them.

Lightfoot responded similarly to Sir James Edward Smith's request for information about the fate of her natural history specimens, writing: "You may depend upon it that her whole Museum will be sold by Auction in Feb^y & March 1786."[4]

Rumors circulated about the Portland auction, focusing on the need for it. Why did the duchess's will stipulate that her natural history and decorative collections should be sold? Why sell off her prints, drawings, and paintings, furniture, decorative items such as snuffboxes, cameos, vases, classical statuary, jewelry, and china? It was understandable that her heirs would sell her natural history specimens, since these were not to everyone's taste. But curiosity was aroused about the other objects in her collection, and particularly the fine and decorative arts, especially the Portland Vase — why were they being sold and not kept in the family? Elizabeth Montagu complained to Elizabeth Carter:

> I have been grieved and provoked at the manner in which the news papers have spoke of the death of a Person who during her life made many happy. They prattle about

the selling her Museum, and her Fortune being useful to her sons Creditors. What disgrace will these vile news papers throw on the best characters of this age in the opinion of those in the next, who shall read them from curiosity after the manners of their Ancestors.[5]

Talk about the family's financial stability spread both in print and in private correspondence, and some wondered if the duchess had mismanaged her finances, greedily buying more than she could afford. Such speculations have continued, with Simon Dewes's undocumented assertion in his biography, *Mrs. Delany* (1940), that the duchess was bankrupt and her family needed the money to pay off her creditors: "There is small comfort, for the Duchess, though she never knew it, died heavily in debt and had, in reality, nothing to leave. She and Mrs. Delany had never bothered about finances. While money was there it was spent. When money was exhausted there was always credit."[6] This notion was taken up and repeated in scholarly and scientific articles, the authors not doubting the assertion that the duchess was bankrupt.[7] If one were to read the duchess's letters to her son, one would not come away with the idea that she did not understand finances; in fact, it becomes clear in her letters to him that she worried about his financial well-being and was quite critical of his costly lawsuits and the expenses he incurred as a result of his political career. She also gave him advice about the running of his estates, Welbeck Abbey in particular, cautioning him against expensive improvements.[8] In urging him to avoid what became hugely expensive lawsuits, she strained their tense relationship to the point that they hardly saw each other.[9]

Furthermore, if the duchess had been in debt, that great gossip Horace Walpole would have told his friends and included any speculation in his letters and account of the auction. He conveyed no rumors about creditors or debt, but he does refer in his journal entry for July 1782 to the younger son, Lord Edward Charles Cavendish Bentinck (1744–1819), as "an idle and worthless younger brother" who "helped to bring his brother, the 3d. D of Portland, into financial straits."[10] Of the third duke, Walpole wrote to a friend in 1782 that he was "overwhelmed by debts without a visible expense of two thousand pounds a year," and that he lived cheaply and privately,

> often in the country, and latterly in Burlington House, lent to him by the Duke of Devonshire, whose sister Portland had married, the Duke of Portland being in too great straits to have a house of his own. His fortune . . . had been noble; but obscure waste, enormous expense in elections . . . and too much compassion for an idle and worthless younger brother . . . had brought him into great distresses.[11]

Agreeing with this characterization of the duke's financial problems, Walpole's friend Lady Mary Coke had noted on 23 December 1777: "Last night it was said at the Princess Amelia's that there had been an execution in the Duke of Portland's house," meaning that the duke had been served with court documents concerning his inability to pay his creditors; in short, he was bankrupt. Lady Mary continued: "I don't affirm this . . . to be true; yet a few days ago I heard he did not pay the Duchess his mother what he had engaged to pay, and this I had from the best authority," which is a reference to a widow's jointure, an annuity to be paid to his mother out of his inheritance as the first-born son. Walpole sums up the Portland ducal fortunes as "The Duke and Lord Edward have both shown how little stability there is in the riches of that family – and *mine* has felt how insecure the permanency of heirlooms!", a reference to his own father's bankruptcy, his nephew's wastrel ways, and the selling of the Walpole family's extensive fine and decorative arts collection, a disaster from which he never really recovered.[12]

At first Walpole believed that the duchess had died intestate, surmising that the bulk of the collection would be sold, but he was corrected by Lady Mary Coke, who wrote that the duchess had made a will:

> [She] has given to the Duke all the best pictures which she had bought and are at Bulstrode. She has likewise given him all her jewels, a cabinet of valuable medals, and a cabinet left by her father full of gems, and other valuable curiosities to which she added. She has left to Lady Wallingford a hundred a year for her life. To Mrs. Delaney two snuff boxes and a picture, the legacy she had told the Duke she desired. She then makes Lady Weymouth, Lady Stamford, and Lord Edward Bentinck [her younger children] executors and residuary legatees. What they will get cannot at present be told; her personal estate must be great, and everything is ordered to be sold. Money I have not heard she has left, but *there are no debts*, and eight thousand pounds due to her from her estates. You may depend on this account as I had it from Lady Bute [a close friend of the duchess].[13]

As Lady Mary asserted, the younger children were to receive the benefit of the auction, while the eldest son, the third duke, would inherit Bulstrode, in addition to Welbeck Abbey, and their revenues, as well as some paintings, jewels, gems, and medals, possibly items Margaret had inherited from her own mother and father.

Many contemporary observers and modern-day scholars have been baffled as to why Mary Delany did not receive, like Lady Wallingford, an annuity, or something more substantial than a few snuffboxes and pictures, though it must be said that one of them, though small, was by Raphael.[14] Delany, though descended from the aristocracy, had few resources as a clergyman's widow and had come to rely on the duchess's largesse, living with her half the year at Bulstrode and the rest of the time in London within a short walk of the duchess's townhouse in Whitehall. With the duchess's death, Mary Delany was bereft of her best friend, as well as the pleasures of residing in ducal splendor at Bulstrode, where she enjoyed a rich and varied social life filled with art projects, scientific activities, and visits from George III and Queen Charlotte, who were charmed by her. Immediately after the duchess's death, as Delany made preparations to leave Bulstrode and to return to her London residence, the third duke, who felt somewhat awkward when he realized how little she was to receive from this will, wanted to make a gesture to acknowledge her intimacy with his mother. As Delany

> got into a chaise to go to her own house, the Duke followed her, begging to know what she would accept of, that belonged to his mother; Mrs. Delany recollected a bird that the Duchess always fed and kept in her room, desired to have it, and felt towards it as you must suppose.

Unfortunately, shortly thereafter, as the story goes, Delany fell ill, and coincidentally the bird died, but the queen came to the rescue by substituting "one of the same sort which she valued extremely (a weaver bird)." Queen Charlotte took the dead bird "with her own hands, and while Mrs. Delany slept, had the cage brought and her own bird [put] into it, charging everyone not to let it go so near Mrs. Delany, as she could perceive the change, till she got enough recovered to bear the loss of her first favourite."[15]

The duchess's old friends, Elizabeth Montagu and her sister Sarah Scott, gossiped about what would become of Delany. Scott wrote: "I see by the papers that the Dowager Duchess of Portland is no more; pray has she done any thing for Mrs. Delany that may add ease and a little more latitude to the days which cannot now be many." Mrs. Montagu wrote that

> her Graces attachment to Mrs. Delany was the most perfect imaginable, and her Grace was not covetous, so I imagine she imparted as much of her wealth as her friends delicacy w^d accept. I do not doubt but her Grace has largely provided for her now, but I have not heard from any one what disposition of her Fortune she has made.[16]

Delany's descendent Lady Llanover, the nineteenth-century editor of her correspondence, argued that it was beneath Delany's dignity and social station to accept an annuity from the duchess. While proponents of this position have asserted that Delany "requested the Duchess not to leave her any money,"[17] skeptics have looked elsewhere for an answer, some believing that Delany's Whitehall residence, called Thatched Cottage, was paid for out of an interest-free loan of £400 made by the duchess "at the time of the death of Revd Delany."[18] Within a month of the duchess's death, as Lightfoot tells us, "The King has just given Mrs Delany a ready furnished House at Windsor, and £300 pr ann: most graciously & royally done indeed." She resided at Windsor, enjoying nearly daily visits from members of the royal family, until her death three years later in 1788.[19] Elizabeth Montagu had a more jaundiced view of these proceedings:

> The Kings bounty had made up the pecuniary loss by the Dss's inattention, but nothing I shd think cd heal the wound such unkindness must give, indeed it is not unkind, but injurious to leave a Friend destitute of such comforts as they had been accustomed to receive, and I cannot but hope some Codicil will be found. I understand this Will was made some years ago.[20]

While it is clear that the duchess's younger children benefited from the sale of her collections, the question remains: why did she stipulate in her will that her natural history collection be sold? Could she not have arranged for it to be given to a scientific institution such as the Royal Society and the British Museum? She had arranged for her father's collection of manuscripts, the Harley papers, to be bought by Parliament for the nation and given to the British Museum, where they now make up an important part of the manuscript collection. Perhaps if she had been a man, she might have been more concerned about her legacy; perhaps her sense of duty to her family trumped her need to protect and valorize the labor of a lifetime of collecting and organizing natural history specimens. But even if she had wanted to sell her natural history collection to keep it together, who would have bought it? Or the shell or the insect collections? The collections were too massive for an individual to purchase, and institutions such as the British Museum were too poor to buy even a few shells. Sir Ashton Lever tried to sell his museum in the 1780s, but then resorted to a lottery as a way of transferring it intact to another person, who agreed to keep it together. It was operated it as a public attraction from 1788 until 1806, when it was sold piecemeal at auction, which was the usual fate of most natural history collections.

THE AUTHORSHIP OF THE AUCTION CATALOGUE

The reason why the auction was held nine months after the duchess's death was that time was needed to prepare the collection for the sale. This involved bringing objects from Bulstrode to Whitehall, though most of the shells and storage cabinets were already housed there (pl. 54).²¹ In addition, the preparations involved breaking down the order that the duchess, with Solander's and Lightfoot's assistance, had imposed on her natural history collections. Drawers in cabinets stored like items, organized by genus and species. The duchess would have kept, for instance, volutes together, and arranged by species. Depending on the number of specimens of each species, a drawer could have contained, for instance, all the *Voluta longata*, but if there were only a few specimens, they would have been kept with other species of the genus *Voluta*. One can make these assumptions because Sir Joseph Banks's shell collection, certainly arranged by Solander, has remained relatively untouched and is organized this way: each genus of mollusks kept together and arranged by species.

Whoever was in charge of the auction, Skinner, the auctioneer, or more likely Humphrey, decided that it was necessary to break up the Linnaean system of arrangement that the duchess had employed to organize her collection, because buyers would not want to purchase lots composed of just one species. The rationale for this is laid out, by way of apology to those invested in Linnaean classification, in the catalogue's preface:

> Some Persons, perhaps, may object to the Promiscuous Assemblage of the various Subjects here exhibited, and be ready to wish that they had been allotted in Order and Method, according to *Genus* and *Species*; and it must be confessed, that such a Proceeding would have proved extremely satisfactory to every true Lover of Science. Such would have been highly pleased to have seen each Article named, and stand in its proper Place. But however desirable such an Attainment might have been to a few *Cognoscenti*, it is very certain that the Majority of the World are not *Methodists*. They love Variety more than Order, and would rather purchase Twenty different Species of *Cones* or *Turbos* in One Lot, than the same Number of *High Admirals* or *Wentletraps*. . . . Whereas, in a methodical Arrangement, it must of Necessity have frequently happened, either that a Multitude of the *same Species* must be sold together in *One Lot*, (which very few would chuse to purchase) or each Individual of that Species, must be disposed *singly*, or in *Pairs*; which would have multiplied the Number of Lots to such a Degree as would extend the Sale to as many *Weeks*, as it consists at present of *Days*.

Putting the emphasis on buyers who were not naturalists or uninterested in taxonomic nomenclature, the auction organizers not only refused Lightfoot's request to sell lots made up of a single species but they also often made up lots that were a mixture of different genera: a lot could consist of a handful of volutes, a handful of turbos, and a handful of cones. To placate the naturalists, however, they did make an attempt to keep the Linnaean names: "all that could be done by the Compiler was only to give *in general* the classical names to such Articles as were known to have any, and to leave the great bulk of Non-descripts to the Examination and Determination of the Curious."[22] The compiler referred to above was Lightfoot, though it is clear from his correspondence that he did not approve of the decision to break up the duchess's collection and to auction it off. In October he wrote to a fellow naturalist:

> The loss of my Noble Friend is still a burden on my mind. To enhance my grief I am appointed to allot and name her fine collection for Sale by Auction next February. This engrosses my thoughts. To see everything I took so much delight in sold to the best Bidder almost rends my heart.[23]

Initially, Lightfoot was an unwilling partner with Humphrey in preparing the collection for auction, a process that involved breaking it up into lots by "Size, Rarity, or Beauty," identifying and describing the shells and other natural history specimens in these lots, providing references for specimens – if possible along with their geographic origin – and even, as some malacologists and historical taxonomists have argued, naming undescribed specimens. By March 1786, after working on the sales catalogue for months, Lightfoot was writing more positively of its scientific value, telling Pennant that it contained geographical information about the specimens, which would aid him in locating the African and Indian shells he was eager to possess.

> [T]he habitat of every rare Shell is therein faithfully recorded, as far as we know, or are authorized to do it. This Catalogue is not of the common Stamp. The classical Names of the various Subjects are religiously attended to. Dr. Solander's new Species are register'd & where figures existed, referred to. Much Pains have been taken in this Respect, in Hope of rendering the Catalogue worthy a Place in the Library of every Naturalist.

Lightfoot explained the identity of the "we" in the above passage when referring to the inclusion of geographical information: "Your request in Regard to the Shells found on the African and Indian shores &c. will be amply comply'd with in the Catalogue now in Hand, I mean as amply & satisfactorily as it can be done by myself and Mr Humphries."[24] Lightfoot and Humphrey collaborated on writing the catalogue, making sure that Solander got credit for naming new species.

The authorship of this anonymous catalogue has been contested. Contemporary newspapers reported Lightfoot as the author, but da Costa introduced the idea that it was Humphrey when a few of his notes on natural history were published posthumously in 1812: "The natural history made by Mr. George Humphrey, and formed or corrected by the Rev. Mr. Lightfoot, her Grace's chaplain." Da Costa's characterization of the working relationship between Humphrey and Lightfoot was taken by Lewis Weston Dillwyn in *A Descriptive Catalogue of Recent Shells* (1817) to mean that Humphrey was the author and responsible for naming and describing new species. This assumption reigned in malacological literature until 1962, when Peter Dance argued that Lightfoot was the author of the catalogue and that many of the new molluscan names could be ascribed to him.[25] More recently, Jean Bowden has suggested that attribution for these new names "should be attributed to them jointly,"[26] "them" referring to Humphrey and Lightfoot, which does not take into account Solander's work on new species in the Portland collection. Dance argued persuasively that Lightfoot himself added some new molluscan names; but Lightfoot did not subsume Solander's work under his and instead made sure that Solander's labor would not go unmarked, including in the list of references: "S. – After one or more names, refers to a Manuscript Copy of Descriptions of Shells, made by the late Dr. Solander, now in the possession of Sir Joseph Banks, Bart. P. R. S.",[27] and he marked each lot that contained Solander's new name for the species with an S. This was as close to publication as Solander got, and the new molluscan names he introduced, augmented in the catalogue by Lightfoot's addition of references and figures, circulated in the work of leading conchologists into the mid-nineteenth century.

Beyond the question of authorship and bestowing credit for new molluscan names, it is clear that the writing of the catalogue was a complex process, requiring the author(s) to negotiate contradictory aims: to describe the dismantled collection in a way that appealed to relatively ignorant buyers, while preserving as much of the scientific information as they could, providing citations to reference materials, geographical origin, Solander's new names, and established Linnaean nomenclature. In the midst of destroying the duchess's collection and dismantling and repackaging it for sale, the catalogue's "Compiler" managed to preserve some, if not much of the knowledge produced by her collecting and classifying practices, which included Solander's and Lightfoot's work on this collection. This was an achievement given the circumstances.

Lightfoot was not alone in his outrage and heartache at seeing the duchess's natural history collections broken down into market-ready, crowd-pleasing, bite-sized lots. Richard Pulteney stayed away from the auction, though some of what was being sold belonged to him – specimens he had loaned to the duchess but was unable to claim after her death. A measure of his grief and anger at what had become of her lifetime's work can be found in the library of the Natural History Museum in the manuscripts "Relative to British Testacea," a collection of Pulteney's notebooks, letters, and lists of shells that William George Maton acquired after he had died. Among these papers are handmade notebooks. One of these contains Latin names for shells in the duchess's collection. Its pages are made up of scrap paper, Pulteney using the back of an advertisement for chocolate and a bill of sale to write on. The hand-lettered title page reads (pl. 55):

CATALOGUS

CONCHYLIORUM

MUSEI PORTLANDICI

A' LINNÆO

non RECENSITORUM;

Quibus affixa sunt

NOMINA TRIVIALIA,

a Viris celeberrimis

SOLANDRO, et LIGHTFOOTIO

confecta.

Haec, per totum Catalogum venalem

disjecta, collegit, et in Methodum

Linnæanam digessit

R. P.

Peter Dance gives the following translation: "A catalogue of the shells in the Portland Museum not taken notice of by Linnaeus: to which are affixed trivial names, given to them by the celebrated gentlemen Solander, and Lightfoot. These, disseminated throughout the sale catalogue, are collected and disposed in the Linnaean Method."[28] This notebook is Pulteney's attempt to undo the chaos that the auction had wrought on the duchess's collection and to restore, if only on paper, the Linnaean order and preserve Solander's labor (pl. 56). It operates as a counter-narrative to the auction catalogue's "Promiscuous Assemblage" and offers testimony to the duchess's goal of describing and publishing every known molluscan species. Pulteney disassembled the auction catalogue's hodge-podge and piecemeal format, extracting from it the information he needed to reassemble what he deemed more useful for him as a scientific naturalist and perhaps as a more appropriate memorial for the duchess: a catalogue of her shells unknown to Linnaeus, but organized using Linnaean classification and taking into account Solander's and Lightfoot's naming of new species. The labor that this act of restoration entailed was enormous; Pulteney must have spent many hours going through the published catalogue of the Portland Museum, extracting and reinterpreting information, translating common names into Linnaean nomenclature, and making decisions about how to classify some specimens.

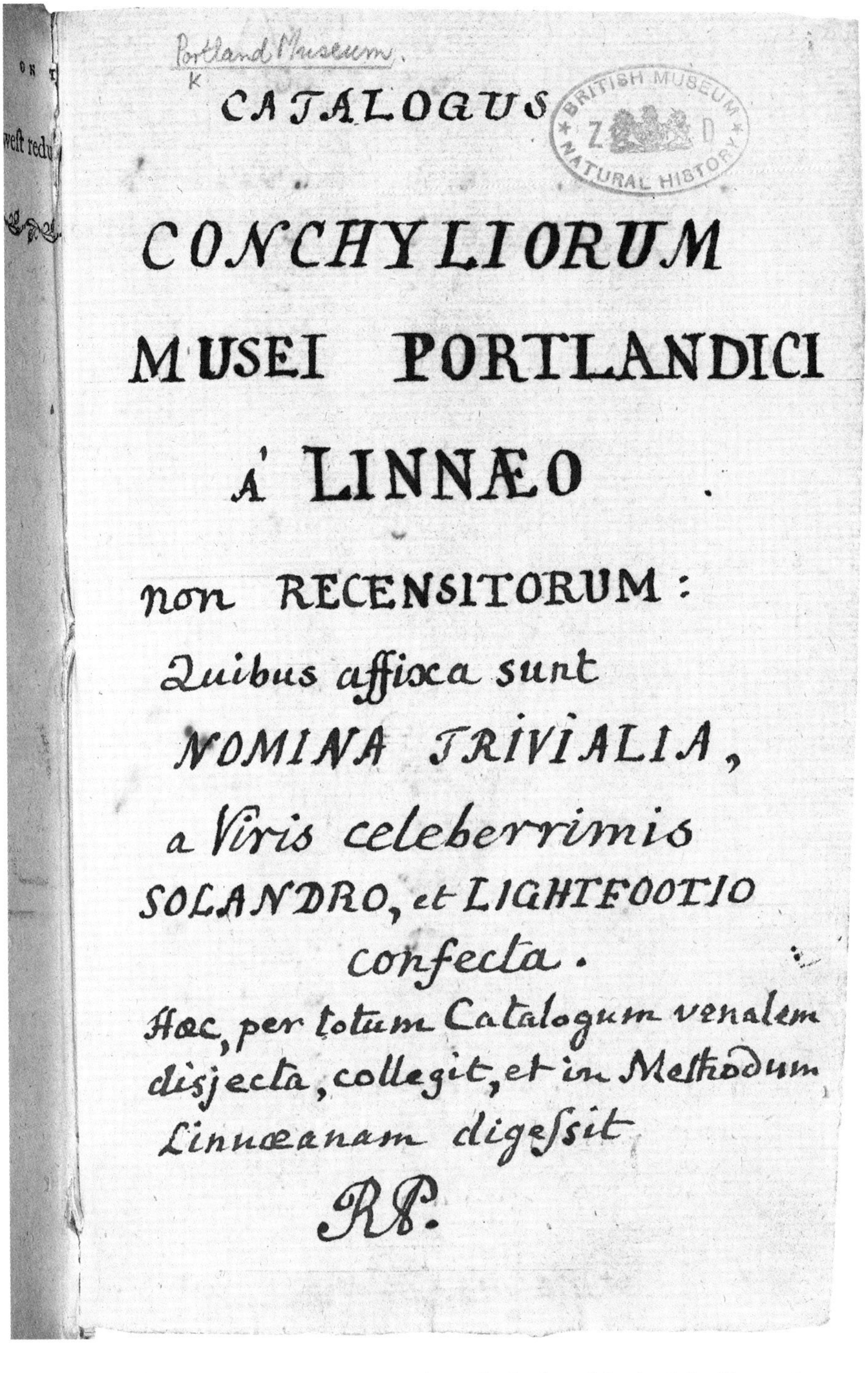

Portland Museum.

CATALOGUS

CONCHYLIORUM

MUSEI PORTLANDICI

à LINNÆO

non RECENSITORUM:
Quibus affixa sunt
NOMINA TRIVIALIA,
a Viris celeberrimis
SOLANDRO, et LIGHTFOOTIO
confecta.
Hæc, per totum Catalogum venalem
disjecta, collegit, et in Methodum
Linnæanam digessit

R.P.

55   Title page of *Catalogus Conchyliorum Musei Portlandici*. Richard Pulteney's handmade booklet. London, Natural History Museum. © Natural History Museum, London

32 aulica   4021. unique, a beautifull
                      red clouded Species of the
                      Wild Music kind.
    Gambaroonica   4024. only one known.
    Dama   4033. or Rau. a most
                      beautifull undescribed
                      Species. from the Coast
                      of Guinea.
    calcarata   4036. unique. of the
                      Coronated division of
                      Volutes.
    pacifica   4039. brought by Capt.
                      Cook from the Reef off
                      Endeavour River.
    maculata   List. 721. 7.
    patula   List. 733. 22 ?
    Auris Zebra   List. 814. 24.
?   acuminata   List. 827. 49 b.
?   torva   List. 826. 48. 49
o   labiata   List. 834. 59.
o   dolabrata Troch   List. 844. 72 a
o   notata   List. 844. 72 b.
o   Butyracea Buc   List. 974. 29.

THE AUCTION

In April of the year after the duchess's death, the newspapers announced an auction to sell her natural history and decorative arts collections for the benefit of her younger children. Lightfoot wrote to Pennant: "The number of Lots will run to 3600 & upwards, & the Sale will continue more than six weeks. Hasten to Town, for you will have enough to do."[29] This lengthy and spectacular auction caught the attention not only of naturalists and collectors but also of the fashionable upper classes and the general, educated public. It was advertised in several newspapers repeatedly throughout the thirty-eight days that it was held, announcing the time (noon to four) and place (the duchess's residence at Privy Garden, Whitehall), as well as giving a brief description of the items to be sold. The *Gazetteer and New Daily Advertiser* ran a large advertisement, announcing: "Catalogues may be had on premises [of the auction]; and of Mr. Skinner and Co. Aldersgate-Street, price 5s. which will admit the bearer during the time of exhibition and sale."[30] In addition to the catalogues sold before and during the auction there were engravings of "a Portrait of the late Dutchess Dowager of Portland, from a Marble Bust, executed by Rysbrack. Sold by G. Humphrey, No. 48, Long-acre" (pl. 57). Humphrey's brother was an illustrator and his sister owned a publishing business, specializing in prints; with these resources at hand, he was able to produce an attractive memento for those who attended the auction and who were curious about what Margaret Cavendish Bentinck had looked like, since very few images of her circulated in public. The illustration cost 1s. 6d., and was "the size of the Catalogue." Some surviving copies of the catalogue still contain this engraving, presumably inserted by a purchaser.[31]

Newspapers printed articles about the auction, as well as advertisements that described in brief the contents of the Portland Museum. An article in the *London Chronicle* stressed its aristocratic lineage, thereby heightening the celebrity quotient:

> This collection, partly founded by her grandfather and father, Robert and Edward, Earls of Oxford, has been continued by her Grace with unremitting industry for above the course of half a century; it is therefore the largest, most various, and perhaps the most curious of any private collection ever yet exhibited.
>
> The curiosities principally consisted of a very large assortment of fossils and minerals, shells and insects, several very curious and illuminated missals, a prayer-book of Queen Elizabeth, consisting of prayers composed as well as written by herself . . .[32]

These announcements must have been effective, because on Monday, 24 April the *Morning Chronicle and London Advertiser* reported on the crowded conditions at the pre-

57 George Vertue, after John Michael Rysbrack, *Margaret Cavendish Bentinck*, 1727, line engraving. London, National Portrait Gallery. © National Portrait Gallery, London

auction viewing held at Whitehall, which lasted ten days, and complained that the auctioneers had not been prepared for the extreme popularity of the event:

> The crowds at the Portland Museum for the last few days have been such as made all the rooms very inconvenient, and yet there was no attempt made to remedy or relieve; no windows open; no extension, as there ought to have been, of the hours of opening.
>
> The hours of opening, as before mentioned, ought to be from ten in the morning to six or seven in the evening, and so they should continue through the whole sale.

The numbers of catalogues already sold is far beyond all expectation; seventeen hundred or more were reported to us on Friday; and computing on the apparent demand for catalogues, it may not be much out of compass to suppose that 500 L. may arise from this part of the business.

The number of people that attended the Duchess of Portland's Muscum on Saturday last (being the last day of viewing) is scarce to be credited; the rooms were at one time so exceedingly hot and crowded, that several ladies fainted. The sale begins this day, and in what room the auctioneer is to exhibit, our correspondent knows not, but is satisfied that the largest room in that house is much too small to contain half the persons who would wish to attend as purchasers. Should the weather continue fine, the square paved yard is by far the most eligible spot, and upon which an old carpet or two might be laid down to prevent the female visitors from catching the cold in the feet.

So crowded was the pre-auction viewing that pickpockets could work the crowd without being caught. The *Public Advertiser* reported on Friday, 21 April that "Pick'd out of a Gentleman's Pocket between One and Two this Afternoon, at the Duchess of Portland's at Whitehall, a plain Gold Watch, Maker's name Emery, with black leather String and Steel Ornaments . . . Reward on Recovery of the said Watch and Seals from the Owner."

Newspaper readers following the auction were, no doubt, curious about who was attending the event and who purchased which items. The *Morning Herald* reported on "a list of Supposed Purchasers of the following lots in this day's sale." This list of purchasers of shells, however, is an amusing fabrication, linking public personages with shells that they supposedly bought; many of the shell names, silly enough on their own, were put into play with real and fictional personalities. For instance, Mr. Eden supposedly bought lot 123, the Devil's Claw; Lord Mulgrave bought lot 128, Chama hippopus, or Bear's Claw; and Lord Thurlow bought lot 132, a Bull's Mouth Helmet; while Lady Worsley bought lot 195, the Orange wide-mouthed Cone; and Chicken Taylor bought lot 198 – Pulletts Egg. These were not the names of the real purchasers, because writing in the margins of the annotated catalogue in the British Library records that Humphrey bought the Pulletts Egg, Dennis the Devil's Claw, and someone with the initials "M Y" bought the wide-mouth cone. Although many of the references in the article are now obscure, some of the jokes are too obvious to miss, at least on one level of meaning: one can see the connection between Chicken Taylor and his pullet's egg, and Eden with his Devil's Claw. Why was Lord Mulgrave, Joseph Banks's friend and a member of the Admiralty Board, linked with a bear's claw, and Edward Thurlow, Lord Chancellor, with a bull's mouth? Perhaps these animal attributes matched their personalities. One can also guess why Lady Worsley was associated with a shell's large orifice: she was rumored to have twenty-seven lovers and was sued in 1782 by her husband, Sir Richard Worsley, an antiquarian, for criminal conversation with an officer in the Hampshire militia.[33] The next day the newspaper continued its joke with another list of "Supposed Purchasers," among them this time was the Prince of Wales, who was reported to have bought "the Imperial Sun."[34] While the *Morning Herald* played with its readers' curiosity about who purchased what for how much, a publisher named Kearsley printed a copy of a priced catalogue with the lengthy title *A Marked Catalogue, containing the lots, what each respectively sold for, and the names of the purchasers of the four thousand two hundred and sixty-three articles. Which constituted the Portland Museum; late the property of the Duchess Dowager of Portland, deceased. Which was sold by auction by Mr. Skinner and Co., etc. Enabling every Connoisseur to know among whom the valuable Curiosities are distributed, and the sum which every lot produced.*

The Portland Museum auction had become an attraction and a form of fashionable entertainment for Londoners and visitors alike. Apart from those who were actively engaged in purchasing items, many people, if not most, came to look at the duchess's things and to watch the spectacle of the bidding itself. The Powys family, for instance, were visiting from the country, spending their usual six weeks in London for the Season, which would have included the theater, public performances, museums, and amusement parks. Mrs. Powys wrote in her diary for 9 May that she had taken her daughter Caroline to the duchess's auction, one of several edifying entertainment venues.

> We went with Caroline to see a very fine Collection to see Des Enfans pictures, ye Great Fish Balloon at the Pantheon and Kensington Gardens. We took Caroline (was too young at eleven) for public places, to see Sir Aston Lever's Museum, the exhibition of pictures, the late Duchess of Portland's Sale of Curiosities and the British Museum. All of which highly entertain'd her, as did Ashleys & Sadlers Wells.

The latter two venues were theater performances, and shared with the mix of natural history and art museums, gardens, and scientific displays a reliance on spectacle to arouse curiosity about art and nature.[35]

Burney's novel *Cecilia* conveys the excitement that the duchess's auction would have afforded spectators.[36] In a scene in which characters discuss going to an auction, Miss Larolles tries to persuade Cecilia to accompany her. Cecilia resists, not seeing much pleasure in it, for as someone who has newly arrived in London, she still retains her country-bred values and is made uncomfortable with the "unbounded extravagance" of the fashionable set:

While they were yet at breakfast, they were again visited by Miss Larolles.

"I am come," cried she, eagerly, "to run away with you both to my Lord Belgrade's sale. All the world will be there; and we shall go in with tickets, and you have no notion how it will be crowded."

"What is to be sold there?" said Cecilia.

"O every thing you can conceive: house, stables, china, laces, horses, caps, every thing in the world."

"And do you intend to buy any thing?"

"Lord, no; but *one likes to see the people's things*."

Cecilia then begged they would excuse her attendance.

"O by no means," cried Miss Larolles, "you must go, I assure you; there'll be such a monstrous crowd as you never saw in your life. I dare say we shall be half squeezed to death [emphasis added].

Cecilia again tries to say no to this excursion, but Miss Larolles counters with what she thinks will persuade her – the fact that this is an auction for the benefit of the Belgrades' creditors. Miss Larolles's excitement over the pleasure of seeing other people's things is heightened by the specter of misfortune that haunts the auction. Not only is the private made public as the Belgrades' objects are put on display and their home opened up for public exhibition, but their bankruptcy enhances the drama of the event by providing spectators with the perverse pleasure of seeing someone whom they might envy suffer ill fortune.

> O but do go, for I assure you it will be the best sale we shall have this season. I can't imagine, Mrs. Harrel, what poor Lady Belgrade will do with herself: I hear the creditors have seized every thing; I really believe creditors are the cruelest set of people in the world! they have taken those beautiful buckles out of her shoes! Poor soul! I declare it will make my heart ache to see them put up. It's quite shocking, upon my word. I wonder who'll buy them. I assure you they were the prettiest fancied I ever saw. But come, if we don't go directly, there will be no getting in.[37]

Miss Larolles's pleasure in Lady Belgrade's humiliation is thinly veiled. Akin to *schadenfreude*, the pleasure she anticipates from such an event puts into play pity, joy, avarice, envy, and the thrill of recognition, along with the exhibitionist pleasure of being seen by the crowds attending the auction. No doubt, the duchess of Portland's auction aroused similar feelings, and the public's pleasure of entering her London residence at Whitehall to examine her things must have been very great, laced most probably with a sense of trespass, especially for those who would never have been invited into such a place during the duchess's lifetime.

Horace Walpole was much interested in the duchess's auction for a variety of reasons, principally because he coveted some of the items, in particular the famous Portland Vase, the illuminated missals, and "the head of Jupiter Serapis, cut out of a green basalts, a most inimitable piece of sculpture, of Egyptian workmanship, from the Barberini cabinet." His copy of the auction catalogue is annotated, naming purchasers and the prices paid for the items. He himself spent £359 3s. 6d.[38] His letters and account of the sale reveal that special combination of delight and envy that auctions inspire in people, conveying the quickly shifting mood of regret and sadness in recognition of misfortune, to delight in the beauty of the objects for sale and envy that he is unable to buy them. His tone wavers much like the fictional Miss Larolles as he announces with what appears to be sadness at the passing of the duchess, then quickly changes register as he delights in the beauty of the objects and eagerly anticipates the prospect of bidding on objects of "virtù" in her collection. He hopes that the duke "will reserve the principal curiosities," not putting them up for auction, "for I should long for some of them, and am become too poor to afford them – besides that it is ridiculous to treat one's self with playthings, when one's eyes are closing."[39] To Walpole's relief, "Several of the most valuable articles in her Collection were not exposed to Sale," including jewels, miniatures, the pictures at Bulstrode, the blue and white china, and an "Ebony cabinet," known as the "Ten thousand pd Cabinet, & had been the legacy of her Father."[40] And although the Portland Vase was listed in the sales catalogue, Sir Joshua Reynolds persuaded the third duke of Portland to buy it back for the family's sake.

Walpole, who had been friends with the duchess for more than thirty years, recalled a scene from 1756 that captures the particular pleasure of viewing dead people's things, a pleasure perhaps aroused by the objects' liminal status as possessions – they hover in that in-between place of belonging for the moment to no one. Foreshadowing the viewing dynamics at the Portland Museum auction thirty years later, Walpole describes his visit to Welbeck Abbey, just after Margaret Cavendish Bentinck's mother, Lady Oxford, had died and before Welbeck Abbey was claimed by her only child and her husband, the second duke of Portland. Walpole roamed through this grand building and fell in love with its new gothic architectural features, its fine library, and pictures:

> I went to Welbeck – It is impossible to describe the bales of Cavendishes, Harleys, Holleses, Veres, and Ogles: every chamber is tapestried with them; nay, and with ten

thousand other fat morsels; all their histories inscribed; all their arms, crests, devices, sculptured on chimneys of various English marbles, in ancient forms (and to say the truth, most of them ugly). Then such a Gothic hall, with pendent fretwork in imitation of the old, and with a chimney-piece extremely like mine in the library! Such water-colour pictures! such historic fragments! In short, such and so much of everything I like, that my party thought they should never get me away.... But it is impossible to tell you half what there is. The poor woman who is just dead [the duchess's mother], passed her whole widowhood, except in doing ten thousand right and just things, in collecting and monumenting the portraits and reliques of all the great families from which she descended, and which centred in her. The Duke and Duchess of Portland are expected there tomorrow, and we saw dozens of cabinets and coffers with the seals not yet taken off. What treasures to revel over![41]

Some of these treasures reappeared at the Portland Museum auction, and Walpole grabbed this chance to purchase a few items that carried with them memories of the duchess, whom he referred to as his "playfellow,"[42] and to satisfy a longing to possess some "relic" from this celebrated aristocratic family, for Walpole was an indefatigable collector of aristocratic memorabilia, even collecting objects that had little intrinsic value other than their aristocratic provenance.[43]

WHO BOUGHT THE DUCHESS'S SHELLS?

Most of the auctioned items were natural history specimens, and these attracted serious collectors as well as natural history dealers. Among the crowds of curiosity seekers and celebrity hounds were naturalists, collectors, and dealers: Emanuel Mendes da Costa, Thomas Martyn, Ingham Foster, Jacob Forster, John Hunter, Thomas Sheldon, Dr. George Fordyce, Dru Drury, Thomas Pennant, George Walker, and George Humphrey.[44] For these people, the auction was serious business. Humphrey attended every day, purchasing at least one or two lots daily until the final two weeks of the sale, when his purchases doubled and tripled, and in the final days he was buying nearly half of all the lots: on the thirty-fifth and thirty-sixth days, for instance, he purchased 40 out of 109 lots and 45 out of 126 lots, respectively.

Another very active buyer at the Portland Museum auction was a Mr. Dillon, an agent who, according to Peter Dance, bought on behalf of Charles Alexandre de Calonne, France's Controller General of Finances, known for his collections of natural history specimens. In two days, for instance, Dillon bought nineteen lots at the Portland auction that went into the Calonne collection, spending £35 on two volutes.⁴⁵ Of Monsieur de Calonne, Humphrey wrote:

> His fortune was ample, and his situation as a minister of France afforded him the means of obtaining early information whenever curious articles were to be disposed of in any country of Europe, whether brought thither by the different circumnavigators, or they were the contents of cabinets collected by preceding amateurs. By the former means he obtained the most rare and curious productions brought from Africa, the East and West Indies, the islands of Ceylon, Amboyna, and Borneo; from China, Peru, New Zealand, and the newly discovered islands in the southern ocean. By the latter means he procured every thing which he did not before possess in greater perfection from the cabinets of Prince Charles of Lorraine, M. Blondel d'Azaincourt, M. de Montriblou, M. de Nanteuil, M. le C. De la Tour d'Auvergne, Mynheer Gevers, the late Duchess Dowager of Portland, and many others.⁴⁶

Calonne's purchases from the Portland Museum are well documented. A decade after purchasing some of the duchess's more valuable shells, he put up his own collection for auction in London, having fled France during the Revolution, and employed Humphrey to organize the sale and prepare the catalogue. This catalogue, *Museum Calonnianum* (1797), indicates which of the shells were purchased from the duchess's collection, with the initials "M.P." followed by the lot numbers in *A Catalogue of the Portland Museum*.⁴⁷ One such shell was a *Murex carinatus* (pl. 58), the subject for one of Edward Donovan's illustrations in his five-volume *The Natural History of British Shells* (1799–1803). This murex had also been illustrated for Pennant's *British Zoology*, Pennant being, according to Laskey,

> the first to figure and describe it, from the Portland cabinet; at the sale of which, it was purchased for the collection of the celebrated Monsieur Calonne, and went to Paris. On the emigration of that gentleman to England, it again came into the market, and became the property of Mr. G. Humphrys, Leicester Square, who has refused a very considerable sum for it.

In a handwritten annotation next to this printed entry is the comment, presumably by Laskey: "it is now in the poss. of Donovan. I offered £5 for it to Humphry's."⁴⁸

Similar narratives about the provenance of the duchess's shells can be found in Donovan's *Natural History of British Shells*. He was quite proud of the Portland shells that he had managed to purchase second-hand from Humphrey and from the Fordyce collection. Of *Lepas dilata* (Plate CLXIV) he wrote (pl. 59):

58 Edward Donovan, "Murex carinatus," Plate CIX of *The Natural History of British Shells* (London, 1799–1803): "It formerly belonged to the late Duchess of Portland." Arizona State University Libraries, SPEC D 756

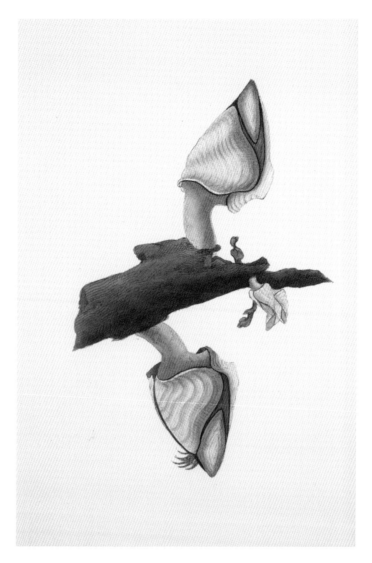

59 Edward Donovan, "Lepas dilata," Plate CLXIV of *The Natural History of British Shells* (London, 1799–1803). Arizona State University Libraries, SPEC D 756

A specimen of this shell, one which we are inclined to think, on pretty good authority, to be the same, or one of the them at least, that was sent by the late Mr. Ellis to the Dutchess of Portland, is at this time in our Cabinet; the late Dr. Fordyce became first possessed of this specimen, and at his death we obtained it, under the title *Lepas sigillantum* of Solander.

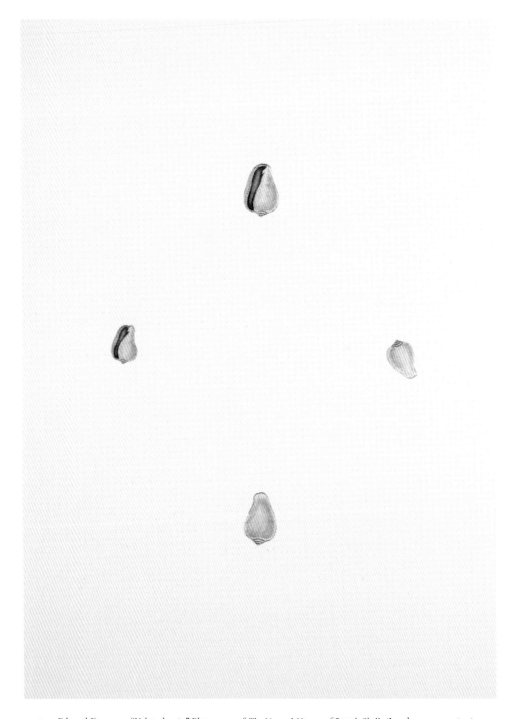

60   Edward Donovan. "Voluta laevis," Plate CLXV of *The Natural History of British Shells* (London, 1799–1803). Arizona State University Libraries, SPEC D 756

Of *Voluta laevis* (Plate CLXV) he boasted that the plate containing illustrations of this Weymouth shell, which had been "dredged up in deep water by some fisherman, and consigned to the cabinet of the late Dutchess of Portland," were figured from specimens, two of which were "originally in the possession of her Grace" (pl. 60). His published descriptions of his shells often contain details about their provenance; for instance, he describes the path that his specimen of *Lepas scalpellum* (Plate CLVI) took into his collection: "dredged up on the coast of Weymouth; a specimen of it affixed to the branches of a coralline that was discovered here, after passing through the collection of the late Dutchess of Portland, and Dr. Fordyce, is at present in our possession." He noted that his *Buccinum glaciale* (Plate CLIV) was found by Agnew, the duchess's gardener, and was in her cabinet; Fordyce bought it at the Portland auction, and when he died in 1802 Charles Bennett, fourth earl of Tankerville, purchased it.[49] Donovan was clearly interested in the provenance of his shells, in particular those that had been owned by the duchess, since this lent them even more value in his eyes. Donovan's *Natural History of British Shells* is very much in the connoisseurial tradition, employing a discourse that mingles disputes about classification with provenance, a discourse readily applied to both natural history and art collecting.

WHERE ARE THEY NOW?

Of those who purchased the duchess of Portland's shells, either first-, second-, or third- hand, the Rev. Clayton Mordaunt Cracherode gains importance as a possible lead in the mystery of their present whereabouts. Guy Wilkins, who was a curator of the invertebrate zoology collections at the Natural History Museum in the 1950s, argued persuasively in a lengthy article that Cracherode had purchased some of the duchess's shells that were in Calonne's collection. Cracherode had a strong working relationship with George Humphrey, who was the interface between Calonne's collection and Cracherode's, for it was he who wrote the *Museum Calonnianum* (1797) and oversaw that auction. In the Natural History Museum's library is a manuscript catalogue of Cracherode's collection, with the prices he paid Humphrey for the shells he bought, several of which were from the Calonne collection. Wilkins suggests that the shells from the Calonne collection are of interest to taxonomists since "in many instances they came from the Portland collection and were named by Solander. Cracherode paid heavily for these shells, the twenty-two marked 'M. C.' in his catalogue costing the not inconsiderable sum of one hundred pounds sixteen shillings."[50] However, of the sixteen shells (out of the total of twenty-two) that Wilkins was able to identify as passing from the Calonne collection into Cracherode's, none is described in the Calonne catalogue as having belonged to the duchess, which means that the likelihood that the Natural History Museum's Cracherode collection contains some of the duchess's shells is minimal.

The one documented shell in the Natural History Museum that is known to have belonged to the duchess was purchased at the Calonne auction, but arrived, not via Cracherode's bequest but through the Broderip collection, which the British Museum bought in 1837 for £1,575.[51] William John Broderip had amassed a huge collection, purchasing some of the Tankerville shells that had come from the Calonne sale. The shell is the lovely *Voluta aulica*, advertised in the Calonne catalogue as belonging to the duchess of Portland (see pl. 5): "273. Aulica – Le Courtisan, ou Le Nuage Rouge – Courtier, or Red Clouded – Voluta Aulica Soland. This beautiful shell is *unique*. Its country is unknown, but presumed to be from some newly discovered island in the South Seas. M. P. 4021."[52]

Could there be shells once owned by the duchess in other museums? A likely candidate is the Hunterian, a museum and art gallery at the University of Glasgow, because it became the repository for William Hunter, a prominent physician, who, like his brother John, the famous anatomist, was a major collector of natural history and medical specimens; John (recipient of the second voyage's Maori head) attended the Portland auction, and William purchased Fothergill's collection before his own death in 1783. Laskey's guide to the Hunterian's collections, *A General Account of the Hunterian Museum* (1813), contains much information about the holdings of shells and their provenance. Typical of his connoisseurial interest, he notes price and origin: "*Genus Ostrea* . . . white Hammer Oyster very rare. This specimen was brought home by Captain Cook from the Coral Reef at Endeavour River, and is very rare. Twelve Guineas have been given for a fine specimen." He refers frequently to Captain Cook, Solander, and the duchess of Portland, and uses them to add luster to the Hunterian's collections. Laskey's guide, however, must be used with caution because some of its information is tainted with the fanciful. It includes a wonderful but apocryphal story about the duchess's shell collection:

> The fact is, when Linné began his System of Nature, he did not consider Conchology as worthy of his notice, (shells being only the covering or exuviae of animals;) till he was struck with their extreme beauty and variety, on seeing and examining the late Duchess of Portland's collection, when he immediately determined on an

arrangement. This he executed in great haste, and consequently it had many errors. Had he lived a few years longer, no doubt he would have improved this division of his system.[53]

Laskey mixed up different narratives and even reversed elements of a truism repeated often by Pulteney, Seymer, and da Costa that it was too bad that Linnaeus never had the opportunity to see the duchess's shell collection because his own was too small to generate enough nomenclature for all the genera and species that came to light after he had formed his system. Although Laskey frequently drops the duchess's name in the context of describing the Hunterian's shells, he does not say exactly which shells belonged to her. Among the Hunterian's shells from the Cook voyages may be more of the duchess's shells than the two *Polymita picta*, Cuban land snails, that are accounted for.[54]

Shells once owned by the duchess must still exist, given that there were thousands of them. Some may be housed in natural history museums, regional museums and local history societies, attics, basements, and secondhand shops; some may be in European or North American collections. Once the link between owner and shell has been severed by mistake, forgetfulness, or indifference, the provenance of shell specimens is impossible to establish. Kathie Way, Collections Manager of the Mollusca department at the Natural History Museum in London, explained to me how even there shells get moved out of collections for scientific purposes. A taxonomist needs to see a shell to help with the task of classifying a specimen; he locates just what he needs, for instance, in one of the drawers of Banks's cabinet; he takes it to his work area, and when he has finished with it he may put it with the specimens of its own kind rather than returning it. He does this because provenance is not a scientific concern; who owned the specimen is a connoisseurial and curatorial question, not a taxonomic one. This kind of disregard for provenance is understandable and reasonable given a scientist's objectives. It is just one of the many ways in which shells can get moved out of their original collection when donated to the museum. Another is that some of these older shells, especially from eighteenth- and early nineteenth-century collections, may have been one of several specimens of the same species, and therefore thought of as superfluous. Museums and natural history societies have been known to sell, swap, or give away extra specimens because of the difficulties of finding the space and time to store and care for them.

A collection is a collection only when it is intact. The ideas and labor that brought the objects together, organized and displayed them, evaporate when the collection is dispersed. All that remains of the duchess's shell collection are three shells, some illustrations, and the auction catalogue. The catalogue conveys the idea that the collection was a "Promiscuous Assemblage,"[55] a site where artificial and natural objects were mingled indiscriminately, taste being the only deciding factor in selection and arrangement. What is left of the duchess's activities that will testify to her passionate engagement with natural history and her specimen hunting expeditions, her efforts to identify, classify, and arrange her shell collection on Linnaean principles, her generosity within the networks of naturalists and her patronage of those involved in the production of conchological books? Apart from a few textual traces found in nineteenth-century shell publications (about which more will be said in the last chapter), all that remains of the duchess's efforts is a clutch of letters in the Linnean Library, which augments and offsets the public record, where she appears in the background of other people's lives – as Mary Delany's friend during the last decades of their lives; as Elizabeth Montagu's correspondent in her youth; as a topic of Walpolian gossip; or fleetingly as Rousseau's botanizing companion – or where she appears in current historical studies as an exemplum of the excesses to which aristocrats were prone, with the purchase of the Portland Vase in the last year of her life and the spectacular auction after her death.

Of the textual traces that remain of this complex life, one potent remnant that still circulates, shaping viewers' understanding of the duchess's collecting practices, is the auction catalogue's prominently placed illustration of what is presumably her collections (see pl. 23). Edward Francis Burney's illustration works well as an invitation to viewers to participate in the auction with its overflowing cornucopia of natural and artificial objects, an image reminiscent of François Boucher's illustration for one of Gersaint's sales catalogues, *Catalogue Raisonné de Coquilles et autre Curiosités naturelles* (1736), which pictures piles of shells on the floor beneath rows of jars filled with specimens and a towering coral specimen. However effective Burney's image was as a sales catalogue illustration, it does the duchess a severe injustice because it portrays her collection in a state of disorder: dismantled and scattered, lying on the floor, waiting to be picked up and purchased by someone who will care for the dislocated objects. The image invites piecemeal purchase, echoing the catalogue's *raison d'être*, and it conveys the idea that she amassed objects without intelligence or reason, that she accumulated without bringing order to bear. This belies the duchess's knowledge of Linnaean taxonomy, her commitment to classification, and her goal to produce scientific knowledge with the publication of a catalogue of her collection. Contemporary and modern-day audiences, unaware of her activities as a naturalist, have assumed on viewing this illustration that she was at best a dilettante and at worst a magpie. Past and current readers of the catalogue who did not know of her scientific activities have been led to believe that she lived in chaotic excess brought on by profligate acquisitiveness, reaffirming the clichéd narratives that circulated about aristocratic women and their proclivities. Because the catalogue and its illustration are widely available, thousands having been published for the auction, the image of the duchess as driven by passionate acquisitiveness dominates the historical record. Only a few scattered contemporary accounts of her naturalist activities survive in print, and, as a result, within a generation of her death, her legacy as a naturalist was nearly forgotten.

CHAPTER SEVEN

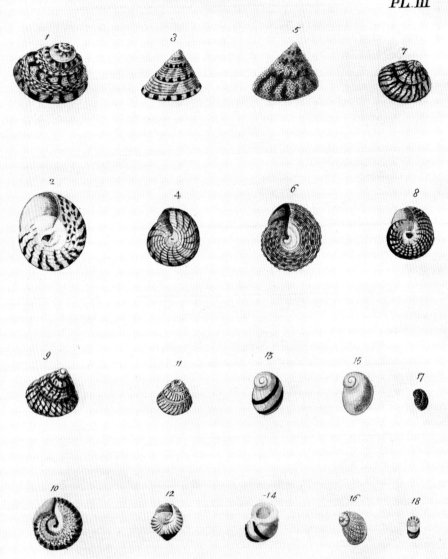

61   Emanuel Mendes da Costa, Plate 3 from *Historia Naturalis Testaceorum Britanniae, or The British Conchology* (London, 1778). Author's photo

LOST LEGACIES: THE DUCHESS'S SHELLS IN THE NINETEENTH CENTURY

Rebecca Stott has pondered why the duchess of Portland "rarely appears as more than a footnote in scientific literature, although the lives of male collectors such as Sir Hans Sloane, John Evelyn, and John Tradescant are well-documented." She suggests that the duchess's absence from the history of science can be attributed to the devastating effect of the auction combined with the complete "disappearance within a generation" of her zoological park, botanical garden, aviary, and even Bulstrode itself, which was badly damaged by fire and torn down in the early nineteenth century.[1] The dispersal of her natural history specimens and destruction of the world she had created at Bulstrode contributed, no doubt, to the erasure of Margaret Cavendish Bentinck's legacy as a naturalist and serious collector of scientifically significant specimens. Her presence was also diminished in the conchological literature published in the generation after her death. In short, her reputation as a naturalist within the scientific literature did not survive beyond the lives of those who knew her personally.

This chapter examines how the duchess's shell collection was represented in print by her contemporaries and by the immediate generation of conchologists after her death, tracing the ways in which her participation in such quintessential Enlightenment scientific practices as collection and classification was elided over time as natural history, along with its goals and practitioners, changed. In general, those who knew her wrote

62 Anonymous, *George Montagu*, oval miniature, oil, 6.9 × 5.6 cm, *circa* 1780–1815. London, Linnean Society. By permission of the Linnean Society of London

about her skills as a naturalist and her wide-ranging knowledge about nature informed by a deep understanding of Linnaean systematics. Those who did not know the duchess personally tended to elide her activities as a collector and naturalist by referring to the collection itself, making the "Portland Museum" stand in for the range of social practices that the duchess deployed as a collector, which had the effect of reducing her agency and expertise to a static corporate object.

The way in which the duchess, her shells, and the Portland Museum are represented in turn-of-the-nineteenth-century shell publications varies subtly but significantly in tone, form, and agenda. What follows here is a textual analysis of shell catalogues in manuscript and published form, specifically those by J. T. Swainson, Richard Pulteney, George Montagu (pl. 62), W. G. Maton and Thomas Rackett, and William Turton.[2] A comparison of how these texts handle the duchess's collecting activities, the shell collection itself, and Solander's work on the collection reveals tensions over a distinction between collectors, on the one hand, and naturalists who were "scientific," to use Montagu's term, on the other. Montagu and Maton, who were of a different generation from the duchess and did not know her personally, were eager to establish themselves as experts, and though amateur naturalists themselves, they took publication as their route to authority and expertise, subtly undermining the work of those collectors and naturalists who did not publish either a catalogue of their own collection or papers in the published transactions of the Royal Society or the Linnean Society. As will be seen, Montagu and Maton reference the duchess's cabinet, Solander's classifications, and the duchess's collecting activities, but they do it in a very different manner from Pulteney and John Timothy Swainson, who, as members of her natural history network, knew the range of activities that her collecting practices entailed.

The process of assessing the duchess's posthumous reputation sheds light on how natural history networks operated at the end of the eighteenth century and how expertise about natural knowledge was constructed. More specifically, the depiction of the duchess in conchological literature, which ranged from handbooks and field guides to reference books and definitive catalogues, reveals the gradual shift in scientific authority from the amateur naturalist to the professional scientist, and the role of publication in this process.

WITHIN THE NETWORK

Margaret Cavendish Bentinck appears in John Timothy Swainson's manuscript notebook, a holograph catalogue of his collection of British Shells (see pl. 31); here she is represented as an individual, not an entity, and as a fellow naturalist eager to share ideas and specimens.[3] It was during his stint as a Customs officer in Margate that Swainson met the duchess, who had summered there in 1782 and 1784.[4] As noted in Chapter Two, they met to examine his shell collection, and perhaps even went shell collecting together. "Mr. Swainson will be a good acquisition," the duchess wrote to Mary Delany; "he shot three or four birds for me yesterday and is gone out to-day trawling, or I shou'd have gone to have seen his collection."[5] Martyn wrote in his *Universal Conchologist* (1784–89):

> For so young a collector, no gentleman has shewn more genius for the study, or been more industrious in exploring new subjects in Conchology, than Mr. John Swainson

of Margate. Success has in some measure already crowned his researches, in his discovering among the sand on the sea-beach at the place of his residence, various new species of minute recent nautili, snails, &c."[6]

Swainson and the duchess also shared interests in birds and insects, and continued their friendship, exchanging specimens, letters, and information, during the winter months. The duchess wrote to Swainson twice in November 1784 about shells that she was sending him and that she had received from him. She also commented on the shells he had sent her, and offered her opinion as to the proper names for them. She was gently suggesting that he may have misidentified a specimen when she wrote: "No. 1. Is it not Buccinum Lapillus young? Da Costa Plate 3, fig.1.2.3.4.", and "No. 5. New it has a resemblance to a land Helix but the Apex of the land shells not so sharp a point." Here again the duchess appears as a naturalist in the same way as Swainson, both eager to talk about the intricacies of identifying shell specimens, sharing opinions and freely disagreeing with each other.[7] She wrote to Swainson: "I am much obliged to you for those in the small Box with the numbers from Walker[.] they seem to be very accurate. I am entirely of your opinion about colours." Not only did Swainson correspond with her and exchange shell specimens (along with insects and birds), he was also the recipient of other kinds of gifts from the duchess, among them "fine copies of da Costa's *Historia naturalis Testaceorum Britanniae* and Pennant's *British Zoology* (the folio edn.)." As his biographer Nora McMillan relates, "of these he was very proud and in his will (1819) he bequeathed them to his daughter Betty (1801–1867)" (see pl. 61).[8]

Swainson depicts the duchess in his notebook as a fellow conchologist, an active collector of shells, and one among several shell experts whom he mentions in the narrative bits of the little notebook. As noted in Chapter Two, this notebook, the product of several years of labor, is a crucial source of information about the duchess's shell hunting activities. *Trochus papillosus* was "found by the D: of Portland at Weymouth & had from her – Da C figures a foreign Spec. – I got it in trawling off Liverpool in Ap. 1808 – r. r.", the "r"s standing for degrees of rarity (r, r.r., r.r.r.). *Bulla patula* was "given me by the D of P found by her at Weym. Da C quotes this as his 'Bubble'," and *Cardium tuberculatum* was "found by dredging off Margate in 1785 – once – therefore r-r-r. – my large specimen given me by G.H. found at Weym. by the D. of Portland, escaped Dr. Solr. Most probably found after his disease [Solander's death from a stroke] – described by Walker." *Tellina lacustris*, Swainson wrote, "had from the D of Port. found in a River near Bulstrode – I got it at Longford in the Colne."[9] The duchess appears in the notebook alongside other noted shell authorities. Da Costa and Pennant, for instance, are invoked as textual entities with which Swainson carries on quarrels over classification, but these quarrels took place only in his head and on paper, since the notebook contains references to 1808, more than a decade after their deaths. Other conchological authorities cited in brief are Montagu, Solander, Boys, and Walker. George Humphrey also appears as someone with whom Swainson went on frequent collecting forays in marshy, wetlands in and around London, where they collected snail shells, both land and freshwater. *Helix contorta* was found "in the ditches in Rotherhithe Marshes with G.H," and *Helix octona* was "found by G.H. near Elms Battersea and by me near Newington Green." *Turbo labiatus* was "found many years ago by G.H.'s sister on the N side of the Serpentine in Hyde Park – & in 1784 I found them with G.H. in an Osier Ground near the Mills in Battersea fields."[10] Clearly, the duchess of Portland was part of the network of naturalists who were engaged in locating and naming British shells, an activity that took on greater significance during the Napoleonic Wars, when travel to Europe was nearly impossible and focus on things British was implicitly patriotic.[11]

Pulteney's published catalogue of Dorset shells also contains information about the duchess's activities as a field collector, information that is dependent on his personal knowledge of her. In other ways, though, his catalogue is unlike Swainson's manuscript: the latter was not intended for publication and operated as a record of where and from whom he acquired shells.[12] In keeping with the conventions of scientific publications of the time, Pulteney presents information about the duchess in a neutral, non-personal way. In his descriptive sections, he is very clear that she was often the first to find certain shells, many of them from the coast at Weymouth. Several times he wrote lines such as: "I do not know that they [Pinnas] were ever found on the English Coast before they were dredged up at Waymouth [sic], under the inspection of the late dowager duchess of Portland, but they are far from common"; "Evidently a new, and a very elegant little shell [*Balanus intertextus*], found at Waymouth, first by the late duchess dowager of Portland. I have seen it common on oysters"; and referring to Phola. Faba.: "First distinguished by the late duchess dowager of Portland at Waymouth, but rarely dredged up."[13]

Each entry in Pulteney's catalogue, as is typical of this genre of natural history writing, contains a list of the published authorities that have described the particular species. In this list, he included the duchess's auction catalogue under the designation "*Mus. Portland.*" [*A Catalogue of the Portland Museum*], and he included Solander's work on her collection under the designation "*Soland. Mus. Portland.*" In addition, he acknowledged Solander's new molluscan names for shells; they are recognized with the notation *Solandri* and most probably refer to Solander's "slips." How did Pulteney know what shells Solander had named? Did he base this on personal knowledge – conversations with the duchess on her visits to Dorset, letters exchanged between them – or did he have access to Solander's catalogue of the Portland museum that he was working on when he died? And did he see Solander's other manuscript papers, his slips, that Sir Joseph Banks acquired after Solander's death? Evidence of Pulteney's first-hand knowledge of Solander's intentions concerning the classification of the duchess's shells includes such statements as: "Agreeable to the plan proposed by Dr. *Solander*, had he lived to publish the Museum Portlandicum, I have separated the Linnaean genus of *Lepas* into two," and "Had Dr. Solander lived to finish his systematic description of the Portland Cabinet, it was his intention to have constituted a new genus, in which these two shells [*Mya praetenuis*], together with the *Solen bullatus*, and other shells in that cabinet, were to have been included."[14] Also, among the miscellaneous Pulteney papers at the Natural History Museum there is a handmade notebook with lists in Pulteney's handwriting of genera and species, which could be his copy of Solander's notes for his revision of Linnaean taxonomy for mollusks. While it cannot be known for certain how Pulteney knew what Solander's intentions were concerning the systematic descriptions of the Portland collection, the fact remains that the duchess, her collection, and Solander's work are clearly given credit in Pulteney's catalogue of Dorset shells. Typical of his insider knowledge is the following passage, which links the duchess as a discoverer of a shell to Solander's naming of it, suggesting a very dynamic relationship between the two: "Nearly allied to the *T. Remies*, but distinguished from it by Dr. *Solander*. Dredged up at Waymouth, under the inspection of the late duchess dowager of Portland." References to Solander and the *Museum Portlandicum* are noted more than thirty-five times in the citation section, along with seven references in the descriptive passages to the duchess as the first to find specific shells. At the risk of seeming to be an overzealous interpreter of genre, style, format, and typography, I suggest that the fact that Solander's name is italicized (while the duchess's is not) is to give Solander's name the weight of a published authority, along with the italicized and abbreviated *Solandri.* and *Soland. Mus. Port.* to indicate his important conchological contributions. Pulteney in this way conferred on Solander the scientific authority that print conveys.

In contrast to Swainson's and Pulteney's first-hand knowledge of the duchess and her collection and Solander's work on it, are two surveys of British conchology: Maton's and Rackett's long article published in the *Transactions of the Linnean Society* in 1807 and George Montagu's weighty, two-volume *Testacea Britannica* (1803) and its Supplement with additional plates (1808). Both of these texts subtly elide Solander's and the duchess of Portland's contributions to conchological knowledge. Unlike Pulteney, these authors make no reference to Solander in their lists of references. Maton and Rackett mention him in their descriptive passages twice only, one passage noting that though Solander may have named a species of *Tellina* first, they are opting for the first *published* mention of the species, as is the current practice in taxonomy:

> We have substituted Gmelin's name of *T. punicea* for Dr. Pulteney's *T. laeta* as his edition of the *Systema Naturae* [Gmelin's updated edition of Linnaeus's classic text] is in general use, and was prior to the publication of the [Pulteney] Dorsetshire Catalogue. This rule we have generally observed in the following pages. The name *T. laeta* was first given to this species by Dr. Solander, in a description of *Testacea* in the Portland Museum, but which he left unfinished, and which was never printed.[15]

Maton and Rackett felt justified in ignoring Solander's contribution – even though his new molluscan names appeared in *A Catalogue of the Portland Museum* and in the descriptive parts of Pulteney's Dorsetshire catalogue – because of his failure to publish. The Portland Museum in the above passage is not italicized, indicating that the authors were referring to the collection rather than the catalogue, which in effect removes the latter from the realm of legitimate scientific publication.

Solander appears more often in Montagu's three volumes, but mostly as someone to argue with in the descriptive passages, his molluscan names criticized for being "obscure." For instance, Solander is taken to task for misidentification: "This shell [*Helix contorta*] was probably mistaken for Linnean *Helix complanata* by Doctor Solander, as we find it by that name in the part of the *Portland* Cabinet now in the possession of Mr. Laskey." Montagu disagrees with Solander's designation over the classification of a *Mya*:

> Doctor Pulteney has made it a *Pholas* we presume under the authority of Doctor Solander. It does not however appear that there ever were sufficient grounds for placing it in the *Pholas* genus. Even Dr. Pulteney himself remarks, that he has seen several of these shells, but had not seen one with the accessory valves, the essential part of the character of that genus.

In such passages, Pulteney is elevated above Solander, Montagu giving him credit as the authority on mollusks and as a source of information about Solander. Montagu's prose style, in particular his use of dependent clauses, removes Solander from the subject position, which has the effect of subordinating him, as in the two following passages: "Doctor Pulteney speaks of it [*Solen fragilis*] as belonging to the *Portland* cabinet, named by Doctor Solander," and "we have followed Doctor Pulteney in the name here prefixed [*Solen antiquates*], and by which it was called by Doctor Solander in the *Portland* cabinet." In Montagu's text, Pulteney's authority simultaneously rests upon and subsumes Solander's.[16]

The duchess of Portland did not fare much better at Montagu's hands. Though he did repeat a few of Swainson's and Pulteney's narratives about her excursions to the Weymouth beaches, her achievements as a naturalist are diminished in his lackluster treatment of her collecting practices. Montagu wrote: "This species [*Mya pubescens*] is not uncommon, of a small size, taken up with sand from *Falmouth Harbour* . . . Dr. Pulteney (who says the shell was first noticed by the late Dutchess Dowager of Portland) speaks of having seen it two inches and a half long."[17] Notice the parentheses that encapsulate and demote her agency. This could be accounted for by the fact that Montagu's information was second-hand, derived primarily from Pulteney's account of Dorset shells and from communication with Swainson. But when compared to his treatment of other collectors, his depiction of the duchess diminishes her role as a naturalist and has the effect of eliding her contribution to conchological knowledge. Furthermore, she is reduced from a person to a textual trace by the repeated use of the phrase "Portland cabinet"; in lieu of mentioning her name or title, Montagu's narrative descriptions substitute the objects in her collection for her activities as a field collector.

Montagu's rhetorical strategy subtly demotes amateur collectors in favor of naturalists who published their findings. For instance, he celebrates field naturalists who classified their own specimens and published catalogues of their collections. Pulteney, as someone who did it all, is depicted in reverential tones:

> To that able naturalist, the late Doctor Pulteney of Blandford, we are indebted for many rare specimens from the coast of Dorsetshire, and for his information and remarks, which were not of less assistance to us than his Catalogue, published in Hutchins's history of that county, and separately for the use of his friends.

Pulteney, "our dear friend," possessed a deep knowledge of British shells: "We were first indebted to our late worthy friend Doctor PULTENEY for a knowledge of this species as an *English* shell." Although Montagu enjoyed arguing over identification, he deferred to Pulteney as the expert on British shells: "We confess this shell [*Solen fragilis*] has hitherto escaped our notice on the various parts of the coast we have examined, but can have no doubt of its being *English* as the Doctor [Pulteney] found it himself on *Studland* beach." Modeling himself on Pulteney, Montagu depicts himself as someone who brings together collecting, taxonomy, and publishing, clearly the ideal mix, as indicated in his preface:

"The species hereafter described, with a few exceptions are in our own cabinet, and have chiefly been collected from their native places by ourselves, or by the hands of a few friends, whose conchological knowledge and scientific researches are too well known to be doubted." The phrase "by ourselves" and the plural noun "we" are Montagu's way of referring to himself.[18]

What emerges in Montagu's text is a split between shell collectors and published naturalists, which has the effect of sidelining Solander who fits into neither category: though an expert taxonomist, he was not a collector per se and died before he could publish his revision of Linnaeus's system of classification and the catalogue for the duchess's museum. While Montagu reserved his highest praise for conchologists who published their findings, he did value greatly the work of field collectors. Boys, the elder Swainson, and Laskey were important sources of information and specimens for him, and they received credit for their contributions to "conchological knowledge." The duchess's contributions, on the other hand, are diminished in Montagu's text, the objects of infrequent and tepid remarks. By contrast, Boys is referred to as "this patron of science" and as "our worthy friend and able conchologist"; Swainson's authority is unquestioned and his generosity praised: "Mr. Swainson, who favoured us with many specimens taken from water plants." Laskey's skills as a shell hunter along the rough Scottish coasts are valued highly: "Mr. H. Boys and Mr. Laskey, two gentlemen whose golden harvest in the field of Northern conchology has not only been amply experienced by the author, but fully demonstrated to the public in these sheets." Laskey's "indefatigable researches on the coast of Scotland" and "a copious catalogue of the *indigena* as well as specimens of shells" have contributed "largely to the elucidation of the subject." Laskey's word in settling disputes about a shell's national origins and local habitats had great weight with Montagu: "We are supported in this opinion by the observation of Mr. Laskey, who has had frequent opportunity of noticing it on the shores of Scotland."[19]

Montagu's praise of Laskey's abilities as a specimen hunter was no doubt informed by Laskey's own self-presentation. He delivered a paper to his local natural history society describing "Testacea collected by me in North Britain":

> It gives me pleasure to acquaint the Members of the Society, that I have added near 50 new species of Testacea to the British Fauna, from their native shores; and also having met with several species, to which much doubt was attached, I have now positively fixed them as inhabitants of the British seas.[20]

Typical of his shell hunting narratives is the following anecdote, concerning a *Helix bidens*:

> This Shell was found in the Cabinet of the late Dr. Poultenay in possn of the Lin. Soc. and was considered as a very doubtful shell in Brit. Collection[.] in 1826 I was fortunate in discovering this shell in a plantation of Sr Alex^r Hope . . . Scotland but with all my diligence for that year & 1827 & 1828 I found only 8 specimens alive . . . I am of the opinion still that [they] are of a foreign growth and were imported here in Moss & that was around the Roots of Plants that was brought from France and other parts.[21]

Laskey, however, also mentions frequently that he was in possession of some of the duchess's shells, something he was clearly proud of: "In my cabinet there is still one of the original shells, of the Portland Museum, which was a Scottish production."[22] Moreover, his descriptions of his and other's shells reveal that he was just as much if not more of a connoisseur than a field collector, suggesting that these two categories are not as mutually exclusive as Montagu implied with his dismissive treatment of the duchess.

Looking back on Montagu's lack of interest in the duchess, nowadays there might be a temptation to suggest that he was unable to admit women into his precious circle of "scientific" conchologists. His collaborative relationship, however, with his long-time companion, partner in publication, and mother of his children, Eliza Dorville, would suggest otherwise. She engraved all of the shell illustrations for Montagu's volumes, in addition to drawing most of those images herself. Of her activities he wrote:

> [S]hould the following sheets be deemed to possess any small share of merit, the public are indebted to the labours of a friend . . . As this friend of science, however, may not undeservedly feel the shafts of the critical artist, it may be right to disarm them, by observing that, the feminine hand of the engraver was self taught, and [who] claims no other merit in the execution, than what results from a desire to further science by a correct representation of the original drawings, taken by the same hand; both equally works of amusement, not labour for emolument. . . . To the naturalist therefore, and not to the artist an appeal is made; if the representations are correct outlines of the objects, the design is accomplished; and we trust science will be considered as having reaped more advantage from such, than from highly finished engravings devoid of correctness and character.

Though he marks her work as the product of feminine accomplishment (rather than artistic expertise), Montagu praises her contribution to science. In the preface to the supplement of 1808, he gives her credit for her skill and attention to the physical details crucial to the taxonomic enterprise: "were it not for some invariable minute distinctions undefinable by the pencil of the artist, they would ever remain confounded."²³

Montagu's dismissive treatment of the duchess can perhaps be explained by a lack of personal knowledge of her work. Of another generation, she was not a member of his conchological network, and, not having had contact with her, he did not understand how much she had contributed to conchological knowledge and specifically to Pulteney's own collection and his naming of shells. Montagu's friends, who constituted his own network, shaped his notion of value. In addition to Pulteney, whose generosity and expertise contributed much to Montagu's catalogue, Montagu repeatedly praised a circle of friends who supported his project by contributing shell specimens, books, and information. Among this group were Boys and his son Henry, who gave his father's shell collection to Montagu after his father's death. Others mentioned respectfully by Montagu are George Walker, who drew the illustrations and did the descriptions of Boys's shells; William George Maton, co-author of *Testaceological Writers* with the Rev. Thomas Rackett; and, of course, Captain James Laskey. Montagu's indifference to the duchess's accomplishments as a conchologist can also be explained by the different agendas that collectors had in the 1790s compared to the 1770s. The duchess's deep interest in acquiring shells from Cook's voyages and other exotic locales in the East and West Indies was typical of the 1770s; but by the 1790s interest in South Pacific shells had run its course with the proliferation of shells procured during later voyages to Australia and New Zealand. Even by the early 1780s Humphrey was complaining to Seymer that the readily available supply of South Pacific shells had reduced the price at which he could sell them. In addition to market saturation, other factors shaped conchologists' objects, some of which were political. The Napoleonic Wars dampened the exchange of ideas and specimens with Europe, and turned British natural history inquiry onto Britain itself – its fauna and flora, its geology and fossil record, and its landscape. By the end of the eighteenth century collectors' interests had shifted from exotic to domestic shells, a shift that carried with it implications about the role of the naturalist and his relation to the field, here defined as Britain. This focus on British shells would have been considered odd, even bizarre, in the 1770s. Da Costa was hesitant to announce his interest in British shells for fear of mockery. Writing to Pulteney in 1776, he could safely share his love of British shells, knowing that he had a fellow enthusiast in Pulteney whose decades-long project of cataloguing the natural history of Dorsetshire included its shells. Pulteney responded: "I cannot flatter you any Gentlemen in this part of the World except Mr. Seymer, Dr. Cuming, & Mr Wilkes that would give Two-pence for all the Shells or the Books thereon in England. And I never open my lips on these subjects in company having too often experienced how they are ridiculed in the Country."²⁴

In the 1770s questions about whether or not a shell was British would have seemed odd or trivial, but by the late 1790s conchology had become specialized, with conchologists cataloguing specimens by region and national boundaries. By the time that Montagu published *Testacea Britannica* in 1803 he could write on this subject with great concern. For instance, the identity of *Turbo tridens* as a British shell was the subject of much anxiety, as the lengthy discussion below indicates:

> In the Boysian cabinet, which contains nothing but what the worthy owner considered as *British*, we find this shell; and in the notes found in one of the drawers, it was referred to under the title of *quadridens*, without any remarks; so that we are still at a loss where to find the native place of it in *England*, for we suspect our late friend Doctor PULTENEY was mistaken in believing it had been found on water plants by the river *Stour*. The banks of that river and neighbourhood have been searched by us and other conchologists in vain.
>
> Mr. Henry BOYS [son of the collector] does not believe it was ever taken about *Sandwich*, but is rather inclined to think those specimens in his father's possession were received from Doctor SOLANDER, out of the Dutchess of PORTLAND's cabinet. In this opinion we are ready to concur, as it stands by the name of *quadridens*, marked by Doctor SOLANDER, and now in the possession of Mr. LASKEY, having been bought at the sale of the PORTLAND cabinet.
>
> It is however with pleasure we are are [*sic*] able to assure our scientific readers that this shell has recently been found to inhabit Scotland, having been discovered together with *Helix lubrica*, in *Carline Park*, near *Leith*, by Mr. Laskey.²⁵

Notice how the duchess's cabinet is mentioned in the context of obscuring the provenance of the shell, alluding to her promiscuous mingling of exotic and domestic shells, and how Laskey comes to the rescue by finding a specimen in Scotland and solving the problem of the shell's national origin. Palpable is Montagu's relief that the shell, though not English, could be labeled British. Such anxieties about where a shell was found convey the complexities in this period of constructing a national identity that was British, a category that included and subsumed the older but still potent and potentially divisive national identities of English, Welsh, Irish, and Scottish.²⁶

By the turn of the century conchology had, in general, changed from being the province of knowledgeable amateurs who as connoisseurs had collected rare specimens. In the 1770s exotic shells, particularly those from the South Pacific, dominated the imagination and desires of collectors due in part to the popularity of the voyages of discovery and Cook's celebrity status. Also, still foremost in the imagination of a collector like the duchess of Portland was the goal of discovering, describing, and naming every molluscan species in the world. This had been Linnaeus's goal, and it became Solander's as he added species, correcting and revising Linnaeus's taxonomic categories, over the course of a decade. However, beginning with the publication of the fourth volume of Thomas Pennant's *British Zoology* (1777), Emanuel Mendes da Costa's *The British Conchology* (1778), and George Walker's *A Collection of Minute and Rare Shells, lately discovered in the sand of the sea shore near Sandwich* (1784), interest in domestic shells grew, and by the 1790s the goal for most British conchologists was no longer to catalogue the world's shells but to locate, collect, and name every shell species in Britain and its coastal regions. Rarity, as a value-enhancing property, still operated but at a much-reduced rate since it was subsumed under the geopolitical directive to name and describe every species within a specific region or national boundary. No longer did a shell need to be rare to be included in a shell publication. Repeated with frequency is the phrase "not uncommon," for instance, in Pulteney's descriptions of Dorset shells; of the *Amonia cepa*, he wrote: "It is sometimes thrown up in great quantities on the sandy beaches all along the coast."[27] This drive to catalogue shells according to national boundaries continued well into the nineteenth century, but by 1819 William Turton, author of *A Conchological Dictionary of the British Islands*, could joke a little about this marriage of patriotism and science: "Every species which has hitherto been found in the British Islands, we have thought it proper to record, leaving to individual collectors their own ideas of purity as to patriotal adoption."[28]

In the process of collecting and identifying British shells, a different kind of conchologist was to emerge. In the place of the patron and connoisseur, who was ultimately, as a collector of exotic shells, a consumer of dealers' or agents' shells, there arose the field collector who was also well versed in taxonomy, knowledgeable about habitat, and learned in conchological authorities. This figure became what is thought of as the nineteenth-century naturalist, who, as either an amateur or professional, collected for scientific reasons, and it was this kind of naturalist who, after the Napoleonic Wars, returned to foreign and remote locales, South America in particular, which proved inviting after the expulsion of the Spanish colonial regimes. Swainson's son, William, was just such a naturalist, collecting specimens in Brazil and Argentina; Hugh Cuming collected on both the Atlantic and Pacific South American coasts, Peru and Chile in particular (as well as in the Philippines); and Charles Darwin gathered shells along the Argentine shore of the Rio Plata where it meets the Atlantic.[29] Part of this construction of the nineteenth-century naturalist involved a shift away from Linnaeus's goal of naming all the species of the globe's biota towards smaller-scale, more circumscribed projects. By 1800 collectors and naturalists tended to focus on a genus or a region. Pulteney's efforts to catalogue the natural history of Dorset was typical of this movement toward more manageable projects, a scaling back from Linnaeus's grand schemes, Solander's equally ambitious attempt to revise Linnaean categories, and the duchess's goal of possessing every molluscan species. As the Preface to *A Catalogue of the Portland Museum* states:

> It was indeed in the Intention of the enlightened Possessor [the duchess] to have had *every unknown* Species described and published to the World; but it pleased God to cut short the Design, not only by the Death of the ingenious Naturalist employed by her for that Purpose; but, in a short Time afterwards, to the great and irreparable Loss of Science, by her *own* also. Had her Life been continued a few Years longer, it is possible that every Subject in this Catalogue would have been properly described and characterised.[30]

It was clear by 1840, when William Swainson published a popular guide to collecting zoological specimens with the misleading title *Taxidermy*, that such goals as the duchess's were no longer possible, nor were they desirable.

> No collector, however zealous his endeavours, or however vast his plans may be, can ever hope to obtain a tenth part of the productions of nature, even in one of her departments. A selection is therefore compulsory . . . A general collection of all the types and species of animals can never be completed. . . . General collections, therefore, of any one kingdom of nature, cannot be recommended, as, independent of the expense of purchasing and the space they will occupy, the time necessary for arranging and preserving the specimens will prevent the naturalist from making any scientific use of his possessions. In proof of this it may be mentioned, that the most extensive collections in modern times have invariably been formed by those who have not benefited the science by their writings: the two occupations, in fact, are incompatible with each other.

Here he is suggesting that putting one's energy into accumulating massive collections will preclude time for writing, which he privileges as more productive for science. Although Swainson was thinking about his contemporaries and was not necessarily referring to the duchess of Portland or Sir Joseph Banks with this sweeping generalization, such a statement is certainly applicable beyond the specific individuals he referenced in the footnote. The extensive collection to which he was referring here was the earl of Tankerville's, which was collected with "the primary regard to size, beauty, and perfection, science having been considered secondary." He urged collectors to specialize in a particular genus or region: "Scientific collections may be either general, particular, typical, local, or economic," and particular collections, focusing on "one or more genera or families," would allow the collector "to investigate in detail." "Collections, formed upon these principles, contribute, above all others, to the effectual advancement of science, while their formation, also, is comparatively easy; the attention of their possessor is not distracted by numberless drawers of unnamed specimens, nor is he tempted to wander from the object it has previously fixed upon."[31] Not only did Swainson discourage his readers from attempting to amass the kind of collection the duchess had possessed, seeing it as antithetical to the aims of science, but he was also critical of the multi-volume, illustrated catalogues such as the duchess had hoped to produce to document her collection. He wrote:

> We are, in truth, heartily set against all these magnificent undertakings: they are absolutely detrimental to science: for they confine the materials of knowledge and of study within narrow limits of the select few (generally wealthy amateurs), to the exclusion of all who cannot expend 400 L. or 500 L. upon a zoological library. We cannot but admire the zeal of their projectors, – for such works never produce profit, and generally bring a heavy loss, – but we think it a zeal misplaced. Now what is most wanted, is a collection of plates of shells . . . within the pecuniary reach of the student.[32]

By the fourth decade of the nineteenth century the duchess's goals as a collector and her adherence to Linnaean classification would have seemed hopelessly antiquated to those who were eagerly embracing the new systems developed in France under the influence of Lamarck and Cuvier. In a rare acknowledgment of the duchess's and Solander's contributions to conchology, however, William Turton, in his *Conchylia Dithyra Insularum Britannicarum* (1822, reissued 1848), added what had been left out of his *Conchological Dictionary of the British Islands* (1819): all references to the duchess's collecting practices and Solander's taxonomic labors. When he was working on the first shell book, the dictionary, "assisted by his daughter" as the title page announced, Turton did not have access to Banks's library and thus did not see Solander's slips, something that irritated him a great deal:

> Although we have not been honoured by the use of the library of the venerable patriarch of natural history, Sir Joseph Banks, without which, we are informed by a

modern compiler on the subject [L. W. Dillwyn], "no writer can hope to attain any tolerable degree of perfection"; we have been not entirely without the means by which this study can be best understood and most faithfully displayed. The whole of what is described, ourselves have seen and accurately examined, with the exception of such as are indistinct and not to be found in cabinets.[33]

By the time that Turton published *Conchylia Insularum Britannicarum* in 1822, he had seen Solander's manuscripts, and perhaps this encouraged him to write one of the few post-1820 references to the duchess and her work as a collector engaged in classification:

> This rare and very singular species [*Pholas papyracea*] appears to have been known to Solander, and the late dutchess of Portland. Mr. [J. T.] Swainson, of Elm-Grove, near Liverpool, who has lately obtained some specimens from Torbay, informs us, that he is in possession of a very small pair, presented to him by the Dutchess, from which he had drawings made by Agnew, her Grace's gardener. And there appears very little doubt but that it is the species alluded to in the manuscripts of Solander, and from these manuscripts quoted in the Portland Catalogue. Mr. Humphrey purchased the lot.[34]

Turton's mention of the duchess in this book is also dependent on the elder Swainson's telling him the story behind one of the shells that the duchess had given him.

By the mid-nineteenth century the duchess of Portland's presence in the literature on mollusks was gone, with the exception of Turton's reissue of *Conchylia Insularum Britannicarum* in 1848. Her work as a naturalist had disappeared and even her legacy as a collector was eclipsed by current collecting preoccupations and new systems of classification based on natural affinities and differences, the new "French" systems of natural arrangement, which stressed the descriptive power of the naming process wherein names capture the organism's "essence."[35] During Solander's time conchologists had relied on Linnaeus's *Systema Naturae*, which continued to be the dominant system for arranging mollusca in Britain through the first two decades of the nineteenth century. As Wilkins notes, "British conchology was entirely Linnaean since the war with France precluded free access to continental literature." In the early 1820s, however, the last three volumes of Lamarck's *Histoire des Animaux sans Vertèbres* "were already available, and this gave the opportunity to start afresh, cataloguing and arranging the Museum collections in accordance with Lamarck's system, which if not absolutely faultless, was considered . . . to be 'at least superior to any other general system extant'."[36] "Progress" precluded looking back to outmoded theories and practices, and the contributions of eighteenth-century conchology, especially those that never took on the form of a publication in a journal or an illustrated shell book, were easily forgotten.

CONCLUSION

This review of the conchological literature produced in the generation after Margaret Cavendish Bentinck's death shows why her contributions as a naturalist disappeared from the scientific record. Many were the forces that worked to exclude her, among them the changing tastes in collecting shaped by war with France and an implicit nationalist agenda; the changing systems of classification, from Linnaeus's focus on the shell to Lamarck's on the animal that produced the shell; and, more subtle and difficult to detect, a lingering distrust and dislike of aristocratic collectors, whose resources far outpaced those of professional scientists and science writers, as well as the culture's too easy dismissal of women's scientific expertise.

The disappearance of the duchess's contributions as a naturalist in the scientific literature can be explained simply by her failure to publish, though she did contribute to and participate in the accumulation and organization of information about the natural world. She was a field collector and could identify her own specimens; she participated in the period's central questions about classification and taxonomy; and she brought together naturalists, taxonomists, and collectors, supporting their work in a variety of ways, including funding collecting expeditions, providing help with identification, giving financial aid to publishing ventures, and opening her collection to other naturalists, who consulted it for their own work. But because she did not publish the catalogue of her collection, her contributions to constructing knowledge of the natural world were overlooked. As this chapter has shown, the privileging of certain kinds of publication, in conjunction with the application of the "Principle of Priority" of the International Code of Zoological Nomenclature, have contributed, in part, to erase the duchess's legacy as a naturalist, in ways that suggest that publication may not be the only or even the best way to access the depth and degree of someone's engagement with the investigation of nature.

Other reasons why the duchess's contributions have been forgotten may have to do with the status of taxonomy within the life sciences and within the discipline of the history of science.[37] Observation, description, identification, and classification methods derived from Enlightenment scientific practices, activities that the duchess, Solander,

Pulteney, and the rest of her shell network employed, are the methods of the current field of systematic biology. Taxonomy is not some quaint remnant from the past but a form of theorized inquiry that has the potential to contribute much to contemporary concerns about biodiversity. As Quentin Wheeler, entomologist and advocate for taxonomy, has argued, because its methods are observational and not experimental, taxonomy based on morphology is today in danger of being overlooked and ignored by funding agencies and universities eager for Federal and corporate money to sustain research.[38] While biologists may argue over the merit of a taxonomy based on observation versus a taxonomy based on genetic codes, less understandable is the comparative lack of interest demonstrated by historians. The study of the history of the experimental sciences has far outpaced interest in recovering and theorizing the observational sciences, with studies on classification and collecting making up a fraction of this relatively neglected field. This neglect can also be explained by the relative dearth of textual traces that record collecting practices. Without experimental science's dedication to inscription, in the form of laboratory notes, journal entries, and published proceedings, natural history's collecting, arranging, and displaying practices may simply be more elusive and resistant to recovery.

After all, a collection, despite its materiality, is ultimately ephemeral because very few collections outlive their collectors. Collections are created by collectors, who orchestrate the acquisition and arrangement of objects; they are brought into existence and kept intact through the expenditure of the collector's energy, expertise, and resources. As such, collections are vulnerable to the corrosive effects of time. Collections that outlive their creators are rare, and those that have succeeded in surviving intact owe their longevity to museums and the efforts of those who arranged for the transfer from collector to museum. Caring for collections of specimens, often bestowed as bequests, has been a role that natural history museums have been playing since the late eighteenth century, and it is one that puts great strain on their limited resources of funds, space, technology, and personnel to maintain these collections properly. Unable to care for and display all the collections that are bequeathed, museums and historical societies sometimes store specimens off-site in warehouses, or loan them semi-permanently to other museums, or even sell them into private hands.

I close by returning to the problem of the instability of an object's value. During the duchess's lifetime, her shells possessed multiple meanings – they were loans and gifts from friends, non-descript specimens found on the seashore at Weymouth, expensive commodities bought from dealers, and, most importantly, rare specimens of exotic species from warm, tropical seas. When she died and they were sold, the shells moved in and out of other collections, losing their identity as once belonging to her, and, with the exception of three shells, disappearing from the public record. Not all the Pacific shells gathered on Cook's voyages have been lost – a few were lucky enough to have a very different fate. The shells that Banks and Solander collected are fairly intact as a collection, snuggled in their tin boxes, nestled into drawers of cabinets designed by Solander for this purpose, and guarded in the basement of the Natural History Museum in London. Other Pacific shells, some of those from the Cracherode Collection, were luckier than the duchess's; they are safely ensconced in the Enlightenment exhibition at the British Museum in beautiful glass-faced cabinets made of polished mahogany. These shells are lucky because they are admired by thousands of people every year; they are well cared for, and are regarded as emblematic of a particular historical moment when Britain was engaged in voyages of exploration and was putting into practice Enlightenment methods of inquiry.

Such displays are rare, because most natural history museums have moved away from displaying natural history collections, burying trays of shells, drawers full of stuffed birds, and jars of pickled fishes in their basement storerooms. Most are unable to care for these collections, with their dwindling resources and changing notions of what constitutes a nature museum. Most natural history museums rarely display their shell collections, regarding them simultaneously as a valuable legacy and a terrible burden. Shells such as these, collected by people like the duchess, have moved out of the categories that they once inhabited – that of rare and valuable possessions, precious gifts, and scientific specimens. For them, the move may be permanent, and they may have fallen into a category out of which it is very difficult to move, that of being suspended somewhere between memorabilia and rubbish, unless some collector or museum curator comes along and rescues them from this fate.[39]

NOTES

ACKNOWLEDGMENTS

1   Rebecca Stott, *Duchess of Curiosities: The Life of Margaret, Duchess of Portland* (Welbeck: Pineapple Press / Harley Gallery, 2006); Alexandra Cook, "Botanical Exchanges: Jean-Jacques Rousseau and the Duchess of Portland," *History of European Ideas* 33 (2007): 142–56; Stacey Sloboda, "Displaying Materials: Porcelain and Natural History in the Duchess of Portland's Museum," *Eighteenth-century Studies* 43/4 (2010): 455–72; Sylvia H. Myers, *The Bluestocking Circle: Women, Friendship, and the Life of the Mind in Eighteenth-century England* (Oxford: Oxford University Press, 1990); Lisa Moore, "Queer Gardens: Mary Delany's Flowers and Friendships," *Eighteenth-century Studies* 39 (2005): 49–70; Mark Laird and Alicia Weisberg-Roberts, eds., *Mrs. Delany and her Circle* (New Haven and London: Yale University Press, 2009).

2   Currently, there may be in private hands correspondence about the duchess's shell collecting that I am not aware of; if so, I hope that this book will be an encouragement to bring such material to light.

INTRODUCTION

1   *Gazetteer and New Daily Advertiser*, Thursday, 27 April 1786 (three days after the sale had begun).

2   Frances Burney, *Cecilia; or, The Memoirs of an Heiress* [1782], ed. Margaret Anne Doody (New York: Oxford University Press, 1999), 31–32. Burney was introduced to the duchess of Portland by their mutual friend Mary Delany, who tried to interest the duchess in Burney's literary talents.

3   Published after the auction was a booklet purporting to list all the buyers and the prices paid, though many of the buyers are listed as "Cash," which is not very illuminating. See *A Marked Catalogue containing the lots; what each respectively sold for, and the names of the purchasers of the four thousand two hundred and sixty three articles which constituted the Portland Museum, late the Property of the Duchess Dowager of Portland* (London, 1786).

4   *A Catalogue of the Portland Museum* (London, 1786), 130, 133–34.

5   For a breakdown of lots into types of items sold, see Nandini Battacharya, *Slavery, Colonialism, and Connoisseurship: Gender and Eighteenth-century Literary Transnationalism* (Burlington, VT, and Aldershot: Ashgate, 2006), 111–12. Sold were 2,104 lots of shells

and coral specimens, 225 lots of insects, and 456 lots of minerals and ores. Each of the shell lots, however, contained several specimens, so that the total number of shells and coral specimens was far greater than the number of lots.

6 Thomas Martyn, *The Universal Conchologist*, 4 vols. (London, 1784–89), 11.

7 Alfred Gell, *Art and Agency: An Anthropological Theory* (New York: Oxford University Press, 1998).

8 Some would argue that shells are particularly feminine or feminized objects not only in qualitative characteristics of fragility and delicacy, but also in terms of shape and smell, with some bivalves resembling women's genitalia. Such thinking is the basis of Linnaeus's system of names for mollusks, all taken from parts of the female anatomy. See Stephen Jay Gould's essay on Emanuel Mendes da Costa's disgust with Linnaean taxonomy in his *Leonardo's Mountain of Clams and Diet of Worms: Essays on Natural History* (New York: Harmony Books, 1998). This linkage of female sexual organs with mollusks was in the eighteenth century – and continues to be – a running joke throughout the literature on shells, a topic that will be discussed in Chapter Four. For a reading of shell grottoes as a libidinal site, see Lisa Moore, "Queer Gardens: Mary Delany's Flowers and Friendships," *Eighteenth-century Studies* 39 (2005): 49–70.

9 Bronislaw Malinowski, *Argonauts of the Western Pacific* [1922] (Prospect Heights, IL: Waveland Press, 1984), and Marcel Mauss, *The Gift: The Form and Reason for Exchange in Archaic Societies* [1925], trans. W. D. Halls (New York and London: Norton, 1990). For the reworking of exchange theory, see, for instance, Jonathan P. Parry, "The *Gift*, the Indian Gift, and the 'Indian Gift,'" *Man* 21 (1986): 453–73; Marilyn Strathern, *The Gender of the Gift: Problems with Women and Problems with Society in Melanesia* (Berkeley and London: University of California Press, 1988); Annette B. Weiner, *Inalienable Possessions: The Paradox of Keeping-While-Giving* (Berkeley and Oxford: University of California Press, 1992); Maurice Godelier, *The Enigma of the Gift* [1996], trans. Nora Scott (Chicago: University of Chicago Press, 1999); Fred R. Myers, ed., *The Empire of Things: Regimes of Value and Material Culture* (Santa Fe and Oxford: School of American Research Press, 2001); and Nicholas Thomas, *Entangled Objects: Exchange, Material Culture, and Colonialism in the Pacific* (Cambridge, MA, and London: Harvard University Press, 1991).

10 This ethnographic approach to social history owes much to Clifford Geertz, *Local Knowledge: Further Essays in Interpretive Anthropology* (New York: Basic Books, 1983).

11 E. C. Spary, "Pierre Pomet's Parisian Cabinet: Revisiting the Invisible and the Visible in Early Modern Collections," in *From Public to Private: Natural Collections and Museums*, ed. Marco Beretta (Sagamore Beach, MA: Science History Publications/Watson Publishing International, 2005), 59–80.

12 Marco Beretta, "Preface," in *From Public to Private*, viii.

13 Beretta, "Preface," in *From Public to Private*, viii.

14 Spary, "Pierre Pomet's Parisian Cabinet," 78. Spary's description recalls the way in which Latour posits the triangulation of hand, eye, and natural object as productive of knowledge about nature. See Bruno Latour, "Visualization and Cognition: Thinking with Eyes and Hands," *Knowledge and Society* 6 (1986): 1–40.

15 Humphrey bought this shell in a lot containing four other shells for a total of £1 2s. 0d. For a description of *Mya declivis*, see http://species-identification.org/species.php?species_group=mollusca&id=980

16 Richard Pulteney to the dowager duchess of Portland, 24 March 1779; Portland Papers, Longleat House, Wiltshire, vol. 14, f. 148.

17 For details concerning Pulteney's loss of specimens after the death of friends, see Chapter Three.

18 Of the *Pinna muricata*, Donovan wrote: "The only British species of Pinna we are acquainted with, is the *P. Muricata* of Linnaeus, or *P. Fragilis* of Pennant, and that is very rare. The latter author describes it from a specimen in the PORTLAND cabinet, which had been fished up at Weymouth, in Dorsetshire. Da Costa says, he has seen a very small one (of the same species) from the coast of Wales. – Both of these are represented in the annexed plate." Edward Donovan, *The Natural History of British Shells*, 5 vols. (London, 1799–1803), vol. 1: pl. 10.

19 See, for instance, Julie Park, *The Self and It: Novel Objects in Eighteenth-century England* (Stanford, CA: Stanford University Press, 2010), and Chloe Wigston-Smith, "Clothes without Bodies: Objects,

Humans, and the Marketplace in Eighteenth-century It-Narratives and Trade Cards," *Eighteenth-century Fiction* 23/2 (2011): 347–80.

20 See Bruno Latour's discussion of the subject/object divide in "Pragmatogonies: A Mythical Account of How Humans and Nonhumans Swap Properties," *American Behavioral Scientist* 37/6 (1994): 791–808, and his article published under the pseudonym Jim Johnson: "Mixing Humans and Non-humans Together: The Sociology of a Door Closer," *Social Problems* 35/3 (1988): 298–310 [special issue: *The Sociology of Science and Technology*].

21 For an excellent overview of this new field, see Robert J. Foster, "Tracking Globalization: Commodities and Value in Motion," in *Handbook of Material Culture*, ed. Chris Tilley et al. (London and Thousand Oaks, CA: Sage, 2006), 285–302. More recently, historians of science have taken up this mode of analysis, tracking the movement of people, instruments, and specimens from one geographical location to another. See the special issue of the *British Journal for the History of Science*, 43/4 (2010), "Circulation and Locality in Early Modern Science," with its introduction by Kapil Raj, 513–17, which seeks to complicate Latourian models of natural knowledge construction by attending to "how experiences of travel, encounter and exchange changed both the knowledges and practices at issue and their bearers, be they stay-at-home or peripatetic savants, their correspondents, missionaries, functionaries, administrators, merchants, assistants, translators or artists" (516).

22 Arjun Appadurai, "Introduction: Commodities and the Politics of Value," in *The Social Life of Things: Commodities in a Cultural Perspective*, ed. Appadurai (Cambridge: Cambridge University Press, 1986), 3–63 (18). Appadurai's introduction, in which he redeploys Simmel's definition of consumption as a form of exchange to argue that objects can move in and out of the "commodity phase," is a crucial anthropological text that promoted this method of following objects as they moved across space and through time. Laying out the theoretical stakes for studying the circulation of commodities, he argues that the meanings of objects emerge in their movement: "all efforts at defining commodities are doomed to sterility unless they illuminate commodities in motion" (16). Rather than thinking of commodities as things produced by alienated labor for exchange, Appadurai broke from "the production-dominated Marxian view of the commodity" to focus "on its total trajectory from production, through exchange/distribution, to consumption" (13).

23 Appadurai ("Introduction" in *The Social Life of Things*, 11–15), building on Igor Kopytoff's work on the social lives of things, has argued that although objects can begin their existence as commodities, they can become something else, including a gift, an heirloom, a museum piece, or an art installation, and, as such, these objects can be described as having been diverted from the "commodity phase" into another phase through various procedures involving "enclaving" and removal from circulation. See also in the same book: Igor Kopytoff, "The Cultural Biography of Things: Commoditization as Process," 64–93.

24 I am using the term "gift economy" here as a heuristic rather than as an accurate description of that particular culture's systems of exchange.

25 On multi-sited ethnography, see George Marcus, "Ethnography in/of the World System: The Emergence of Multi-Sited Ethnography," *Annual Review of Anthropology* 24 (1995): 95–117.

26 Mary Louise Pratt, *Imperial Eyes: Travel Writing and Transculturation* (New York and London: Routledge, 1992). For studies of the ways in which a region's biota was taken up by the imperial project, see also Lucile Brockway, *Science and Colonial Expansion: The Role of the British Botanic Gardens* (New York: Academic Press, 1979); Richard Drayton, *Nature's Government: Science, Imperial Britain, and the "Improvement" of the World* (New Haven and London: Yale University Press, 2000); Londa Schiebinger, *Plants and Empire: Colonial Bioprospecting in the Atlantic World* (Cambridge, MA, and London: Harvard University Press, 2004); and my *Colonizing Nature: The Tropics in British Arts and Letters, 1760–1820* (Philadelphia: University of Pennsylvania Press, 2005).

27 See James Delbourgo and Nicholas Dew, eds., *Science and Empire in the Atlantic World* (New York: Routledge, 2008); Neil Safier, *Measuring the New World: Enlightenment Science and South America* (Chicago: University of Chicago Press, 2008); Daniela Bleichmar and Peter C. Mancall, eds., *Collecting across Cultures: Material Exchanges in the Early Modern Atlantic World* (Philadelphia: University of Pennsylvania Press,

2011); and Daniela Bleichmar, *Visual Empire: Botanical Expeditions and Visual Culture in the Hispanic Enlightenment* (Chicago and London: University of Chicago Press, 2012).

28 Thomas, *Entangled Objects*, 126. See also Spary's call to complicate Latour's actor-network model with careful historical examination of the processes by which natural history specimens were collected and carried to the centers of calculation: "The work put into rendering objects of natural history both mobile and immutable must be investigated by historians, but so also must the work done in enforcing particular interpretations both on voyages and at home" (E. C. Spary, *Utopia's Garden: French Natural History from the Old Regime to Revolution*, Chicago: University of Chicago Press, 2000, 97).

29 On tensions between professional naturalists and amateur collectors and their competing claims to authority over natural knowledge, see Steven Shapin, "The Image of the Man of Science," in *The Cambridge History of Science*, vol. 4: *Eighteenth-century Science*, ed. Roy Porter (Cambridge: Cambridge University Press, 2003).

1 THE DUCHESS, NATURAL HISTORY, AND CULTURES OF COLLECTING

1 For an overview of the duchess of Portland's collecting, see the booklet that accompanied the Harley Gallery exhibition on her collections: Rebecca Stott, *Duchess of Curiosities: The Life of Margaret, Duchess of Portland* (Welbeck: Pineapple Press / Harley Gallery, 2006). Her collection was rivaled by those of Sir Joseph Banks and Sir Ashton Lever, but both these contained artifacts, of which the duchess had very few. See Adrienne L. Kaeppler, *Holophusicon: The Leverian Museum, an Eighteenth-century English Institution of Science, Curiosity, and Art* (Altenstadt: ZKF, 2011). For contexts in which such collections were made, see Arthur MacGregor, *Curiosity and Enlightenment: Collectors and Collections from the Sixteenth to the Nineteenth Centuries* (New Haven and London: Yale University Press, 2007), and Kim Sloan, ed., *Enlightenment: Discovering the World of the Eighteenth Century* (Washington, DC: Smithsonian Books / British Museum Press, 2003).

2 Lady Margaret also inherited her mother's family name, which was Cavendish, for, as Kate Retford avers, "it was a stipulation of the second Duke of Newcastle [Lady Margaret's great-grandfather] in his will that all his heirs should be called Cavendish" (325): "Patrilineal Portraiture? Gender and Genealogy in the Eighteenth-century Country House," in *Gender, Taste, and Material Culture in Britain and North America, 1700–1830*, ed. John Styles and Amanda Vickery (New Haven and London: Yale University Press, 2006), 315–44. However, I refer to Margaret, duchess of Portland, as Margaret Cavendish Bentinck, as she has been named recently in various articles, including the *Oxford Dictionary of National Biography*, although I tend to refer to her more often as "the duchess," since this is how her correspondents, even Mary Delany, referred to her. She signed her letters MC Portland, and Solander referred to her as "MCP." To refer to her as Margaret Cavendish may be most correct, but this can lead to confusion with the poet and philosopher who was her great-grandfather's second wife, who was childless and therefore not a blood relation.

3 Horace Walpole, *The Duchess of Portland's Museum*, ed. W. S. Lewis (New York: Grolier Club, 1936), 5–10.

4 In 1741 the duchess's husband wrote a letter to Sloane asking him if he would show his wife and Mrs. Pendarves (Mary Delany) his "Cabinet of Curiosities" (William Bentinck, duke of Portland, to Sir Hans Sloane, Whitehall, Tuesday, 2 'O the Clock, Sloane MS 4058, f. 5, British Library). For a discussion of material culture exchanges between the duchess and Elizabeth Montagu, see Elizabeth Eger, "Paper Trails and Eloquent Objects: Bluestocking Friendship and Material Culture," *Parergon* 26/2 (2009): 109–38.

5 Margaret Cavendish Bentinck's children: Elizabeth (1735–1825), viscountess of Weymouth and marchioness of Bath; Henrietta (1737–1827), countess of Stamford; William Henry (1738–1809), third duke of Portland; Margaret (1739–1756); Francis (1741–1743); Edward Charles (1744–1819).

6 Mary Delany to Ann Dewes, Bulstrode, 20 November 1753, in Mary Delany, *The Autobiography and Correspondence of Mary Granville, Mrs. Delany: With Interesting Reminiscences of King George the Third and*

Queen Charlotte, ed. Lady Llanover, 6 vols. (London: Bentley, 1861–62), 3: 241.

7   Duchess of Portland to Elizabeth (Robinson) Montagu, Bulstrode, 30 June 1738; quoted by Eger, "Paper Trails and Eloquent Objects," 128.

8   Duchess of Portland to Ann Granville, Bulstrode, 24 August 1737, in Delany, *Correspondence and Autobiography*, 1: 618.

9   Walpole, *The Duchess of Portland's Museum*, 5–10. In *An Essay on Prints* (London, 1802) William Gilpin, a visitor to Bulstrode, was most likely referring to the duchess's efforts to collect every Hollar print: "A fourth caution, which may be of use in collecting prints, is, not to rate their value by their scarceness.... Yet, absurd as this false taste is, nothing is more common; and a trifling genius may be found, who will give ten guineas for Hollar's shells, which, valued according to their merit (and much merit they certainly have), are not worth more than twice as many shillings" (169).

10 Peter Dance's work on shells is invaluable to understanding the history of shell collections. See *Shell Collecting: An Illustrated History* (Berkeley, CA: University of California Press, 1966) and the revised version, *A History of Shell Collecting* (Leiden: Brill, 1986). He includes a section on the duchess. Walpole, *The Duchess of Portland's Museum*, 5–10.

11 An exception is Stacy Sloboda's important article on the duchess's china collection: "Displaying Materials: Porcelain and Natural History in the Duchess of Portland's Museum," *Eighteenth-century Studies* 43/4 (2010): 455–72.

12 See "Introduction: The Material of Visual Cultures", in *Material Collecting, 1740–1920: The Visual Meanings and Pleasures of Collecting*, ed. Alla Myzelev and John Potvin (Farnham: Ashgate, 2009): 1–17 (1).

13 Exceptions to this lack of interest in natural history collecting within the history of science are Paula Findlen's masterful study of late Renaissance Italian naturalists, *Possessing Nature: Museums, Collecting, and Scientific Culture in Early Modern Italy* (Berkeley and London: University of California Press, 1994); Marjorie Swann's engaging examination of seventeenth-century English collecting, *Curiosities and Texts: The Culture of Collecting in Early Modern England* (Philadelphia: University of Pennsylvania Press, 2001); and Susan Scott Parrish's *American Curiosity: Cultures of Natural History in the Colonial British Atlantic World* (Chapel Hill: University of North Carolina, 2006), on the colonial dynamics of naturalists' networks.

14 Robert E. Kohler ("Finders, Keepers: Collecting Sciences and Collecting Practice," *History of Science* 95, 2007: 428) attributes this neglect to the lingering effects of older historical models, "the grand narratives of scientific progress," in which collecting was portrayed as "what naturalists did before they became scientists and built labs and gardens and learned to experiment, measure, and model." He notes with irony that "we have long since renounced the grand narrative; yet its implicit bias against collecting seems to live on." See also his *All Creatures: Naturalists, Collectors, and Biodiversity, 1850–1950* (Princeton and Oxford: Princeton University Press, 2006), and *Landscapes and Labscapes: Exploring the Lab–Field Border in Biology* (Chicago and London: University of Chicago Press, 2002).

15 E. C. Spary, "Luminous Learning," *History Workshop Journal* 57 (2004): 271–75. For an exception to the lack of interest in the material culture of collecting, see Mary Terrall's wonderful article, "Following Insects Around: Tools and Techniques of Eighteenth-century Natural History," *British Journal for the History of Science* 43/4 (2010): 573–88. For work on how the physical arrangement of natural history specimens stressed their materiality, see Emma Spary, "Scientific Symmetries," *History of Science* 92 (2004): 1–46. See also Spary, "Codes of Passion: Natural History Specimens as Polite Language in Late 18th-century France," in *Wissenschaft als kulturelle Praxis, 1750–1900*, ed. Peter Hanns Reill and J. Schlumbohm (Göttingen: Vanderhoeck & Ruprecht, 1999), 105–35, and Anke te Heesen and Emma C. Spary, eds., *Sammeln als Wissen* (Göttingen: Wallstein, 2001).

16 Susan M. Pearce, "Collecting Reconsidered," in *Interpreting Objects and Collections*, ed. Pearce (London and New York: Routledge, 1994), 193–204 (204).

17 David Elliston Allen has noted: "The majority of rich collectors ... had little thought of science ... when the ... new, thrusting spirit of enquiry ... faltered and finally petered out, the motive of science ... tended to deteriorate into a mindless hoarder's lust ... the main functions of the wealthy collector were to prime the subject with

money and build up a wide selection of the raw material for later systematists to study. Natural History, in the true sense, gained . . . only incidentally" (*The Naturalist in Britain* [1978], 30; quoted by R. I. Vane-Wright and Harold W. D. Hughes, *The Seymer Legacy: Henry Seymer and Henry Seymer Jnr. of Dorset, and their Entomological Paintings with a Catalogue of Butterflies and Plants, 1755–1783*, Tresaith: Forrest Text, 2005, 29).

18 For polite science, see Alice N. Watters, "Conversation Pieces: Science and Politeness in Eighteenth-century England," *History of Science* 35/2 (1997): 121–54; and Geoffrey V. Sutton, *Science for a Polite Society: Gender, Culture, and the Demonstration of Enlightenment* (Boulder, CO: Westview Press, 1995).

19 For the country house as site of natural history craftwork, see "Defining Femininity: Women and the Country House," chapter 5 in Dana Arnold, *The Georgian Country House* (Stroud: Sutton, 1998), 79–99; for grottoes, see 94–99.

20 Elizabeth Montagu to Mrs. Donnellan, [Fall] 1749, [Tunbridge Wells]; Matthew Montagu, ed., *The Letters of Mrs. Elizabeth Montagu*, 3 vols. (Boston, MA, 1825), 2: 167.

21 *Lloyd's Evening Post and British Chronicle*, issue 734, Friday, 26 March 1762.

22 *Gazetteer and New Daily Advertiser*, issue 11196, Friday, 18 January 1765.

23 *Daily Advertiser*, issue 14187, Friday, 7 June 1776.

24 For the link between disembodiment and scientific value, see "Introduction: The Body of Knowledge," in *Science Incarnate: Historical Embodiments of Natural Knowledge*, ed. Christopher Lawrence and Steven Shapin (Chicago and London: University of Chicago Press, 1998), 1–19.

25 Lynch and Law wrote (319–20): "'Natural order is discovered and organized through the use of texts. . . . [Bird-watching] requires an active consultation of texts as part of the embodied performance of a socially organized activity. . . . Bird-watchers use optical equipment, field guides, and lists in a reflexive way, as they go back and forth between the textual categories in hand and the proverbial bird-in-the-bush" (Michael Lynch and John Law, "Pictures, Texts, and Objects: The Literary Language Game of Bird-Watching," in *The Science Studies Reader*, ed. Mario Biagioli, New York and London: Routledge, 1999, 315–41).

26 Amanda Vickery, *Behind Closed Doors: At Home in Georgian England* (New Haven and London: Yale University Press, 2009), 242.

27 Pamela H. Smith's and Paula Findlen's collection of essays, *Merchants and Marvels: Commerce, Science, and Art in Early Modern Europe* (New York and London: Routledge, 2002), raises important questions about the connections between science, art, and commerce, as does David Philip Miller and Peter Hanns Reill, eds., *Visions of Empire: Voyages, Botany, and Representations of Nature* (Cambridge: Cambridge University Press, 1996).

28 For Seymer's life and his work as an illustrator of beautiful butterfly watercolors, see Vane-Wright and Hughes, *The Seymer Legacy*.

29 Dr. John Doran, *A Lady of the Last Century (Mrs. Elizabeth Montagu): Illustrated in her Unpublished Letters* (London: Richard Bentley, 1873), 326, 335.

30 For advice on how to preserve a bird, see Hannah Robertson, *The Young Ladies School of Arts* (York, 1777).

31 For the relationship between decorative art practices and the study of natural history in this period, see the essays by Lisa Ford, John Edmondson, Mark Laird, Janice Neri, and Maria Zytaruk in *Mrs. Delany and her Circle*, ed. Mark Laird and Alicia Weisberg-Roberts (New Haven and London: Yale University Press, 2009).

32 W. Hugh Curtis, *William Curtis, 1746–1799* (Winchester: Warren and Son, 1941), 73.

33 Ruth Hayden, *Mrs. Delany: Her Life and her Flowers* [1980] (New York: New Amsterdam Books / British Museum Press, 1992), 158.

34 *Letters from Mrs. Delany . . . to Mrs. Frances Hamilton from the Year 1779 to the Year 1788* (London: Longman, Hurst, Rees, Orme, and Brown, 1820), xix.

35 Dru Drury, *Illustrations of Natural History: wherein are exhibited upwards of two hundred and forty Figures of Exotic Insects, according to their different Genera*, 3 vols. (London, 1770–82), 1: xiii and 10. For more details about the coloring of illustrations, see my chapter "Butterflies, Spiders, and Shells: Coloring Natural History Illustrations in Late Eighteenth-Century Britain," in *The Materiality of Color: The Production, Circulation, and Application of Dyes and Pigments, 1400–1800*, ed. Andrea Feeser, Maureen Daly Goggin, and Beth Fowkes Tobin (Farnham, Surrey: Ashgate, 2012), 265–80.

36 Emanuel Mendes da Costa, *Historia Naturalis Testaceorum Britanniae; or, The British Conchology* (London, 1778), 221–22.

37 Da Costa, *British Conchology*, vii, viii.

38 Mary Julia Young, *The East Indian; or, Clifford Priory*, 4 vols. (London, 1799).

39 Mary Delany to Bernard Granville, Bulstrode, 10 October 1774, in *Autobiography and Correspondence*, 5: 40.

40 Robert John Thornton, *The Religious Use of Botany* (London, 1824), 4.

41 Mary Delany to Mary Dewes, Bulstrode, 4 October 1768, in *Autobiography and Correspondence*, 4: 173.

42 Stott, *Duchess of Curiosities*, 38.

43 Vickery, *Behind Closed Doors*, 152. Though Vickery mentions the duchess of Portland only in passing, she is the sole scholar to portray her as engaged in doing the work of classifying her specimens: "The duchess funded natural history expeditions and drew experts *to work alongside her* on the collections" (emphasis added).

44 John Timothy Swainson, "MS Catalogue of British Shells in the Possession of J. T. Swainson"; Zoology Library, Natural History Museum, London, Mollusca 1, n. p.

45 Jean K. Bowden, *John Lightfoot: His Work and Travels* (Kew and Pittsburgh: Royal Botanic Gardens/Hunt Institute for Botanical Documentation, 1989), 8–10.

46 For Mary Delany's involvement in natural history, see *Mrs. Delany and her Circle*. For Delany's life and accomplishments, see Hayden, *Mrs. Delany*, who is to be credited for calling attention to the achievements of her ancestor, and for inspiring such recent work on Delany as *Mrs Delany and her Circle*. See Janice Farrar Thaddeus's careful reworking of the Delany created by her Victorian great-niece Lady Llanover in the *Autobiography and Correspondence* in "Mary Delany, Model to the Age," in *History, Gender, and Eighteenth-century Literature*, ed. Beth Fowkes Tobin (Athens and London: University of Georgia Press, 1994), 113–40. For women's friendships, see Lisa Moore, "Queer Gardens: Mary Delany's Flowers and Friendships," *Eighteenth-century Studies* 39 (2005): 49–70; Betty Rizzo, *Companions without Vows: Relationships among Eighteenth-century British Women* (Athens and London: University of Georgia Press, 1994); and Sylvia H. Myers, *The Bluestocking Circle: Women, Friendship, and the Life of the Mind in Eighteenth-century England* (Oxford: Oxford University Press, 1990).

47 Mary Delany to Miss Port, Bulstrode, 1 August 1779, in Delany, *Autobiography and Correpondence*, 5: 448–49

48 Quoted in Hayden, *Mrs. Delany*, 155.

49 Jean Bowden (*John Lightfoot*, 10) gives "Thomas" as Agnew's first name, but on his drawings the initial he uses looks more like a "J" or an "I," and the Natural History Museum's library catalogue uses "J."

50 Ann B. Shteir, *Cultivating Women, Cultivating Science: Flora's Daughters and Botany in England, 1760–1860* (Baltimore and London: Johns Hopkins University Press, 1996), chapter 2: "Women in the Polite Culture of Botany," 47–50. For a discussion of aristocratic women's leisured activities, see also Vickery, *Behind Closed Doors*, chapter 5: "Rooms at the Top."

51 The history of collecting in the early modern period has been addressed by several scholars most ably, in particular Paula Findlen in her *Possessing Nature: Museums, Collecting, and Scientific Culture in Early Modern Italy* (Berkeley and London: University of California Press, 1994) and her astute analysis of natural history collecting and the construction of scientific networks. Among the other most influential studies are Oliver Impey and Arthur MacGregor, eds., *The Origin of Museums: The Cabinet of Curiosities in Sixteenth- and Seventeenth-century Europe* (Oxford: Clarendon Press, 1985); Krzysztof Pomian, *Collectors and Curiosities: Paris and Venice, 1500–1800* [1987], trans. Elizabeth Wiles-Portier (Cambridge: Polity Press, 1990); Thomas DaCosta Kaufmann, *The Mastery of Nature: Aspects of Art, Science, and Humanism in the Renaissance* (Princeton: Princeton University Press, 1993); Swann, *Curiosities and Texts*; and the essays in *The Cultures of Natural History*, ed. Nicholas Jardine, James A. Secord, and E. C. Spary (Cambridge: Cambridge University Press, 1996). Also, John Elsner and Roger Cardinal, eds., *The Cultures of Collecting* (Cambridge, MA: Harvard University Press, 1994), is an excellent collection of essays that provides a range of readings (semiotic, anthropological, psychoanalytic, and art historical) of various kinds of collecting practices.

52 Pomian, *Collectors and Curiosities*, 58–59.

53 Katie Whitaker, "The Culture of Curiosity," in *The Cultures of Natural History*, 75–90 (75).

54 For Mrs. Barrington, Lady de Clifford, and Lady Amelia Hume, see Sir James Edward Smith, *Exotic Botany*, 2 vols. (London, 1804–05); for Lady Bute, see Lisa Ford, "A Progress in Plants: Mrs. Delany's Botanical Sources," in *Mrs. Delany and her Circle*, 204–23; for Lady Shelburne, see Vickery, *Behind Closed Doors*, 145–56.

55 Mary Delany to Rev. John Dewes, Bulstrode, 9 July 1778, and Delany to Miss Port, Bulstrode, 1 August 1779, in Delany, *Autobiography and Correspondence*, 5: 363–64 and 448–49.

56 Emily J. Climenson, ed., *The Diaries of Mrs. Philip Lybbe Powys* (London: Longmans, 1899), 120–21. For aristocratic expenditure and spectacular waste, see Georges Bataille, "The Notion of Expenditure," in *Visions of Excess: Selected Writings, 1927–1939*, trans. Allan Stoekl (Minneapolis: University of Minnesota Press, 1985).

57 Elizabeth Montagu to the duchess of Portland, 3 September 1745, Tunbridge-Wells; Montagu, *The Letters of Mrs. Elizabeth Montagu*, vol. 2: 116.

58 Shteir, *Cultivating Women, Cultivating Science*, 47.

59 Elizabeth Montagu to her cousin Gilbert West, Sandleford, 25 October 1753; Montagu, *The Letters of Mrs. Elizabeth Montagu*, vol. 2: 269.

60 Emanuel Mendes da Costa to Richard Pulteney, 20 July 1776; Natural History Museum, Pulteney correspondence.

61 Jane Austen, *Emma* [1815] (New York: Bedford/St. Martins, 2002), 290.

62 Anon., *An Essay in Defense of the Female Sex* (1696), 91.

63 Donald F. Bond, ed., *The Tatler*, 3 vols. (Oxford: Clarendon Press, 1987), 3: 134 [no. 216, 26 August 1710]. For this literary tradition of ridiculing virtuosi, see Claude Lloyd, "Shadwell and the Virtuosi," *PMLA* 44 (1929): 472–96; Joseph M. Levine, *Dr. Woodward's Shield: History, Science, and Satire in Augustan England* (Berkeley: University of California Press, 1977); and Swann, *Curiosities and Texts*, chapter 2.

64 Hannah Cowley, *The Belle's Stratagem* (1780), quoted in Nandini Battacharya, *Slavery, Colonialism, and Connoisseurship: Gender and Eighteenth-century Literary Transnationalism* (Burlington, VT, and Aldershot: Ashgate, 2006), 105.

65 For the complex and contradiction discourse on curiosity, see Barbara Benedict, *Curiosity: A Cultural History of Early Modern Inquiry* (Chicago: University of Chicago Press, 2001). For the connection between curiosity and collecting, see Chapter Four.

66 Emanuel Mendes da Costa, *The British Conchology*, vi.

67 Benedict, *Curiosity*.

68 Craig Ashley Hanson, *The English Virtuoso: Art, Medicine, and Antiquarianism in the Age of Empiricism* (Chicago and London: University of Chicago Press, 2009), 8.

69 Among these are Paula Findlen's masterful study of early modern Italian naturalists, Marjorie Swann's engaging examination of seventeenth-century English collecting, Susan Scott Parrish's important book on the colonial dynamics of naturalists' networks between Britain and its North American colonies, and Matthew Eddy's study of the career of Dr. John Walker as a mineralogist, which attends to the ways in which Walker amassed and organized his mineral collection: *The Language of Mineralogy: John Walker, Chemistry, and the Edinburgh Medical School, 1750–1800* (Aldershot: Ashgate, 2008).

70 An exception to this is Spary's work on shell collecting, as well as Bettina Dietz's "Mobile Objects: The Space of Shells in Eighteenth-century France," *British Journal of the History of Science* 39 (2006): 363–82. See Spary's "Scientific Symmetries," *History of Science* 92 (2004): 1–46, and her "Codes of Passion."

71 Sigmund Freud, *On Sexuality: Three Essays on the Theory of Sexuality and Other Works*, ed. James Strachey, vol. 7 (Harmondsworth: Penguin, 1977), 351–408. See also Jean Baudrillard, "The System of Collecting," in Elsner and Cardinal, eds., *The Cultures of Collecting*, 7–24.

72 See, for instance, Pierre Bourdieu, *Distinction: A Social Critique of the Judgement of Taste*, trans. Richard Nice (London: Routledge, 1984), and Jean Baudrillard, "The Art Auction: Sign Exchange and Sumptuary Value," in *For a Critique of the Political Economy of the Sign*, trans. Charles Levin (St. Louis, MO: Telos Press, 1981), 112–22.

73 For bird metaphors applied to the duchess of Portland, specifically "bowerbird mentality," see David Elliston Allen, *The Naturalists in Britain: A Social History* [1976] (Princeton: Princeton University Press,

1995), 25; and for "magpie's enthusiasm," see Hayden, *Mrs. Delany*, 106.

74 Basing her analysis of the duchess's collecting practices on the auction catalogue's illustration, Stacey Sloboda argues: "Reading this image in tandem with other contemporary textual and visual accounts reveals that disorder, pleasure, and curiosity facilitated by the commercial and sociable exchange of objects were not just functions of the auction, but also key to understanding the logic of the collection itself" ("Displaying Materials," 455). For a very thoughtful and nuanced response to the way in which *A Catalogue of the Portland Museum* (London, 1786) represents the duchess's collecting practices, see *Promiscuous Assemblage, Friendship, and the Order of Things* (New Haven: Yale Center for British Art, 2009), the catalogue of Jane Wildgoose's amazing installation, an exuberant "Celebration" of the friendship between Mary Delany and the duchess of Portland.

75 For "chaotic jumble," see Sloboda, "Displaying Materials," 460; for "a clutter of shells in a jumble of unsorted boxes," see Judith Pascoe, *The Hummingbird Cabinet: A Rare and Curious History of Romantic Collectors* (Ithaca: Cornell University Press, 2006), 64. To be fair, Sloboda is much more thoughtful than most cultural historians about the duchess's collecting practices. She quotes Montagu, who in 1742 wrote to the duchess about her "Closet," saying: "What cunning confusion, and vast variety, and surprising Universality, must the head possess that is but worthy to make an inventory of the things in that closet" (quoted on p. 463). It is unclear if this phrase is a description of the duchess's "head" or her closet because Montagu's verbal style was to stretch a metaphor to its breaking point, when it disintegrated into nonsense. Montagu's suggestion was that there was so much stuff in the closet that only a genius would be able to see order in it. This may have been a fair assessment of the duchess's state of collecting in 1742 before she began to study Linnaean classification seriously and focus her attention on conchology.

76 Mary Delany to Mary Dewes, Bulstrode, 3 September 1769, in *Autobiography and Correspondence*, 4: 238.

77 Mark Laird, "Introduction (2): Mrs Delany and Compassing the Circle: The Essays Introduced," in *Mrs. Delany and her Circle*, 29–39 (37). For more on the productive nature of this "agreeable confusion," see Maria Zytaruk, "Mary Delany: Epistolary Utterances, Cabinet Spaces, and Natural History," in *Mrs. Delany and her Circle*, 131–49.

78 Linnaean taxonomy has been characterized as an imperial imposition of one system onto the world's biota. See, for instance, Londa Schiebinger, *Plants and Empire: Colonial Bioprospecting in the Atlantic World* (Cambridge, MA, and London: Harvard University Press, 2004).

79 W. S. Lewis, "Introduction," in Walpole, *The Duchess of Portland's Museum*, v.

80 See John Brewer, *The Pleasures of the Imagination: English Culture in the Eighteenth Century* (Chicago: University of Chicago Press, 1997), chapter six: "Connoisseurs and Artists."

81 Pearce, "Collecting Reconsidered," 204.

82 Susan Stewart, *On Longing: Narratives of the Miniature, the Gigantic, the Souvenir, the Collection* (Baltimore and London: The Johns Hopkins University Press, 1984), 136, 81.

83 Susan M. Pearce, *Museums, Objects, and Collections: A Cultural Study* (Washington, DC: Smithsonian, 1992), 81, 72.

84 Pascoe, *The Hummingbird Cabinet*, 6. See also Clive Wainwright, *The Romantic Interior: The British Collector at Home, 1750–1850* (New Haven and London: Yale University Press, 1989).

85 Pearce, *Museums, Objects, and Collections*, 84.

86 Pearce, *Museums, Objects, and Collections*, 84.

87 Baudrillard, "The System of Collecting," 22.

88 Pearce, *Museums, Objects, and Collections*, 81. See Walter Benjamin, "Unpacking my Library: A Talk about Book Collecting," in *Illuminations*, ed. Hannah Arendt (New York: Schocken Books, 1968), 59–67.

89 Pearce, "Collecting Reconsidered," 202.

90 Stewart, *On Longing*, 143, 151.

91 "It is invariably *oneself* that one collects" (Baudrillard, "The System of Collecting," 12).

92 James H. Bunn, "The Aesthetics of British Mercantilism," *New Literary History* 11 (1980): 303–21.

93 Stewart, *On Longing*, 133.

94 For a discussion of China's history of consumption and the rise of non-European modernities, see Craig Clunas, "Modernity Global and Local: Consumption and the Rise of the West," *American Historical Review* 104 (1999): 1497–1511.

95 Mark Laird, "Mrs. Delany's Circles of Cutting and Embroidering in Home and Garden," in *Mrs. Delany and her Circle*, 166.

96 Shteir, *Cultivating Women, Cultivating Science*, 47. See also Vickery, *Behind Closed Doors*, 152–53.

97 For a defense of the duchess as a naturalist, see my essay "The Duchess's Shells: Natural History Collecting, Gender, and Scientific Practice," in *Material Women, 1750–1950: Consuming Desires and Collecting Practices*, ed. Maureen Daly Goggin and Beth Fowkes Tobin (Aldershot: Ashgate, 2009), 244–63.

98 See Alexandra Cook, "Botanical Exchanges: Jean-Jacques Rousseau and the Duchess of Portland," *History of European Ideas* 33 (2007): 142–56.

99 Montagu, *Testacea Britannica*, 1: x–xi.

100 See H. S. Torrens, "Mawe, John (1766–1829)," in *Oxford Dictionary of National Biography* (Oxford: Oxford University Press, 2004); online edn., 2008 [accessed 25 January 2009].

101 Thomas Martyn, *The Universal Conchologist*, 4 vols. (London, 1784–89), 1: 10–12.

102 Margaret C. Jacob and Dorothée Sturkenboom, "A Woman's Scientific Society in the West: The Late Eighteenth-century Assimilation of Science," *Isis* 94/2 (2003): 221.

103 See Bruno Latour, "Visualization and Cognition: Thinking with Eyes and Hands," *Knowledge and Society* 6 (1986): 1–40; and Michael Lynch and John Law, "Pictures, Texts, and Objects: The Literary Language Game of Bird-watching," in *The Science Studies Reader*, ed. Mario Biagioli (New York: Routledge, 1999), 315–41, a revised version of "Lists, Field Guides, and the Descriptive Organization of Seeing: Birdwatching as an Exemplary Observational Activity," *Human Studies* 11/2–3 (1988): 271–304. I have been reminded by my taxonomist colleagues that the text–eye–specimen *techné* continues to be how taxonomists do the work of classification, even with the inroads of the genetic testing of specimens.

104 According to John Bowyer Nichols, Tunstall's bird collection alone had cost him more than £3,000 (*Illustrations of the Literary History of the Eighteenth Century*, London, 1828, 5: 514).

105 Fortunately, much of Pulteney's massive correspondence has been preserved and is held in the libraries of the Linnean Society and the Natural History Museum, London. This study of the duchess's shell collection is completely dependent on the generosity of these two institutions. For an overview of the extent of Pulteney's correspondence, see Robert H. Jeffers, "Richard Pulteney, MD, FRS (1730–1801) and his Correspondents," *Proceedings of the Linnean Society of London* 171 (1959–60): 15–26.

106 Martyn, *Universal Conchologist*, 1: 20.

107 Aylmer Lambert, "Account of Henry Seymer," MS, Linnean Society.

108 It is important to note that the duchess was not the only woman who was an active specimen hunter. In addition to her friends Lady Bute, Mary Delany, and Mrs. LeCoq and those listed in Thomas Martyn's Preface to *The Universal Conchologist*, there were Mary Anning, the famous shell and fossil hunter of Lyme Regis, and a Miss Pocock, who pops up with regularity in Edward Donovan's volumes on *The Natural History of British Shells* as someone who found many new molluscan species along the coast of Wales. Many more women collected shells for a range of purposes, and I hope that other scholars are working on bringing their activities to light.

109 The literature on the relationship between science and the Enlightenment is vast and beyond the scope of this book; but very helpful in addressing the crucial issues concerning this relationship as cultural context can be found in the introduction to the collection of essays edited by William Clark, Jan Golinski, and Simon Schaffer: *The Sciences in Enlightened Europe* (Chicago: University of Chicago Press, 1999). See also Jan Golinski, *Science as Public Culture: Chemistry and Enlightenment in Britain, 1760–1820* (Cambridge and New York: Cambridge University Press, 1992); John Gascoigne, *Joseph Banks and the English Enlightenment* (Cambridge and New York: Cambridge University Press, 1994); Dorinda Outram, *The Enlightenment* (Cambridge and New York: Cambridge University Press, 1995); and Geoffrey Sutton, *Science for a Polite Society: Gender, Culture, and the Demonstration of Enlightenment* (Boulder, CO: Westview Press, 1995). For early modern concepts of order in nature, see Lorraine Daston and Katharine Park, eds., *Wonders and the Order of Nature, 1150–1750* (New York: Zone Books, 1998). For the British engagement with Enlightenment natural history, see Sloan, *Enlightenment*, and K. G. W. Anderson et al., eds., *Enlightening the British: Knowledge, Discovery and the Museum in the Eighteenth Century*

(London: British Museum Press, 2003), both published in conjunction with the opening of the British Museum's Enlightenment Gallery. See also MacGregor, *Curiosity and Enlightenment*. For in-depth study of Sir Joseph Banks's career, see John Gascoigne, *Science in the Service of Empire: Joseph Banks, the British State and the Uses of Science in the Age of Revolution* (Cambridge: Cambridge University Press, 1998). For a lucid explanation of natural classification as scientific praxis, see Peter Dear, *The Intelligibility of Nature* (Chicago and London: University of Chicago Press, 2006), in which the author suggests that "Classification was not just cataloguing, the imposition of arbitrarily selected ordering principles; it was not mere 'natural history' in the usual sense of the straightforward description of nature's contents. Eighteenth-century taxonomists generally aimed at being natural philosophers who could provide the conceptual understanding that would raise the status of their work and establish it as philosophically meaningful, not simply as useful" (53). See also Tore Frängsmyr, J. L. Heilbron, and R. E. Rider, eds., *The Quantifying Spirit in the Eighteenth Century* (Berkeley: University of California Press, 1990).

2 THE DUCHESS AND SHELL COLLECTING

1 Dru Drury, Letter-Book, 1761–83; Natural History Museum, London, SB F D.6. For Henry Smeathman (1742–1786), see Starr Douglas and Felix Driver, "Imagining the Tropical Colony: Henry Smeathman and the Termites of Sierra Leone," in *Tropical Visions in an Age of Empire*, ed. Felix Driver and Luciana Martins (Chicago: University of Chicago Press, 2005), 91–112.

2 Sir James Edward Smith, *A Selection of the Correspondence of Linnaeus, and other Naturalists, from the Original Manuscripts*, 2 vols. (London: Longman, Hurst, Rees, Orme, and Brown, 1821), 2: 44–45.

3 *Mrs. Montagu, "Queen of the Blues": Her Letters and Friendships from 1762 to 1800*, ed. Reginald Blunt from material left to him by Montagu's great-great-niece, Emily Climenson, 2 vols. (London: Constable, 1923), 1: 259. Thomas Shaw was the author of *Travels or Observations relating to Barbary and the Levant* (Oxford, 1738), and a frequent visitor to Bulstrode (duchess of Portland to Pulteney, Whitehall, 6 April 1781). For more on Shaw, see *Mrs. Montagu, "Queen of the Blues"* (1: 288–89), in which Montagu refers to him as the "huge godfather of all shell fish." Katherine H. Porter's doctoral dissertation, "Margaret, Duchess of Portland" (Cornell University, 1930, 313), quotes from a letter written by the duchess (in a private collection) in which she bemoans the loss of a ship coming from Madras because it was carrying South Asian land and sea shells for her.

4 The Rev. John Lightfoot to the duchess of Portland, no date; Portland Papers, Longleat House, Wiltshire, vol. 14, f. 187: "I have the Pleasure to inform you that by a letter from Mr Smith of Chelsea (the Purchaser of the Linnean Treasures), who had lately heard from Dr. Sibthorpe that the Doctor has made for you a complete Collection of the Land & Water shells of Switzerland and waits only for a fair Opportunity to convey them to England . . . that you will shortly receive considerable Acquisitions to your magnificent Collection, which is indeed almost beyond Improvement."

5 David Elliston Allen, "Naturalists in Britain: Some Tasks for the Historian," XVI in his *Naturalists and Society: The Culture of Natural History in Britain, 1700–1900* (Aldershot: Ashgate / Variorum, 2001), 97.

6 David Elliston Allen, *The Naturalist in Britain: A Social History* [1976] (Princeton: Princeton University Press, 1994), 25. See also Allen, "Tastes and Crazes," in *The Cultures of Natural History*, ed. N. Jardine, J. A. Secord, and E. C. Spary (Cambridge: Cambridge University Press, 1996), 394–407 (397). For feminist work on women and science, see Ann B. Shteir, *Cultivating Women, Cultivating Science: Flora's Daughters and Botany in England, 1760–1860* (Baltimore and London: Johns Hopkins University Press, 1996), and Londa Schiebinger, *Nature's Body: Gender in the Making of Modern Science* (Boston, MA: Beacon Press, 1993). See also Sally Festing, "Rare Flowers and Fantastic Breeds: The 2nd Duchess of Portland and her Circle," *Country Life* (12 June 1986): 1684–86; and "Grace without Triviality: The 2nd Duchess of Portland and her Circle" (19 June 1986): 1772–74. For women and collecting, see Rémy Saisselin's study of nineteenth-century French consumption, which suggests that "women were perceived as mere buyers of bibelots . . . men were col-

lectors. Women bought to decorate and for the sheer joy of buying, but men had a vision for their collections, and viewed their collections as an ensemble with a philosophy behind it" (*Bricabracomania: The Bourgeois and the Bibelot*, New Brunswick: Rutgers University Press, 1984, 68). Susan Pearce, a leading scholar in museum studies, critiques this prevalent notion concerning the difference between men's and women's collecting practices in *Museums, Objects, and Collections: A Cultural Study* (Washington, DC: Smithsonian Institution Press, 1993). See also Russell W. Belk and Melanie Wallendorf, "Of Mice and Men: Gender Identity in Collecting," in *The Material Culture of Gender: The Gender of Material Culture*, ed. Kenneth L. Ames and Katharine Martinez (Winterthur, DE: Henry Francis du Pont Winterthur Museum, 1997), 7–27.

7 Some of these practices – cleaning, sorting, and identifying shell specimens – were once considered to contribute to knowledge about nature, but they are no longer recognized as belonging to the purview of science, surviving into the present as a form of amusement for amateur collectors and naturalists. Remnants of these eighteenth-century shell activities can be found today in such venues as the Philadelphia Shell Show.

8 "This species was first described in the transactions of the Royal Society of London, by the Rev. Mr. Lightfoot, chaplain to the late Duchess of Portland. He says it was found adhering to the leaves of the *Iris Pseudacorus*, in waters near Beaconsfield in Buckinghamshire, by Mr. Agneu, the Duchess of Portland's Gardener" (Edward Donovan, *The Natural History of British Shells*, 5 vols. (London, 1799–1803), vol. 5, plate CL, n. p.).

9 John Lightfoot, in *Philosophical Transactions of the Royal Society* 76 (1786): 160–70 (167).

10 George Montagu, *Testacea Britannica*, 2 vols. (London, 1803), 2: 430: "Lightfoot, who has given a good figure of this shell, in the *Philosophical Transactions* for 1786, says, it was found near Bullstrode."

11 John Timothy Swainson, "MS Catalogue of British Shells in the Possession of J. T. Swainson"; Zoology Library, Natural History Museum.

12 Nineteenth-century naturalists, William Swainson for instance, were far more articulate about techniques and materials of shell collecting than eighteenth-century writers. Compared to da Costa's *Elements of Conchology*, Swainson's *Taxidermy* is full of explicit instructions and visual language to enable the reader to apprehend and put into practice what he reads. See William Swainson, *Taxidermy: Bibliography and Biography* (London: Longman, Orme, Brown, Green, & Longmans, 1840). This detailing of the material world is typical of nineteenth-century writing, novels especially, and stands in marked contrast to the eighteenth-century's lack of specifics about the materiality of ordinary life. Cynthia Wall argues that the lack of material detail in the eighteenth-century novel suggests that there was no need to be specific since the readership and author had a picture of the world in common. Readers belonged to a smaller, closer-knit group of people who would know what a drawing room would contain, so that when Burney or Austen mentioned it, there was no need to explain that there were chairs, sofas, rugs, paintings, prints, china cabinets, writing desks, etc., a practice quite unlike mid-nineteenth-century fiction, with its long paragraphs full of descriptive detail of all the objects that might be housed in a parlor. She argues that by the mid-nineteenth century the readership was so wide that assumptions about space and things were no longer held in common. Applying Wall's thesis to natural history writing, in particular the kind that sought to instruct new naturalists in the techniques of the avocation, it can be argued that one of the signs that natural history had become widely popular was that people relied on books for instruction rather than face-to-face and small-scale interactions with other naturalists, which had been the dominant mode of instruction in the art and science of natural history collecting in earlier periods. See Cynthia Sundberg Wall, *The Prose of Things: Transformations of Description in the Eighteenth Century* (Chicago and London: University of Chicago Press, 2006).

13 Swainson, *Taxidermy*, 220–21.

14 Swainson, *Taxidermy*, 20–21.

15 Anon. [George Annesley, earl of Montmorris?], "Short Instructions for Collecting Shells" ([1815?]), 4.

16 Anon., "Short Instructions for Collecting Shells," 5.

17 Swainson, "MS Catalogue of British Shells," n.p. Proud that many of his British shells were "principally of my own collecting," Swainson found many

of his land and freshwater mollusks when he was stationed as a Customs officer in London (Transcript of J. T. Swainson's will, Swainsoniana, Natural History Museum, MSS 89 f s 1). "Swainsonia" refers to Nora F. McMillan's collection of documents concerning William Swainson and his father, John Timothy, which she collected in preparation to write a biography of the former, a task she was unable to complete. The senior Swainson and George Humphrey (G. H. in the manuscript) would go out together to Battersea, then undeveloped wetlands, in search of shells. Referring to his *Turbo bidens*, he wrote that he "found them with G. H. in an Osier Ground near the Mills [an inn] in Battersea fields." He also notes that he found a *Sabella dentaleformis* "in a pond in the Garden of an Inn before the ascent into Barnet Midd[lesex] on the left hand in 1783," a *Nerita fluviatilis* "in the Thames adhering to Stones & old Tiles above Kingston," and a *Helix contorta* "in the ditches in Rotherhithe Marshes with G. H." Swainson's "MS Catalogue of British Shells" records where he found shells, as well as where his friends found the shells that they then gave to him. For example, next to *Turbo bidens* is the comment: "found many years ago by G. H.'s sister on the N side of the Serpentine in Hyde park," and next to *Helix octona*: "found by G. H. near Elms Battersea."

18 William Swainson (*Taxidermy*, 23) wrote: "the coasts of Exmouth, Sandwich, and Weymouth are particularly productive."

19 Montagu, *Testacea Britannica*, 1: x.

20 Duchess of Portland to Thomas Pennant, Whitehall, 26 February 1778; Warwickshire County Record Office, Pennant Correspondence, CR 2017/TP 172/2.

21 Duchess of Portland to Mrs. Delany, Weymouth, 23 August 1776; Mary Delany, *The Autobiography and Correspondence of Mary Granville, Mrs. Delany: With Interesting Reminiscences of King George the Third and Queen Charlotte*, ed. Lady Llanover, 6 vols. (London: Bentley, 1861–62), 5: 253–54.

22 Lightfoot to the duchess of Portland, Uxbridge, 31 August 1771; Portland Papers, Longleat House, vol. 14, f. 153.

23 Duchess of Portland to Richard Pulteney, 6 July 1782; Pulteney Correspondence, Linnean Society, London.

24 Duchess of Portland to Richard Pulteney, Margate, 21 August 1784; Pulteney Correspondence, Linnean Society.

25 Duchess of Portland to Mary Delany, Margate, 31 July 1784, in Delany, *Autobiography and Correspondence*, 6: 223.

26 Angélique Day, ed., *Letters from Georgian Ireland: The Correspondence of Mary Delany, 1731–68*. (Belfast: Friar's Bush Press, 1991), 249 (6 September 1732). "Do you not wish yourself extended on the beach gathering shells, listening to Phill [Miss Donnellan] while she sings at her work, or joining in the conversation, always attended with cheerfulness?" (246). For a discussion of Victorian codes of propriety concerning shell and seaweed collecting on the shore, see David Elliston Allen, "Tastes and Crazes," in *The Cultures of Natural History*, 394–407.

27 Delville, 24 March 1759; Delany, *Autobiography and Correspondence*, 3: 542–43.

28 Swainson, *Taxidermy*, 21.

29 Anon., "Short Instructions for Collecting Shells," 3.

30 Swainson, *Taxidermy*, 21–22.

31 Emanuel Mendes da Costa, *Elements of Conchology; or, An Introduction to the Knowledge of Shells* (London, 1776), 67.

32 Edward Donovan, *Instructions for collecting and preserving various subjects of natural history* (London, 1794), 59.

33 Anon., "Short Instructions for Collecting Shells," 4.

34 Nora F. McMillan, "John Timothy Swainson, the Second (1756–1824), the Zoologist," *Archives of Natural History* 14/3 (1987): 289–96 (295).

35 Da Costa to Richard Pulteney, October 1776; Natural History Museum, MSS PUL A, f. 16.

36 I owe this observation to the participants in an art history seminar at the University of South Florida in 2010, who were upset to hear that the duchess had dredged for shells since such practices can be destructive to the environment.

37 Da Costa, *Elements of Conchology*, 64.

38 Swainson, *Taxidermy*, 22.

39 Apparently the duchess's advice on dredging was helpful because Cordiner included a section on marine life in his natural history and antiquarian tour of Scotland, having had "access to rich stores of all manner of productions [which] are dredged up with

ease from the bottom of these seas . . . frequently seen in all their healthful, animated state." See Charles Cordiner, *Remarkable ruins and romantic prospects of North Britain, with ancient monuments and singular subjects of natural history*, 2 vols. (London, 1788–95), 1: iii.

40 She wrote in response to Pennant's query about dredging: "the Dss of P is extreamly obliged to Mr Pennant for Mr Cordiners Letter she will make an enquiry if there are different ways of dredging & will let him know but that she will not be able to learn till she gets to Weymouth" (duchess of Portland to Pennant, Whitehall, 26 June 1778; Warwickshire County Record Office, Pennant Correspondence, CR 2017/TP149/5b). The duchess, however, was not always successful in dredging up new specimens: "I . . . am very sorry to inform [you] of the very indifferent success I have had in dredging that I fear there is not any you wish for. I have put up a few which I shall bring to Blandford with me next Thursday" (duchess of Portland to Richard Pulteney, Weymouth, 25 August 1777; Pulteney Correspondence, Linnean Society).

41 Delany, *Autobiography and Correspondence*, 1: 565.

42 Swainson, *Taxidermy*, 25.

43 Of just such a cabinet of shells from the late eighteenth century, which surfaced intact in London in the 1980s, S. Peter Dance wrote: "They are redolent of the eighteenth century; at that time it was more important to bring out the ornamental aspects of conchology than to worry about the niceties of scientific arrangement and documentation. . . . Thus it is no surprise to see so many shells stripped of their natural coating and highly polished. We should expect several shells to be sectioned so that their inner structure is revealed – and here they are" (*Description of a Cabinet of Shells: A Remarkable Survival from the Time of the Voyages of Captain Cook, 1768–1780*, London: Jonathon Harris, 1983, a privately printed, limited edition describing a cabinet filled with shells "virtually untouched since the beginning of the nineteenth century").

44 Donovan, *Instructions*, 61.

45 Hannah Robertson, *The Young Ladies School of Arts* (York, 1777), 10–12.

46 Delville, 17 June 1750; Delany, *Autobiography and Correspondence*, 2: 557.

47 Delville, 8 June 1750; Delany, *Autobiography and Correspondence*, 2: 551.

48 "A Catalogue of Shells chiefly from the South Seas sent down to Mr. Seymer from Mr. Geo. Humphreys"; Natural History Museum, "Mss Relative to British Testacea," vol. 2: f. 156v.

49 Natural History Museum, "Mss Relative to British Testacea," vol. 2: f. 156.

50 Henry Seymer to Richard Pulteney, 10 o'clock p.m., 23 December 1770; Pulteney Correspondence, Linnean Society.

51 Da Costa, *Elements of Conchology*, 71.

52 Swainson, *Taxidermy*, 25.

53 Mary Delany to Lady Andover, Bustrode, 24 November 1768, in Delany, *Autobiography and Correspondence*, 4: 192. Patty pans are little shallow pans that were generally used to bake small cakes and pies. In this instance, the duchess must have used up all the cook's patty pans to sort her specimens and turned to the deck of playing cards as a substitute. Linnaeus, Solander, and Banks all used little tin boxes to store shells. For these metal specimen boxes, see Kathie Way, "The Linnaean Shell Collection at Burlington House," in *The Linnaean Collections* (Linnean Special Issue, no. 7), ed. B. Gardiner and M. Morris (London: The Linnean Society, 2007), 37–46.

54 Mary Delany to Miss Hamilton, 19 March and 2 February 1781, in Delany, *Autobiography and Correspondence*, 6: 4, 9.

55 I am grateful to Ilya Tëmkin for sharing this information with me.

56 Guy L. Wilkins, "The Cracherode Shell Collection," *Bulletin of the British Museum (Natural History) Historical Series* 1/4 (1957): 121–84 (132).

57 Swainson, *Taxidermy*, 95.

58 *A Catalogue of the Portland Museum, lately the property of the Duchess Dowager of Portland* (London, 1786), 78.

59 Lot 118 in *A Catalogue . . . of Natural History which will be Sold by Auction by Mr King . . . April 30 1804* (Barker & Son, Covent Garden). This cabinet matches the description of lot 2443 in *A Catalogue of the Portland Museum*, which was bought by Humphrey.

60 Mary Hamilton's Diary, 17 December 1783, in Delany, *Autobiography and Correspondence*, 3: 182–83.

61 Mary Delany to Miss Hamilton, 19 March 1781, in Delany, *Autobiography and Correspondence*, 6: 9.

62 On 5 March 1779; British Library, Add. MS 45,875, f. 19v. For a photograph of a late eighteenth-century shell cabinet, see Dance, *Description of a Cabinet of Shells*.

63 Delany, *Autobiography and Correspondence*, 2: 471 (11 July 1747); quoted in Day, *Letters from Georgian Ireland*, 250.

64 Swainson, *Taxidermy*, 94.

65 Swainson, *Taxidermy*, 90.

66 Swainson, *Taxidermy*, 77 and 81.

67 Swainson, *Taxidermy*, 72.

68 For the importance of aesthetics and the role of taste in French eighteenth-century natural philosophy and natural history collecting, see E. C. Spary, "Rococo Readings of the Book of Nature," in *Books and the Sciences in History*, ed. Marina Frasca-Spada and Nick Jardine (Cambridge and New York: Cambridge University Press, 2000), 255–75.

69 Henry Seymer to Richard Pulteney, 10 November 1770; Pulteney Correspondence, Linnean Society.

70 Seymer to Pulteney, 10 November 1770; Pulteney Correspondence, Linnean Society.

71 Seymer to Pulteney, 17 March 1771, Pulteney Correspondence, Linnean Society.

72 Da Costa, *Elements of Conchology*, 71–72.

73 Da Costa, *Elements of Conchology*, 72.

74 Humphrey bought three lots (#3412, #3903, and #3961) of very rare shells (Arca, Voluta, and Venus, respectively) at the Portland auction with their epidermises (for 10s., £1 1s., and £4. 10s., respectively).

75 Mrs. Le Cocq seems to have been more than an expert specimen hunter since her letter to the duchess demonstrates a zoologist's concern with anatomy: "I am exceedingly sorry this shell is broke, as I think I never met with the sort before; it had the fish in it, and I think it is as wonderful a thing of the kind as ever I saw. After I had taken it out I was endeavouring which way to preserve it, but found I could not succeed, as it was of such a fleshy substance. On opening it I found the inside to resemble a fowl's; it had a distinct heart, liver, and this that I have sent, which I take to be the gizzard, as your Grace will see by a little shell going into the inside." This letter was published along with a description and illustration of this "gizzard" by George Humphrey, who bought the item for another collector, and then published a brief article about the Bulla's gizzard in the *Transactions of the Linnean Society* 2 (dated 1 December 1789).

76 Da Costa, *Elements of Conchology*, 6–10.

77 Da Costa, *Elements of Conchology*, 10.

78 Swainson, *Taxidermy*, 77, 81–82.

79 Elizabeth Montagu to the duchess of Portland, Sandleford, 2 August 1743; *The Letters of Mrs. Elizabeth Montagu*, ed. Matthew Montagu, 3 vols. (Boston, MA, 1825), 2: 80–81.

80 Elizabeth Montagu to the duchess of Portland, 23 August [1747]; *Elizabeth Montagu: The Queen of the Bluestockings*, ed. Emily J. Climenson, 2 vols. (New York: Dutton, 1906), 1: 245.

81 Grottoes interested the duchess a great deal. For the grotto that Mrs. Delany designed at Bulstrode, see Janice Neri, "Mrs Delany's Natural History and Zoological Activities: 'A Beautiful Mixture of Pretty Objects'," in *Mrs. Delany and her Circle*, ed. Mark Laird and Alicia Weisberg-Roberts (New Haven and London: Yale University Press, 2009), 172–87. For a history of grottoes, see Barbara Jones, *Follies and Grottoes*, second edn. (London: Constable, 1974)

82 Mary Delany to the Right Hon. Viscountess Andover, 21 January 1771, in Delany, *Autobiography and Correspondence*, 4: 326.

83 Alicia Weisberg-Roberts, "Introduction: Mrs Delany from Source to Subject," in *Mrs. Delany and her Circle*, 10.

84 Mary Delany to Mary Dewes, Bulstrode, 17 September 1769, in Delany, *Autobiography and Correspondence*, 4: 240.

85 Thomas Pennant, *British Zoology*, 4 vols. (London, 1776–77), 4: 25.

86 Ilya Tëmkin, "Linnaeus, Solander, and Conchology" (unpublished paper), 3.

87 Thomas Martyn, *The Universal Conchologist*, 4 vols. (London, 1784–89), 1: 4.

88 Pennant, *British Zoology*, 4: 117.

89 Emanuel Mendes da Costa, "Mytilus discors," in *Historia Naturalis Testaceorum Britanniae; or, The British Conchology* (London, 1778), 221.

90 The Marine Species Identification Portal's glossary contains all the above-listed terms with the exception of gutter, wing, mouth, and slope: http://

species-identification.org/species.php?species_group=mollusca&menuentry=woordenlijst [accessed 21 September 2011].

91 S. Lovén, "On the species of Echinoidea described by Linnaeus in his work Museum Ludovicae Ulricae" [1887], quoted in Tëmkin, "Linnaeus, Solander, and Conchology," 2.

92 Caroli Linnæi, *Systema Naturae: A Photographic Facsimile of the First Volume of the Tenth Edition: Regnum Animale* [1758] (London: Trustees of the British Museum, 1956), 707.

93 Tëmkin, "Linnaeus, Solander, and Conchology," 1.

94 Linnaeus's impact on British natural history is too complex to deal with here. For a sample of various approaches to this topic, see Frans A. Stafleu, *Linnaeus and Linnaeans: The Spreading of their Ideas in Systematic Botany, 1735–1789* (Utrecht: Oosthoek, 1971); Harriet Ritvo, *The Platypus and the Mermaid, and Other Figments of the Classifying Imagination* (Cambridge, MA: Harvard University Press, 1997); Lorraine Daston, "Type Specimens and Scientific Memory," *Critical Inquiry* 31 (2004): 153–82; Lorraine Daston and Peter Gallison, *Objectivity* (New York: Zone Books, 2007), in particular chapter 2: "Truth-to-Nature"; M. D. Eddy, "Tools for Reordering: Commonplacing and the Space of Words in Linnaeus's *Philosophia Botanica*," *Intellectual History Review* 20/2 (2010): 227–52.

95 See Wilfrid Blunt, *Linnaeus: The Compleat Naturalist* [1971] (Princeton: Princeton University Press, 2001). See also Lisbet Koerner, *Linnaeus: Nature and Nation* (Cambridge, MA: Harvard University Press, 1999), in which she wrote: "Linnaeus at once abhorred, and was sexually attracted to, adult women. Arguably, his disgust for women's bodily functions went beyond the ordinary misogyny of his era" (154). See also Schiebinger, *Nature's Body*, and Sam George, *Botany, Sexuality and Women's Writing, 1760–1830; From Modest Shoot to Forward Plant* (Manchester: Manchester University Press, 2007).

96 Stephen Jay Gould, "The Clam Stripped Bare by her Naturalists, Even," in his *Leonardo's Mountain of Clams and the Diet of Worms* (New York: Harmony Books, 1998), 77–98 (79).

97 Da Costa, "Preface," in *Elements of Conchology*.

98 Da Costa to Richard Pulteney, 3 October 1778; Natural History Museum, MSS PUL A, f. 35v. For other contemporary critiques of Linnaean taxonomy, see Londa Schiebinger, "Naming and Knowing: The Global Politics of Eighteenth-century Botanical Nomenclatures," in *Making Knowledge in Early Modern Europe: Practices, Objects, and Texts, 1400–1800*, ed. Pamela H. Smith and Benjamin Schmidt (Chicago: University of Chicago Press, 2007), 90–108.

99 Gould, "The Clam Stripped Bare by her Naturalists, Even," 98.

100 Richard Pulteney to da Costa, 8 October 1778; Natural History Museum, MSS PUL A, f. 45v.

101 William George Maton, MD, and the Reverend Thomas Rackett, "An Historical Account of Testaceological Writers," *Transactions of the Linnean Society*, 7 (1803): 119–244 (200).

102 Duchess of Portland to Richard Pulteney, 5 July 1777; Pulteney Correspondence, Linnean Society.

103 Duchess of Portland to Richard Pulteney, Whitehall, 9 April 1772; Pulteney Correspondence, Linnean Society.

104 In response to his query about a particular shell he had sent her, she wrote: "the only long shaped patella I recollect is not the same as the *Patella Unguis* of Lin. & which I have sent you" (duchess of Portland to Richard Pulteney, Whitehall, 9 April 1772; Pulteney Correspondence, Linnean Society).

105 Duchess of Portland to Richard Pulteney, Whitehall, 9 April 1772; Pulteney Correspondence, Linnean Society.

106 Duchess of Portland to Richard Pulteney, 8 June 1772; Pulteney Correspondence, Linnean Society.

107 Richard Pulteney to the duchess of Portland, 23 March 1772; Pulteney Correspondence, Linnean Society.

108 Natural History Museum, "Mss Relative to British Testacea," vol. 2: f. 181.

109 Henry Seymer to Richard Pulteney, 29 March 1771; Pulteney Correspondence, Linnean Society.

110 Henry Seymer to Richard Pulteney, Fall 1772?; Pulteney Correspondence, Linnean Society.

111 Richard Pulteney to the duchess of Portland, 24 March 1779; Portland Papers, Longleat House, vol. 14, f. 148.

112 "Dr. P.s duplicates of Shells"; Natural His-

tory Museum, "Mss Relative to British Testacea," vol. 2: f. 148. For the importance of lists, see Valentina Pugliano, "Specimen Lists: Artisanal Writing or Natural History Paperwork?", *Isis* 103/4 (2012): 716–26.

113 "I recollect you have a Mya which was intirely new to me, if you have no Objection to sending *that shell* up for Dr *Solander* to examine I will take care to return it very safely to you again." (duchess of Portland to Richard Pulteney, Bulstrode, March 1778; Pulteney Correspondence, Linnean Society). Early on in their relationship Seymer complained "the D*s* knowledge in Natural history, is not so much founded on her own judgement, & study of the subjects, as from what she is told by others, who she imagines are capital judges" (Henry Seymer to Richard Pulteney, 28 November 1775; Pulteney Correspondence, Linnean Society). Countering Seymer's statement (perhaps uttered in a pique) is Fabricius's assertion that she was "a true expert in natural history." Johann Christian Fabricius (1745–1808), a very accomplished Danish naturalist and student of Linnaeus, described her ability to distinguish slight variations within specimens of the same species as so acute that "Solander probably allowed himself to be led into introducing new species based on distinguishing features that were perhaps only slight variations" (J. C. Fabricius, 10 September 1782, Letter 19, in *Briefe aus London vermischten Inhalts* (Dessau & Leipzig, 1784). Fabricius's brief description suggests that the duchess was deeply involved in the process of classifying her specimens and was not the passive recipient of others' advice as Seymer suggests. I am grateful to Geoff Hancock, Curator of Entomology at the Hunterian, for bringing Fabricius's comments to my attention.

114 Richard Pulteney to the duchess of Portland, 13 August 1777; Natural History Museum, "Mss Relative to British Testacea," vol. 2: f. 134.

115 The duchess reported on some of Solander's changes to the Linnaean system.

116 Duchess of Portland to Richard Pulteney, Bulstrode, March 1778; Pulteney Correspondence, Linnean Society. Wondering how long it would take to have her collection "properly ranged," she guessed that "it is a work that will take up a great deal of time as Dr. Solander is not less than five hours at a time as the six first familys took up nine days & I leave it to you to judge how long it will be before it is finish'd but there is some progress made" (duchess of Portland to Richard Pulteney, Whitehall, 19 March 1778; Pulteney Papers, Linnean Society).

117 These descriptions are called Solander's "slips," now housed in the Natural History Museum. He had been using the duchess's collection to work on his own project, which would have been a book revising Linnaeus's categories. These slips were consulted by malacologists throughout the nineteenth century and even into twentieth, though Solander's failure to publish them has resulted in the diminishment of his legacy as a taxonomist. For these observations, I am indebted to Ilya Tëmkin's unpublished paper "Linnaeus, Solander, and Conchology." The authorship of the Portland catalogue has been much disputed: George Humphrey's authorship was espoused by Guy L. Wilkins, "A Catalogue and Historical Account of the Banks Shell Collection," *Bulletin of the British Museum (Natural History) Historical Series* 16 (1955): 71–119; S. Peter Dance argues for Lightfoot's involvement in the process, "The Authorship of the Portland Catalogue," *Journal of the Bibliography of Natural History* 4 (1962): 30–34; and E. Allison Kay championed Lightfoot as the one responsible for the final product in "The Reverend John Lightfoot, Daniel Solander, and the Portland Catalogue," *Nautilus* 79 (1965): 10–19. For a fuller treatment of these issues, see Chapter Six.

3 PATRONAGE, BROKERS, AND NETWORKS OF EXCHANGE

1 Lot 3928, *A Catalogue of the Portland Museum, lately the property of the Duchess Dowager of Portland* (London, 1786), 184.

2 Respectively, lots 3832, 3831, and 2116, *A Catalogue of the Portland Museum*, 178 and 96.

3 See Introduction, note 10. See also Pierre Bourdieu's description of the complexity and social instability of non-monetary exchanges in "Symbolic Capital," a section of *Outline of a Theory of Practice* [1972], trans. Richard Nice (Cambridge: Cambridge University Press, 1972, reprinted 1989), 171–83.

4 Harold Perkin, *The Origins of Modern English Society* (London: Routledge & Kegan Paul, 1985), 49. For the importance of gift giving and the way in which

patronage and scientific exchange operated within early modern court cultures, see Mario Biagioli, "Galileo's System of Patronage," *History of Science* 28 (1990): 1–62, and Biagioli, *Galileo, Courtier: The Practice of Science in the Culture of Absolutism* (Chicago: University of Chicago Press, 1993); Paula Findlen, "The Economy of Exchange in Early Modern Italy," in *Patronage and Institutions: Science, Technology, and Medicine at the European Court, 1500–1750*, ed. Bruce T. Moran (Rochester, NY: Boydell Press, 1991), 5–24, and Findlen, *Possessing Nature: Museums, Collecting, and Scientific Culture in Early Modern Italy* (Berkeley and London: University of California Press, 1994). For an analysis of the disintegration of late eighteenth-century French patronage within natural history circles, see E. C. Spary, *Utopia's Garden: French Natural History from the Old Regime to Revolution* (Chicago: University of Chicago Press, 2000), in particular the chapter entitled "Patronage, Community, and Power."

5 As Mario Biagioli and Paula Findlen have argued, the relationship between patronage and scientific practice in the early modern period was intertwined, presenting scientists with multiple social and economic challenges involving patrons' expectations, degrees of engagement, and economic limitations. As Emma Spary has suggested, by the end of the eighteenth century patronage systems had been sufficiently eroded by the social and economic forces of modernization to render the role of patrons minimal. However, this movement from a quasi-feudal and hierarchical social structure to a capitalized, commercialized economy structured around the contract was uneven across Europe, with France taking the lead in state funding for science (the Revolution doing away with aristocratic patronage), leaving Britain well behind in continuing to rely on the patronage of wealthy amateurs. See Biagioli, *Galileo, Courtier*; Findlen, *Possessing Nature*; and Spary, *Utopia's Garden*.

6 Findlen further elaborates on the nature of this relationship: "Friendship, the security of knowing another, emerged from the cycle of letters and visits that organized the life of a collector" (*Possessing Nature*, 131).

7 Duchess of Portland to J. T. Swainson, Bulstrode, 29 November 1784. These letters were found by Nora F. McMillan, who transcribed them, and left the transcriptions to the Zoology Library of the Natural History Museum, London, along with her hundreds of pages of notes and research on John Timothy Swainson, which were gathered initially in preparation for a biography of his son.

8 Duchess of Portland to Richard Pulteney, 10 January 1772; Pulteney Correspondence, Linnean Society, London.

9 Duchess of Portland to Richard Pulteney, 24 December 1771; Pulteney Correspondence, Linnean Society.

10 Duchess of Portland to Richard Pulteney, 16 January 1772; Pulteney Correspondence, Linnean Society.

11 Duchess of Portland to Richard Pulteney, Whitehall, 4 May 1772; Pulteney Correspondence, Linnean Society.

12 Duchess of Portland to Richard Pulteney, Bulstrode, 27 June 1780; Pulteney Correspondence, Linnean Society.

13 Duchess of Portland to Richard Pulteney, Whitehall, 8 June 1772; Pulteney Correspondence, Linnean Society.

14 Duchess of Portland to Richard Pulteney, Bulstrode, 23 September 1778; Pulteney Correspondence, Linnean Society.

15 Henry Seymer to Richard Pulteney, no date; letters 77 and 78 of the Pulteney Correspondence, Linnean Society.

16 Duchess of Portland to Mary Delany, Weymouth, 19 July 1772; Mary Delany, *The Autobiography and Correspondence of Mary Granville, Mrs. Delany: With Interesting Reminiscences of King George the Third and Queen Charlotte*, ed. Lady Llanover, 6 vols. (London: Bentley, 1861–62), 1: 444. For more information on Seymer, his interest in butterflies, and lovely reproductions of his artwork, see R. I. Vane-Wright and Harold W. D. Hughes, *The Seymer Legacy: Henry Seymer and Henry Seymer Jnr. of Dorset, and their Entomological Paintings with a Catalogue of Butterflies and Plants, 1755–1783* (Tresaith: Forrest Text, 2005).

17 Much of Seymer's correspondence is in private hands and not easily accessed, which means that there might be letters from the duchess that I have not seen.

18 Duchess of Portland to Richard Pulteney, 8 June 1772; Pulteney Correspondence, Linnean Society.

19   Richard Pulteney to George Montagu, 17 June 1800; Natural History Museum, London, "Mss Relative to British Testacea," vol. 2: f. 104.

20   Richard Pulteney to George Montagu, 17 June 1800; Natural History Museum, "Mss Relative to British Testacea," vol. 2: f. 104.

21   See Spary on botanical networks and their interplay with colonial dynamics in "Acting at a Distance," chapter 2 in *Utopia's Garden*. Susan Scott Parrish, *American Curiosity: Cultures of Natural History in the Colonial British Atlantic World* (Chapel Hill: University of North Carolina, 2006), an examination of the dynamics of transatlantic natural history networks, reveals tensions in the exchanges between metropolitan and colonial spheres, but argues that ultimately, despite implicit hierarchies between metropolitan elites and colonial naturalists and specimen hunters, the knowledge produced on the peripheries of the empire was valued at the metropolitan core.

22   Emanuel Mendes da Costa will be examined in greater detail in Chapter Five. His troubled career as a naturalist has been described by P. J. P. Whitehead, "Emanuel Mendes da Costa (1717–91) and the *Conchology; or, Natural History of Shells*," *Bulletin of the British Museum (Natural History) Historical Series* 6/1 (1977): 1–24; Stephen Jay Gould, "The Clam Stripped Bare by her Naturalists, Even," in his *Leonardo's Mountain of Clams and the Diet of Worms* (New York: Harmony Books, 1998), 77–98; and G. S. Rousseau and David Haycock, "The Jew of Crane Court: Emanuel Mendes Da Costa (1717–91), Natural History, and Natural Excess," *History of Science* 38 (2000): 127–70. Da Costa's Jewish heritage and his disgrace – caught embezzling funds from the Royal Society in his role as clerk – have captured the attention of these writers to the exclusion of an examination of the complexity of his post-incarceration career as a specimen broker and the author of important conchology books.

23   For regimes of value, see Arjun Appadurai, "Introduction: Commodities and the Politics of Value," in *The Social Life of Things: Commodities in a Cultural Perspective*, ed. Appadurai (Cambridge: Cambridge University Press, 1986), 3–63, and Fred R. Myers's introduction to *The Empire of Things: Regimes of Value and Material Culture* (Santa Fe and Oxford: School of American Research Press, 2001), 3–64.

24   Henry Seymer to Richard Pulteney, 15 July 1782; Natural History Museum, Pulteney Correspondence, MSS PUL A, f. 79.

25   Henry Seymer to Richard Pulteney, 24 April 1782; Natural History Museum, Pulteney Correspondence, MSS PUL A, f. 76.

26   Henry Seymer to Richard Pulteney, 29 March 1771; Pulteney Correspondence, Linnean Society.

27   Henry Seymer to Richard Pulteney, 11 June 1782; Natural History Museum, Pulteney Correspondence, MSS PUL A, f. 71.

28   Henry Seymer to Richard Pulteney, 3 May 1772; Pulteney Correspondence, Linnean Society.

29   These terms are, of course, taken from Marcel Mauss, *The Gift: The Form and Reason for Exchange in Archaic Societies* [1925], trans. W. D. Halls (New York and London: Norton, 1990).

30   Henry Seymer to Richard Pulteney, 26 December 1772; Pulteney Correspondence, Linnean Society.

31   Henry Seymer to Richard Pulteney, 17 June 1773; Pulteney Correspondence, Linnean Society.

32   Henry Seymer to Richard Pulteney, 30 July 1773; Pulteney Correspondence, Linnean Society.

33   See Biagioli's astute analysis of gift exchange within patron/client hierarchies: "the beginning of a permanent patronage relationship with a great patron was marked by the patron's *acknowledging but not reciprocating the client's gift*" (*Galileo, Courtier*, 47). Seymer, though a landed gentleman, did not seem to recognize fully what was required of him in this initial encounter with the duchess. Though he knew enough to give her a lavish gift, he was surprised and angered by her response, which was acknowledgment without reciprocating, exactly what Biagioli describes. The misunderstandings and messiness of this encounter are symptoms of the dissolution of patronage as a coherent social system in the 1770s.

34   Duchess of Portland to the duke of Portland, 29 October 1773; University of Nottingham Library, PWF 886.

35   Duchess of Portland to Richard Pulteney, Whitehall, 3 May 1777; Pulteney Correspondence, Linnean Society.

36   Richard Pulteney to the duchess of Portland, 7 May 1777; Pulteney Correspondence, Linnean Society.

37   Nora F. McMillan, "John Timothy Swainson

38 Da Costa to Richard Pulteney, 25 April 1776; Natural History Museum, MSS PUL A, f. 6.

39 Richard Pulteney to da Costa, 15 June 1776; Natural History Museum, MSS PUL A, f. 7.

40 Da Costa to Richard Pulteney, 15 June 1776; Natural History Museum, MSS PUL A, ff. 7–7v.

41 Richard Pulteney's draft of letter to da Costa, 18 June 1776; Natural History Museum, MSS PUL A, f. 10.

42 On Da Costa's role as an intermediary, there is a letter to him in the Pulteney correspondence from Humphrey, the dealer, outlining his plan to send da Costa's friend (unnamed) a regular supply of shells with the goal of building up a solid collection (Humphrey to da Costa, Long Acre, 31 January 1782; Natural History Museum, MSS PUL A, f. 27). This letter may have been passed on to Pulteney from da Costa.

43 For a thoughtful analysis of da Costa's character and the degree to which his Sephardic background limited his access to the kind of support he required as a naturalist, see Rousseau and Haycock, "The Jew of Crane Court."

44 Da Costa, King's Bench, to Solander, 23 September 1771; British Library, Add. MS 28,542, f. 240.

45 William George Maton, MD, and the Rev. Thomas Rackett, "An Historical Account of Testaceological Writers," *Transactions of the Linnean Society*, 7 (1803): 119–244 (201).

46 John Marra to Sir Joseph Banks, 30 July 1775 ("From on board his Majesty's Sloop Resolution, 1776 [5]"); British Library, Add. MS 33,977, f. 53:

47 Mrs. Delany and the duchess paid a visit to Mr. Deard's, a Pall Mall shop that was selling "a curious collection of shells" for £300, but their visit did not end in a purchase because the price was much too high for the relatively small number of shells to tempt either of them (Mary Delany to Ann Dewes, Suffolk Street, 9 May 1754, in Delany, *Autobiography and Correspondence*, 3: 271).

48 See *Natural History Auctions, 1700–1972: A Register of Sales in the British Isles*, compiled by J. M. Chalmers-Hunt (London: Sotheby Parke Bernet, 1976), which lists at least a dozen auctioneers who specialized in selling natural history specimens.

49 Richard Pulteney to Joseph Banks, 25 July 1774; Natural History Museum, Dawson Turner Copy, 1: 72.

50 In his article on Cook's shells, Peter Dance wrote: "Seymer acquired his specimens from dealers, but the PORTLAND and LEVER collections, though enriched by shells bought from dealers or acquired at auctions, benefited from gifts of shells from leading participants in the voyages" (369). For more on this topic, see Peter Dance, "The Cook Voyages and Conchology," *Journal of Conchology* 26 (1971): 354–79.

51 Duchess of Portland to Richard Pulteney, Bulstrode, 29 June 1780; Pulteney Correspondence, Linnean Society.

52 Duchess of Portland to Richard Pulteney: 24 December 1771; Whitehall, 16 January 1772; Whitehall, 9 April 1772; Pulteney Correspondence, Linnean Society.

53 Duchess of Portland to Richard Pulteney, March 1778; Pulteney Correspondence, Linnean Society.

54 Delany, *Autobiography and Correspondence*, 3: 262.

55 W. S. Lewis, "Introduction," in Horace Walpole, *The Duchess of Portland's Museum*, ed. Lewis (New York: Grolier Club, 1936), vi; *The Yale Edition of Horace Walpole's Correspondence*, ed. W. S. Lewis, 48 vols. (New Haven: Yale University Press, 1937–83), 9: 114.

56 "What additions have you made to your shells? mine has received very few additions from the South seas & an irreparable loss of a large collection from Madras of sea shells & a Box of Land & River shells all sunk & gone in the General [Barker?] that I am almost inclined to give up any future hopes of getting any having so often been disappointed" (duchess of Portland to Richard Pulteney, Whitehall, 6 April 1781; Pulteney Correspondence, Linnean Society).

57 Daniel Solander to Linnaeus, 16 November 1761, in Edward Duyker and Per Tingbrand, ed. and trans., *Daniel Solander: Collected Correspondence, 1753–1782* (Melbourne: Miegunyah Press, 1995), 190.

58 Duchess of Portland to Richard Pulteney, Margate, 21 August 1784; Pulteney Correspondence, Linnean Society.

59 Mary Delany and the duchess of Portland visited Banks in London in 1771: "We were yesterday at Mr Banks to see some of the fruits of his travels, and were delighted with paintings of the Otaheitie plants,

quite different from anything the Dss *ever* saw, so they must be very new to me!" quoted in Ruth Hayden, *Mrs. Delany: Her Life and her Flowers* [1980] (New York: New Amsterdam Books/British Museum Press, 1993), 114–15.

60 See Chalmers-Hunt, *Natural History Auctions*.

61 Delany, *Autobiography and Correspondence*, 3: 271.

62 Richard Pulteney copied letters and catalogues originally written by other people; these documents are therefore often without clear attribution, some seeming to have been sent originally to Seymer from Humphrey and other London correspondents, and Seymer passing them onto Pulteney, who copied them, and probably sent them back to Seymer, all of which makes for some confusion. Such is the case with this letter, which seems to have been written originally by Martyn and is dated 9 December 1780. Natural History Museum, "Mss Relative to British Testacea," Pulteney MSS 89, vol. 2: f. 159.

63 Swainson, "Bibliography of Zoology," in *Taxidermy*, 210–20.

64 In volume 6 of Solander's slips, on f. 115 for *Ostrea peregrina*, the following is written: "Habitat in Oceano pacifico. J. R. Forster. In Oceano Indico."

65 "A Catalogue of Shells chiefly from the South Seas sent down to Mr. Seymer from Mr. Geo. Humphreys 18 September 1775." Natural History Museum, "Mss Relative to British Testacea," vol. 2: f. 157.

66 "Mr Humphreys's Shells out of the Resolution and Discovery sent down to Mr. S. [Seymer] Dec. 3 1780" (copied in Richard Pulteney's hand); Natural History Museum, "Mss Relative to British Testacea," Pulteney MSS, vol. 2: f. 161.

67 Duchess of Portland to Richard Pulteney, Margate, 21 August 1784; Pulteney Correspondence, Linnean Society.

68 Delany's correspondence also provides information on auctions, some of which she attended with the duchess. For instance, she went to Pitts's auction with the aim of purchasing some shells (Mary Delany to Ann Dewes, Bolton Row, 22 February 1755, in Delany, *Autobiography and Correspondence*, 3: 332).

69 Duchess of Portland to her son, the duke of Portland, Bulstrode, 2 August 1772; Bulstrode, 14 August 1772; University of Nottingham Library, pwF 872 and pwF 874.

70 Mary Delany to Bernard Granville, Spring Garden, 27 April 1758, in Delany, *Autobiography and Correspondence*, 3: 495–96.

71 The West Collection sale catalogue (1773), now housed in the British Library, lists everything sold at auction with the prices and names of some of the buyers written in the margins. *A Catalogue of the large and capital Collection of Pictures . . . of James West . . . sold by auction, by Mess. Langford's* (London, 1773), 7.

72 Richard D. Altick, *The Shows of London* (Cambridge, MA: Harvard University Press, 1978), 23.

73 Chalmers-Hunt searched the contents of more than sixty libraries and museums (mostly in Britain) in an effort to locate eighteenth-century sales catalogues for auctions featuring natural history specimens, primarily shells, corals, fossils, minerals, insects, mammals, and reptiles (J. M. Chalmers-Hunt, *Natural History Auctions*, 59–67).

74 S. Peter Dance, "Shells," in Chalmers-Hunt, *Natural History Auctions*, 45–51.

75 Henry Seymer to Richard Pulteney, 28 November 1775; Pulteney Correspondence, Linnean Society.

76 Henry Seymer to Richard Pulteney, 28 November 1775; Pulteney Correspondence, Linnean Society. Seymer's misspelling of Ingham Foster's name – Forster – has led to some confusion with Jacob Forster, a mineral dealer. P. J. P Whitehead ("Some Further Notes on Jacob Forster [1739–1806], Mineral Collector and Dealer," *Mineralogical Magazine* 39, 1973: 361–63) argues that Seymer meant Jacob Forster in his article. However, I agree with Peter Dance (*A History of Shell Collecting*, Leiden: Brill, 1986, 76) that it was Ingham Foster who bought shells on Seymer's behalf.

77 Henry Seymer to Richard Pulteney, 28 November 1775; Pulteney Correspondence, Linnean Society.

78 Henry Seymer to Richard Pulteney, 28 November 1775; Pulteney Correspondence, Linnean Society.

79 Duchess of Portland to Richard Pulteney, Margate, 21 August 1784; Pulteney Correspondence, Linnean Society.

80 Henry Seymer to Richard Pulteney, 3 May 1772; 23 October 1775; 19 August 1774; Pulteney Correspondence, Linnean Society.

81 Henry Seymer to Richard Pulteney, 28 November 1775; duchess of Portland to Richard Pulteney, Whitehall, 9 April 1772; Pulteney Correspondence, Linnean Society.

82 For Thomas Martyn's activities as a dealer, see Dance, *A History of Shell Collecting*, chapter 4.

83 Solander to the Board of British Museum, 1 April 1779; 30 July 1779. *Diary and Occurrence-Book of the British Museum*, 2 April 1773 to April 1782 (signed Daniel Solander), British Library, Add. MS 45,875, ff. 20 and 21.

84 Henry Seymer to Richard Pulteney, 28 November 1775; Pulteney Correspondence, Linnean Society.

85 Emanuel Mendes da Costa to Richard Pulteney, 20 July 1776; Natural History Museum, MSS PUL A, f. 14v.

86 Fortunately, Da Costa's habit of writing down the auction prices was a practice shared by others, and many such annotated sales catalogues survive today, including, more importantly for this project, the sale catalogue of the duchess's auction. The British Library possesses several of these catalogues; I assume they were a part of Banks's collection of natural history books.

4 PACIFIC SHELLS, COOK'S VOYAGES, AND THE QUESTION OF VALUE

1 Daniel Solander to Sir Joseph Banks, 14 August 1775; National Maritime Museum, Greenwich, Banks Correspondence, Dawson Turner Copy, 1: 95–96.

2 Quoted by Arthur MacGregor in *The Origin of Museums: The Cabinet of Curiosities in Sixteenth- and Seventeenth-century Europe*, ed. Oliver Impey and Arthur MacGregor (Oxford: Clarendon Press, 1985), 155–56.

3 S. Peter Dance wrote: "Many of the officers and seamen who sailed with Cook returned with collections of shells and other 'curiosities.' But few of them intended to keep such things; they wanted to sell them. Because such material was almost unknown in Europe, dealers were willing to pay well for it; they knew they would make a handsome profit when they hawked it in turn to wealthy collectors." See "The Cook Voyages and Conchology," *Journal of Conchology* 26 (1971): 354–79 (366). See also P. J. P. Whitehead, "A Guide to the Dispersal of Zoological Material from Captain Cook's Voyages," *Pacific Studies* (1978): 52–93.

4 For such artifacts, see, for instance, Adrienne L. Kaeppler, ed., *Cook Voyage Artifacts in Leningrad, Berne, and Florence Museums* (Honolulu: Bishop Museum Press, 1978); Adrienne L. Kaeppler, *"Artificial Curiosities": An Exposition of Native Manufactures Collected on the Three Pacific Voyages of Captain James Cook, RN* (Honolulu: Bishop Museum Press, 1978); and Stephen Little and Peter Ruthenberg, eds., *Life in the Pacific of the 1700s: The Cook/Forster Collection of the Georg August University of Göttingen*, 3 vols. (Honolulu: Honolulu Academy of the Arts, 2006), which contains photographs of George Humphrey's handwritten "Catalogue of Manufactures, Mechanical Performances, and other Inventions of the Natives of the New-discovered, or but Seldom Visited Countries in the Pacific Ocean" (1782).

5 See Jeremy Coote on the timing of and motivations behind these gifts to the Ashmolean Museum, Oxford, as well as a complete list of the items from the Forster Collection held at the Pitt Rivers Museum, Oxford. Jeremy Coote, et al., "Curiosities Sent to Oxford: The Original Documentation of the Forster Collection at the Pitt Rivers Museum," *Journal of the History of Collections* 12 (2000): 177–92.

6 Daniel Solander to Sir Joseph Banks, 5 September 1775; National Maritime Museum, Banks Correspondence, Dawson Turner Copy, 1: ff. 98–99.

7 See Michael E. Hoare, *The Tactless Philosopher: Johann Reinhold Forster (1729–98)* (Melbourne: Hawthorn Press, 1976), 157–58.

8 For the gentlemanly code of conduct and questions of identity, rank, status, and the construction of truth about nature, see Steven Shapin, "A Scholar and a Gentleman: The Problematic Identity of the Scientific Practitioner in Early Modern England," *History of Science* 29 (1991): 279–327; Shapin, *A Social History of Truth: Civility and Science in Seventeenth-century England* (London and Chicago: University of Chicago Press, 1994); and Shapin, "The Image of the Man of Science," in *The Cambridge History of Science*, vol. 4: *Eighteenth-century Science*, ed. Roy Porter (Cambridge: Cambridge University Press, 2003), 159–83; Adrian Johns, *The Nature of the Book: Print and Knowledge in the Making* (Chicago and London: University of Chicago Press, 1998), in particular chapter 7; and Mario Bia-

gioli, *Galileo, Courtier: The Practice of Science in the Culture of Absolutism* (Chicago: University of Chicago Press, 1993). For a consideration of the tensions between amateurs and professionals in the construction of knowledge about nature, see Ann B. Shteir, *Cultivating Women, Cultivating Science: Flora's Daughters and Botany in England, 1760–1860* (Baltimore and London: Johns Hopkins University Press, 1996); John Gascoigne, *Science in the Service of Empire: Joseph Banks, the British State, and the Uses of Science in the Age of Revolution* (Cambridge: Cambridge University Press, 1998); and my discussion of Sir James E. Smith's career as President of the Linnean Society in chapter 6 of *Colonizing Nature: The Tropics in British Arts and Letters, 1760–1820* (Philadelphia: University of Pennsylvania, 2005).

9   See Spary's call to complicate Latour's actor-network model with careful historical examination of the processes by which natural history specimens were collected and carried to the centers of calculation: "The work put into rendering objects of natural history both mobile and immutable must be investigated by historians, but so also must the work done in enforcing particular interpretations both on voyages and at home" (E. C. Spary, *Utopia's Garden: French Natural History from the Old Regime to Revolution*, Chicago: University of Chicago Press, 2000, 97). For examples of this kind of historical work on networks, see also David Philip Miller, "Joseph Banks, Empire, and Centers of Calculation," 21–37; David Mackay, "Agents of Empire: The Banksian Collectors and Evaluation of New Lands," 38–57; and Alan Frost, "The Antipodean Exchange: European Horticulture and Imperial Designs," 58–79, all three essays in *Visions of Empire: Voyages, Botany, and Representations of Nature*, ed. David Philip Miller and Peter Hanns Reill (Cambridge: Cambridge University Press, 1996).

10   "Experimental Gentlemen" is John Marra's phrase in his letter to Banks of 30 July 1775; British Library, Add. MS 33,977, f. 53. Marra's name as the author of a book on the voyage suggests that he wrote a book about his experiences. However, it is doubtful that he wrote the *Journal of the Resolution Voyage* (1775), since the letter he wrote to Banks suggests someone who is semi-literate (no sense of grammar, punctuation or spelling or how to complete a thought in sentence form), and the book reads like the prose of a literary hack mimicking Hawkesworth's grand literati style. The book contains no first-person account of the collecting of specimens, neither does it describe personal events such as his flogging for insubordination, all of which leads one to suspect that he may have authorized the use of his name to sell the work of a hack. For Marra's punishment for insubordination, see Robert Cooper, "A Journal of the Proceedings of His Majesty's Sloop Resolution, on a voyage upon Discovery towards the South Pole"; The National Archives, Adm/ 104, August 1773, f. 30v.

11   Quoted in Robert McCracken Peck, "Alcohol and Arsenic, Pepper and Pitch: Brief Histories of Preservation Techniques," in *Stuffing Birds, Pressing Plants, Shaping Knowledge: Natural History in North America, 1730–1860*, ed. Sue Ann Prince (Philadelphia: American Philosophical Society, 2003), 34.

12   J. R. Forster, "Short Directions for Lovers and Promoters of Natural History," in *A Catalogue of the Animals of North America* (London, 1771), 39.

13   John Mawe, "A Short Treatise . . . to encourage the collecting of Natural History" (1804), 4; in facsimile edition compiled by Jeffrey D. Stilwell, *The World's First Shell Collecting Guide* (Perth: Western Australia Museum, 2003), 36.

14   John Mawe, *The Voyager's Companion; or, Shell Collector's Pilot* (London, 1821), xii; in facsimile edition compiled by Stilwell: *The World's First Shell Collecting Guide*, 48.

15   The word "negro" was not only a racialized category but also the term used euphemistically as a substitute for the word "slave" (Mawe, *The Voyager's Companion*, 10). Slavery was not abolished in the British West Indies until 1833, and in Brazil in 1888.

16   Mawe, *The Voyager's Companion*, 14–16.

17   Much scholarly attention has been bestowed on Pacific artifacts collected by Cook's crew, with such anthropologists as Adrienne L. Kaeppler leading the way. Most notably and recently, a stunningly beautiful and intelligently contextualized exhibition of the artifacts collected by the Forsters has been organized by the Honolulu Academy of Arts and the University of Göttingen. For more information, see the beautifully produced catalogue: *Life in the Pacific of the 1700s*.

18   Thomas Edgar, "A Journal of a Voyage undertaken to the South Seas on discoveries by His Majesty's Ships Resolution and Discovery. Captain James

Cook & Chas. Clerke, Esqrs. Commanders, Kept by Thomas Edgar, "Master" of the Discovery"; British Library, Add. MS 37,528, f. 16v.

19 The National Archives, Clerke's log of the Adventure, Adm 55/103, f. 181.

20 J. R. Forster, *The Resolution Journal of Johann Reinhold Forster, 1772–1775*, ed. Michael E. Hoare, 4 vols. (London: Hakluyt Society, 1982), 4: 652.

21 Cook, *Journal of H.M.S. Endeavour*, Success Bay, 15 January 1769; National Library of Australia, Manuscript 1, p. 28: http://nla.gov.au/nla.ms-ms1-s28r [accessed 5 October 2008].

22 Jacques-Julien Labillardière, *An Account of the Voyage in Search of La Pérouse . . . during the Years 1791, 1792, 1793, and 1794*, 2 vols. (London, 1800), 1: 212 [French original].

23 Some early English settlers to North America regarded clams and mussels found on the beaches and rocks of the New England coastline as inappropriate food for humans, though they knew Indians ate them. The Puritans of the Massachusetts Bay Colony let their hogs run loose on the beaches to feed on clams and mussels, a practice that upset the local native inhabitants. See William Cronon, *Changes in the Land: Indians, Colonists, and the Ecology of New England* (New York: Hill and Wang, 1983).

24 Nicholas Thomas, *Cook: The Extraordinary Voyages of Captain James Cook* (New York: Walker, 2004), 51–52.

25 Quoted in Thomas, *Cook*, 55.

26 Thomas, *Cook*, 54.

27 John Hawkesworth, *An Account of the Voyages . . . By Commodore Byron, Captain Wallis, Captain Carteret, and Captain Cook*, 3 vols. (London: Straham and Cadell, 1773), 3: 498.

28 Cook, *Endeavour Journal*, crossed out section for 29 April 1770; transcription of National Library of Australia, Manuscript 1, p. 228: http://nla.gov.au/nla.cs-ss-jrnl-cook-17700429 [accessed 5 August 2008].

29 Cook's autograph log and journal of the *Endeavour* voyage; British Library, Add. MS 27,886, f. 88v.

30 This passage was written by Tobias Furneaux as he commanded the *Adventure*, the ship that was to accompany the *Endeavour* but became separated and returned home to England a year earlier. Cook included Furneaux's narrative in his published volume, *A Voyage towards the South Pole*, 2 vols. (London: Straham and Cadell, 1777), 1: 117.

31 Hawkesworth, *An Account of the Voyages*, 2: 347.

32 Hawkesworth, *An Account of the Voyages*, 3: 441.

33 Anne Salmond, *The Trial of the Cannibal Dog: Captain Cook in the South Seas* (London: Penguin, 2004), 140; Hawkesworth, *An Account of the Voyages*, 2: 392–95 (22 January 1770).

34 "Horeta as told to John White," quoted in Salmond, *Trial of the Cannibal Dog*, 131–32.

35 Arthur Bowes-Smyth, *Journal of Voyage of Penrhyn*, 10 February 1788; British Library, Add. MS 47,966, f. 46v.

36 Arthur Bowes-Smyth, *Journal of Voyage of Penrhyn*, 25 July 1788; British Library, Add. MS 47,966, f. 67. This passage refers to Otaheite, a destination he visited after the ship had delivered its cargo of convicts.

37 Arthur Bowes-Smyth, 27 September 1788; quoted in *Journal of Arthur Bowes-Smyth* [based on British Library MS X 800/ 31062], ed. Paul J. Fidlon and R. J. Ryan (Sydney, 1979), 120.

38 Bowes-Smyth, *Journal of Voyage of Penrhyn*, 28 January 1788; British Library, Add. MS 47,966, f. 42.

39 Bowes-Smyth, *Journal of Voyage of Penrhyn*, 23 January 1788; British Library, Add. MS 47,966, f. 41.

40 Hawkesworth, *An Account of the Voyages*, 3: 521.

41 Sydney Parkinson, *A Journal of a Voyage to the South Seas, in His Majesty's Ship, the Endeavour* (London, 1773); reprinted by Australiana Facsimile Editions (Adelaide: Libraries Board of South Australia, 1992), 138.

42 J. C. Beaglehole, ed., *The Endeavour Journal of Joseph Banks, 1768–1771*, 2 vols. (Sydney: Angus and Robertson, 1962), 223. See also Guy L. Wilkins, "A Catalogue and Historical Account of the Banks Shell Collection," *Bulletin of the British Museum (Natural History), Historical Series* 16 (1953): 71–119.

43 *The Resolution Journal of Johann Reinhold Forster*, 2: 248; 2: 287.

44 *The Resolution Journal of Johann Reinhold Forster*, 2: 250.

45 *The Resolution Journal of Johann Reinhold Forster*, 2: 250.

46 *The Resolution Journal of Johann Reinhold Forster*, 2: 287.

47 George Forster, *A Voyage Round the World*, 2 vols. (London, 1777), 1: 263 (referring to Tahiti) and 1: 456–57 (referring to Tonga).

48 Forster, *A Voyage Round the World*, 2: 118.

49 Forster, *A Voyage Round the World*, 1: 263.

50 Forster, *A Voyage Round the World*, 1: 453.

51 Forster, *A Voyage Round the World*, 1: 357.

52 Edgar, "A Journal of the Voyage . . . Kept by Thomas Edgar, Master of the Discovery"; British Library, Add. MS 37,528; f. 57v.

53 Edgar, "A Journal of the Voyage," f. 43v.

54 Forster, *A Voyage Round the World*, 2: 299.

55 Edgar, "A Journal of the Voyage," f. 38v.

56 Clerke quoted in Edgar, "A Journal of the Voyage," f. 55v.

57 Edgar, "A Journal of the Voyage," f. 38v.

58 Forster, *A Voyage Round the World*, 1: 436.

59 Forster, *A Voyage Round the World*, 1: 436. *Museum Humfredianum: A Catalogue of the Large and Valuable Museum of Mr. George Humphrey* (London, 1779) mentions a shell specimen with a hole in it, made by an islander, so that it could be threaded with string and worn around the neck.

60 Edgar, "A Journal of the Voyage," f. 22.

61 Edgar, "A Journal of the Voyage," f. 31v.

62 Peter Dance ("The Cook Voyages and Conchology," 370) states: "Whatever the scientific value of the shells brought back by Cook's men, the overriding incentive to collect them was the prospect of making money."

63 See Hoare's discussion of tensions between the Forsters and the crew in *The Tactless Philosopher* (1976).

64 *The Resolution Journal of Johann Reinhold Forster*, 4: 647, 555–57.

65 Nicholas Thomas, *Entangled Objects: Exchange, Material Culture, and Colonialism in the Pacific* (Cambridge, MA: Harvard University Press, 1991), 141.

66 Forster, *A Voyage Round the World*, 2: 419–20.

67 Forster, *A Voyage Round the World*, 2: 420.

68 *The Resolution Journal of Johann Reinhold Forster*, 2: 251.

69 William Wales wrote: "Before we reached New Zealand the first time, there was scarce a man in the ship with whom he [Forster] had not quarrelled with on one pretence or another" (*Remarks on Mr. Forster's Account of Captain Cook's Last Voyage Round the World*, London, 1778, 6–7).

70 Logbook and Journal of Captain James Cook in the "Resolution" during his Second Voyage Round the World, 5 October 1773; British Library, Add. MS 27,886, f. 148.

71 See Thomas, *Cook*, 142. See also Bernard Smith, "Cook's Posthumous Reputation," in *Captain James Cook and his Times*, ed. Robin Fisher and Hugh Johnston (Canberra: Australian National University Press, 1979), 159–85.

72 Anon. "To the Memory of Captain James Cook," in *A Voyage to the Pacific Ocean*, 3 vols. (London, 1784), 1: lxxxviii.

73 Witnesses described the severity of the punishments Cook meted out to Pacific islanders during the third voyage for their supposed thefts, the master of the *Discovery* writing of his treatment of a "thief": "as Captain Cooke said that he might be known here after, as well as to deter the rest from theft or using us ill when on Shore – this was by scoring both his arms with a common knife by one of our Seamen. Longitudinally and transversely, into the Bone – This the man bore with all the Fortitude imaginable, as indeed they all did their punishments" (Edgar, "A Journal of the Voyage," f. 45).

74 *Omai* was written by John O'Keeffe, the music was composed by William Shield, and John Webber, artist from the third voyage, may have painted the scenes. Webber's drawings were used by Loutherbourg for the costume and set designs.

75 The Honolulu Academy of Art possesses French wallpaper (*circa* 1805) and upholstery fabric (copper-plate print on plain weave cotton, *circa* 1785) depicting Cook's encounters with Pacific islanders. Dickens's comfortable Victorian parlor in *Bleak House* contains framed engravings of all sorts of subjects, including "the death of Captain Cook" (Charles Dickens, *Bleak House* [1853], Oxford: Oxford University Press, 2008, 79). For the popularity of Cook-themed decorative items, see Rüdiger Joppien, "Artistic Bequest of Captain Cook's Voyages," in *Captain James Cook and his Times*, 187–210; and Rüdiger Joppien and Bernard Smith, *The Art of Captain Cook's Voyages*, vol. 3: *The Voyage of the* Resolution *and* Adventure (New Haven and London: Yale University Press / Australian Academy of the Humanities, 1985). For wallpaper and ceramics, see, respectively, Roger Collins, "The Inside Story: Dufour Wallpaper," *Bulletin of New Zea-*

land *Art History* 9 (1985): 5–13, and M. K. Beddie, ed., *Bibliography of Captain Cook, RN, FRS, Circumnavigator* (Sydney: Mitchell Library, 1970).

76 George Tobin, *Captain Bligh's Second Chance: An Eyewitness Account of his Return to the South Seas*, ed. Roy Schreiber (London: Chatham, 2007), 117.

77 For an astute discussion of how Pacific islanders were collectors too, see Jennifer Newell's "Collecting from the Collectors: Pacific Islanders and the Spoils of Europe," a part of the Göttingen University's Cook-Forster exhibition, made available by the National Museum of Australia online at: www.nma.gov.au/online_features/cook_forster/background/collectors.

78 Arthur Bowes-Smyth, *Journal of Voyage of Penrhyn*; British Library, Add. MS 47,966.

79 *Museum Humphredianum*, 120.

80 *Museum Calonnianum* (London, 1797), iii–v, 44.

81 For a fuller treatment of late eighteenth-century British museums, see Richard D. Altick, *The Shows of London* (Cambridge, MA: Harvard University Press, 1978); Edward P. Alexander, "William Bullock: Little-Remembered Museologist and Showman," *Curator* 28 (1985): 117–47; J. C. H. King, "New Evidence for the Contents of the Leverian Museum," *Journal of the History of Collections* 8 (1996): 167–86; Clare Haynes, "A 'Natural' Exhibitioner: Sir Ashton Lever and his *Holosphusikon*," *British Journal of Eighteenth-century Studies* 24 (2001): 1–14; and Adrienne L. Kaeppler, *Holophusicon: The Leverian Museum: An Eighteenth-century English Institution of Science, Curiosity, and Art* (Altebstadt, Germany: ZKF Publishers, 2011).

82 *A Companion to the Museum (late Sir Aston Lever's)*, (London, 1790), 6–7.

83 *Leverian Auction Catalogue* (1806), 47–58. This catalogue is in the Bishop Museum, Honolulu, and has a hand-lettered rather than a printed titled page. This one and those written by Boulter and Bullock were collected in the 1930s and 1940s in London by A. W. F. Fuller, who also collected Pacific Island artifacts (and Sarah Stone's drawings of them), many of which are now in the Bishop Museum.

84 *The World* (25 June 1790), 1; quoted in J. C. H. King, "New Evidence for the Contents of the Leverian Museum," *Journal of the History of Collections* 8 (1996): 173.

85 *Museum Boulterianum* (Yarmouth, 1793), 12, 17.

86 *Museum Boulterianum*, 76–78.

87 *A Companion to the London Museum . . . by William Bullock* (1814), 1. For Bullock's showmanship, see Susan Pearce, "William Bullock: Collections and Exhibitions at the Egyptian Hall, 1816–25," *Journal of the History of Collections*, 20 (2008), 17–35.

88 Captain J. Laskey, *A General Account of the Hunterian Museum* (Glasgow, 1813), 100. It is uncertain what the "J" stands for. James is the most frequently cited.

89 *A Catalogue of the Portland Museum, lately the property of the Duchess Dowager of Portland* (London, 1786), 190.

90 Charles Smith, *Auctions* (New York: Free Press, 1989), 40.

91 J. R. Forster to Sir Joseph Banks, Paddington Green, 26 September 1778; Natural History Museum, Banks Correspondence, Dawson Turner Copy, 1: 213.

92 George Humphrey to Henry Seymer, 18 September 1775; Natural History Museum, London, "Mss Relative to British Testacea," vol. 2: f. 157.

93 Edgar, "A Journal of a Voyage," f. 74v.

94 Edgar, "A Journal of a Voyage," f. 78v.

95 Edgar, "A Journal of a Voyage," f. 22.

5 PATRONAGE, PUBLICATION, AND THE ILLUSTRATED CONCHOLOGY BOOK

1 David Allen notes that two important aspects of natural history benefited from the patronage and activities of wealthy amateur natural history collectors: one was the improvement in the techniques of taxidermy, and the other was "in supporting the production of sumptuous illustrated works on the subject" (David Elliston Allen, *The Naturalist in Britain: A Social History* [1976], (Princeton: Princeton University Press, 1994), 30).

2 Most observers, contemporary and modern, have described the duchess of Portland as a patroness of natural history. For instance, John Lightfoot prefaced his *Flora Scotica* (1777) with a dedication "To Her Grace The Most Nobel Margaret Cavendishe, Duchess Dowager of Portland, That Great and Intelligent Admirer and Patroness of Natural History in Gen-

eral," and in the 1950s Guy Wilkins of the Natural History Museum in London wrote: "For many years she was the leading patroness of natural history in England, and particularly devoted to conchology" (Guy L. Wilkins, "A Catalogue and Historical Account of the Banks Shell Collection," *Bulletin of the British Museum (Natural History) Historical Series* 16 (1955): 71–119 (88)). Typically, historians describe the duchess in passing as "a leading patron of natural history in England," as do Edward Duyker and Per Tingbrand, ed. and trans., in *Daniel Solander: Collected Correspondence, 1753–1782* (Melbourne: Miegunyah Press, 1995), 420.

3 For the kinds of disagreements and ill feeling generated by the collaborative nature of publishing illustrated books, see Elisabeth A. Fraser, "Books, Prints, and Travel: Reading in the Gaps of the Orientalist Archive," *Art History* 31/3 (2008): 342–67. For disputes that arose within the publication process, as well as print culture's participation in the construction of scientific knowledge, see Adrian Johns, *The Nature of the Book: Print and Knowledge in the Making* (Chicago and London: University of Chicago Press, 1998).

4 George Walker, *Testacea Minuta Rariora; or, Minute and Rare Shells . . . discovered by William Boys* (London, 1784), ii.

5 George Montagu, *Testacea Britannica*, 2 vols. (London, 1803), Preface, xii. Solander sent Boys and Walker some of the duchess's specimens that he had already classified to help them with naming Boys's shells. Montagu also wrote in his *Supplement to Testacea Britannica* (London, 1808), i–ii: "By the same means we have also ascertained some of the obscure names which had been given by Doctor Solander in the Portland Museum as the identical specimens received by Mr. Boys from the Doctor [Solander] were marked. Of the accuracy of these names we find full confirmation by a lot of land and fresh-water shells, which were bought at the sale of that museum, now in the possession of Mr. Laskey, with their original titles affixed; and who obligingly indulged us with them for comparison."

6 Walker, *Minute and Rare Shells*, ii.

7 Da Costa wrote to Pulteney in a state of confusion about some shells sent to him to be described and drawn for his *British Conchology* (1778): "I have a letter from the D^r in which he says he does not explain himself whether they are by way of *loan* or *present*[.] I wish you would let me know how it is intended" (Da Costa to Richard Pulteney, 25 February 1777; British Library, Da Costa Correspondence, Add. MS 28,541, f. 74).

8 Duchess of Portland to Thomas Pennant (the duchess's emphasis), no date; Warwickshire County Record Office, Correspondence of Thomas Pennant (1726–1798), CR 2017/TP 149/3.

9 Peter Brown to Thomas Pennant, no date; Warwickshire County Record Office, Correspondence of Thomas Pennant (1726–1798), CR 2017/TP186/1. Apparently, the duchess liked Brown's work enough to keep his drawings in her own collection. Lot 2656 of *A Catalogue of the Portland Museum, lately the property of the Duchess Dowager of Portland* (London, 1786) is described as "Peter Brown drawing of Trochus Solaris, from New Zealand, in Sir Aston Lever's Museum" (119).

10 In da Costa's correspondence with Pulteney held at the Natural History Museum, he describes preparations for the plates of *British Conchology*, mentioning Peter Brown as the artist and engraver for this book.

11 Thomas Pennant, *British Zoology*, fourth edn., 4 vols. (Warrington, 1776–77), 4: 127.

12 William Hudson to Thomas Pennant, 18 June 1778; Warwickshire County Record Office, Correspondence of Thomas Pennant (1726–1798), CR 2017/TP149/1A–1B.

13 William Hudson to Thomas Pennant, 24 June 1778; Warwickshire County Record Office, Correspondence of Thomas Pennant (1726–1798), CR 2017/TP149/8 and 149/8b; duchess of Portland to Pennant, 26 June 1778, CR 2017/ TP149/5 and 5b; Hudson to Pennant, no date, CR 2017/ TP149/2.

14 Thomas Pennant to [William Eyres?], no date; Warwickshire County Record Office, Correspondence of Thomas Pennant (1726–1798), CR 2017/TP53.

15 Pennant, *British Zoology*, 4: 72. With thanks to Professor Taylor Corse of Arizona State University for the English translation. Apparently, in antiquity these shellfish were considered to be aphrodisiacs.

16 Duchess of Portland to Thomas Pennant, Whitehall, 26 June 1778; Warwickshire County

Record Office, Correspondence of Thomas Pennant (1726–1798), CR 2017/TP149/5.

17 This particular copy of the first installment of *Conchology; or, The Natural History of Shells* (1770–72), located in the Zoology Library at the Natural History Museum, was owned by Pulteney.

18 "Preface," *Conchology Number III*, i (New York, American Museum of Natural History). Since there is much variation between copies in the series, the location of copies examined is placed after the citation.

19 P. J. P. Whitehead's carefully researched and richly detailed article, "Emanuel Mendes da Costa (1717–91) and the *Conchology; or, The Natural History of Shells*," *Bulletin of the British Museum (Natural History) Historical Series* 6/1 (1977): 1–24, has been tremendously helpful in understanding da Costa's situation and relationship to Humphrey. I would like to suggest, however, that while the writing was da Costa's, the book was a collaboration, and Whitehead's focus on who was the "true author" (2), while perfectly understandable given the demands made by taxonomic codes for naming the "author," such an approach underestimates the collaborative nature of the production of natural history books.

20 W. G. Maton assumes that da Costa is the author of *Conchology*, placing his description of this work under da Costa's name in his catalogue (William George Maton, MD, and the Reverend Thomas Rackett, "An Historical Account of Testaceological Writers," *Transactions of the Linnean Society*, 7, 1803: 119–244), 200–01, whereas L. W. Dillwyn in *A Descriptive Catalogue of Recent Shells* (London, 1817), ix, assigns dual authorship: "Supposed to be the joint work of E. M. Da Costa and George Humphreys." For the arguments as to authorship of *Conchology*, see: C. Davies Sherborn, "Note on the 'Museum Humfredianum', 1779" *Annals and Magazine of Natural History* 16 (1905): 262–64; Sherborn, "The Museum Humfredianum, 1779," *Geology Magazine* 2 (1905): 379–81; Tom Iredale, "On Humphrey's Conchology," *Proceedings of the Malacological Society of London* 11 (1915): 307–09; Iredale, "Book Notes," *Proceedings of the Malacological Society of London* 15 (1922): 78–92; J. Wilfrid Jackson, "A Letter from George Humphrey to William Swainson, 1815," *Journal of Conchology* 20 (1937): 332–37; Whitehead, "Emanuel Mendes da Costa," 1–24.

21 *Museum Humfredianum: A Catalogue of the Large and Valuable Museum of Mr. George Humphrey* (London, 1779), 104, 112, 110. Sherborn thinks that Humphrey did not use Latin names before the Calonne catalogue, a position that has implications concerning the disputed authorship of *A Catalogue of the Portland Museum*.

22 Whitehead ("Emanuel Mendes da Costa," 2) argues: "[T]he conclusion is reached here that the true author was da Costa and not Humphrey. The latter saw the work through the press and acted as editor, but it was actually written by da Costa as an unrepentant debtor in the King's Bench Prison."

23 The sale catalogue, *Museum Humfredianum*, not only lists *Conchologie* as "Humphrey's," but also lists for sale: "The Copper Plates of the said Work, with the Remainder of the Impression of the Cuts and Letter-Press" and "*Ten Original Drawings of Shells in their natural Colours*, upon Vellum (five of which are unpublished) by Brown (to be sold separately)" (167–68). It is intriguing that Humphrey would have the copper plates of *Conchology* and some of Brown's drawings since these would have been valuable to da Costa in producing *British Conchology* (1778); this suggests that da Costa and Humphrey continued to collaborate, Humphrey lending da Costa the plates and drawings for his book of 1778, and selling them only once it was published.

24 Da Costa to William Borlase; British Library, Add. MS 35,230, f. 31v.

25 Duchess of Portland to Richard Pulteney, Whitehall, 9 April 1772; Pulteney Correspondence, Linnean Society.

26 *Conchology, Number III*, 25 (American Museum of Natural History).

27 Da Costa, *British Conchology*, 12.

28 See Iredale, "On Humphrey's Conchology," 309. His use of the word "anonymous" in *The British Conchology*'s list of reference books is merely in keeping with scientific practice of citing the publication's author as printed on the title page.

29 The measure of hostility that professional writers felt about patrons can be found in Dr. Johnson's frequent attacks on them in the form of definitions: a patron is "Commonly a wretch who supports with insolence, and is paid with flattery."

30 George Humphrey to da Costa, St. Martin's Lane, Wednesday morning, February 1770; British

Library, Add. MS 28,538, f. 224. The duchess owned the shell figured in Plate 4, Fig. 9.

31 George Humphrey to da Costa, February 1770; British Library, Add. MS 28,538, f. 225. Solander made quite a bit of money as a guide to the British Museum's collections.

32 An artist, J. Wicksteed, gained access to the British Museum and perhaps drew the limpet mentioned below in *A Catalogue of the Portland Museum* there or borrowed the duchess's specimen. "Patella tricarinata, L., the triple-ridged Limpet. Of this curious species there is only one other specimen known, which is in the British Museum. Its native place we have not yet discovered, but have reason to think it is an inhabitant of New Guinea, *Humphrey's Conch*. pl. 4 fig. 9" (lot 3601, *A Catalogue of the Portland Museum*, 165).

33 Da Costa's letter of application, February 1770; British Library, Add. MS 28,538, f. 225v. Modern scholars may recognize in da Costa's letter their own attempts to get permission to view, copy, and publish materials from archives, museums, and libraries. Fortunately, the British Museum, Natural History Museum, and the British Library no longer seem to require the cringing and abjection that da Costa seemed to think the proper method to obtain permission to use these marvelous collections.

34 Humphrey's brother was a draftsman and engraver, and eventually became a printer; he was enlisted along with other artists, including Pierre Brown, to draw shell specimens, and Humphrey had to ask permission for his brother "to be with me to draw any particular shell" (British Library, Add. MS 28,538, f. 225). The duchess owned some drawings by William Humphrey, *A Catalogue of the Portland Museum* describing lot 2657 as "two portraits, drawings after Holbein, by Mr. Humphreys, very fine" (119).

35 Da Costa correspondence, British Library, Add. MS 28,542, f. 182v. This letter is da Costa's copy of Seymer's letter addressed to Mr. George Humphrey, Jr., which was dated 16 February 1771. It is not certain that Seymer knew at this point that Humphrey and da Costa were collaborating on this series, but the fact that Humphrey gave da Costa the letter to copy and put it in his files indicates that he thought that da Costa needed to attend to Seymer's advice about their mutual project.

36 Da Costa to J. C. Sepp, London, 3 May 1771; British Library, Add. MS 28,542, f. 179–179v.

37 George Humphrey to da Costa, noon, Wednesday, 6 March 1771; British Library, Add. MS 28,538, f. 230.

38 George Humphrey to da Costa, 2 October 1770; British Library, Add. MS 28,538, f. 228.

39 George Humphrey to da Costa, 6 March 1771; British Library, Add. MS 28,538, f. 231. Along with Chemnitz, Friedrich Heinrich Wilhelm Martini was one of the authors of *Neues Systematisches Conchylien-Cabinet* (1769–95), and he was responsible for vols. 1–3 (1769).

40 George Humphrey to da Costa, 12 April 1771; British Library, Add. MS 28,538, f. 233.

41 Plates 5 and 7 contain images of the duchess's shells. See the folllowing descriptions from *A Catalogue of the Portland Museum* for clues: "A very fine specimen of Patella umbraculum, or Umbrella Limpet, from China, extremely scarce. *Humphrey's Conch*. Pl 5 fig. 5 which was taken from this shell" (lot 1301, *A Catalogue of the Portland Museum*, 178). "A very large Patella Ungarica, L. or Fool's Cap Limpet; and another scarce shell of the same genus, *Humphrey's conchology*, pl. 5 fig. 15" (lot 1965, *A Catalogue of the Portland Museum*, 89). "An extremely fine pair of a species of perforated patella, the only two that are known, named Macroschisma, *Humph Conch*. pl. 7. fig. 3" (lot 1601, *A Catalogue of the Portland Museum*, 71).

42 Emanuel Mendes da Costa, *Elements of Conchology; or, An Introduction to the Knowledge of Shells* (London, 1776), 51.

43 Maton and Rackett, "An Historical Account of Testaceological Writers," 200–01.

44 See, for instance, Edme François Gersaint, *Catalogue Raisonné de Coquilles et autre Curiosités naturelles* (Paris, 1736). For French auction catalogues, see E. C. Spary, "Rococo Readings of the Book of Nature," in *Books and the Sciences in History*, ed. Marina Frasca-Spada and Nick Jardine (Cambridge: Cambridge University Press, 2000), 255–75.

45 Da Costa to Daniel Solander, King's Bench, 23 September 1771; British Library, Add. MS 28,542, f. 240.

46 In the Natural History Museum's general library there are three letters from the duchess to

Swainson; they are transcripts that Nora McMillan made, and now they are the only copies of these letters. One letter is published in her article: Nora F. McMillan, "John Timothy Swainson the Second, 1756–1824: Zoologist," *Archives of Natural History* 14/3 (1987): 289–96. These letters "were pasted inside the da Costa volume" (292), but this was destroyed during the Second World War. The duchess gave Swainson "fine copies of da Costa's Historia naturalis Testaceorum Britanniae and Pennant's British Zoology (the folio ed.); of these he was very proud and in his will (1819) he bequeathed them to his daughter Betty (1801–1867)" (292).

47   "I fear I may have lost me your greatly Honourd friendship respect" (Da Costa to William Borlase; British Library, Add. MS 35,230, ff. 29–31v). In this letter he also asks for the loan of some fossil specimens "being destitute of specimens as my collections were dispersed & sold," a reference to the compulsory sale of his books and specimens in an attempt by the Royal Society to recoup lost revenues due to his embezzling of the Society's funds.

48   Dru Drury to Dr. Pallas, 28 February 1768, reprinted in J. D. A. Cockerell, "Dru Drury, an Eighteenth Century Entomologist," *Scientific Monthly* (January 1922): 67–82; quoted in Iredale, "Book Notes," 86.

49   See G. S. Rousseau and David Haycock, "The Jew of Crane Court: Emanuel Mendes da Costa (1717–91), Natural History and Natural Excess," *History of Science* 38 (2000): 127–70.

50   See the description of this project in the Natural History Museum's online catalogue entitled: "A coloured set of the seventeen plates to Mendes da Costa's 'Historia naturalis Testaceorum Britanniae', 1778, and six others."

51   E. [Edward] Donovan, "Illustration of British Zoology. On May 1, 1799 will be published, the first number of an entirely new, complete, and elegant work." London, 1799: iii–iv (Eighteenth Century Collections Online, Gale Group, Document Number: CW3324535882). Donovan's advertisement states that "I. This work shall be elegantly printed in OCTAVO, on a fine WOVE ROYAL PAPER, and HOT-PRESSED. II. Three Plates shall be given in each Number, with Descriptions: they will be printed on WHATMAN'S DRAWING PAPER, and correctly finished in Colours from perfect SPECIMENS of the SHELLS – Prce 2s. 6 d. each Number. III. This Work shall be completed in SIXTY MONTHLY NUMBERS. An Index for the systematic Arrangement will be given at the End of each VOLUME, containing TWELVE NUMBERS" (title page).

52   Maton and Rackett, "An Historical Account of Testaceological Writers," 201.

53   Peter Dance, *A History of Shell Collecting* (Leiden: Brill, 1986), 71. Thomas Martyn, *The Universal Conchologist*, 2 vols. (London, 1784). The publishing history of this book is quite complex; see William Healey Dall, "Thomas Martyn and the *Universal Conchologist*," *Proceedings of the United States National Museum* 29 (1905): 415–32.

54   Thomas Martyn, *The Universal Conchologist*, second edn., 4 vols. (London, 1789), 6.

55   "One of the principal inducements however to this performance was, his becoming about that time possessed of a considerable number of new species, found amongst different collections of Shells which he had recently purchased of several officers then lately returned from the Pacific Ocean" (Martyn, *The Universal Conchologist* (1789), 26).

56   Martyn, "Introduction," in *The Universal Conchologist* (1789), 14, 11.

57   Thomas Martyn to Henry Seymer, 9 December 1780; Natural History Museum, Pulteney Correspondence; quoted in Dance, *A History of Shell Collecting*, 70.

58   Dance, *A History of Shell Collecting*, 73.

59   Martyn, *The Universal Conchologist* (1789), 4.

60   Martyn, *The Universal Conchologist* (1789), 26.

61   Martyn, *The Universal Conchologist* (1789), 34.

62   Martyn, *The Universal Conchologist* (1789), 26–28.

63   George Montagu solved the problems of expense and necessary expertise needed to illustrate his volumes of *Testacea Britannica* in an equally creative (and exploitative?) way: his mistress and long-time companion, Eliza [Elizabeth] Dorville, etched all of the shell drawings and drew many of the illustrations herself.

64   Maton and Rackett, "An Historical Account of Testaceological Writers," 203–04.

65   See the Linnean Society's online catalogue's entry for Huddesford's edition of Lister's shell book, which notes: "3 letters from Huddesford to Forster"

Provenance: "Martin Lister/Ingham Forster/ Will. Huddersford/Duchess Dowager of Portland."

66 Da Costa to Richard Pulteney, London, 22 October 1776; Natural History Museum, MSS PUL A.

67 Dance (*A History of Shell Collecting*, 73) says that this edition was dedicated to the duchess because it "had been produced under her patronage."

68 Da Costa to I. Foster; British Library, Add. MS 28,539, ff. 115–16.

69 See A. MacGregor, "William Huddesford (1732–1772): His Role in Reanimating the Ashmolean Museum, his Collections, Researches and Support Network," *Archives of Natural History* 34/1 (2007): 47–68.

70 *A Catalogue of the Portland Museum*, iii.

71 Pulteney, Natural History Museum, "MSS Relative to British Testacea," vol. 2: ff. 101–101v.

72 Lewis Weston Dillwyn, *A Descriptive Catalogue of Recent Shells Arranged according to the Linnaean Method*, 2 vols. (London, 1817), 1: 527.

73 Edward Donovan, *The Natural History of British Shells*, 5 vols. (London, 1799–1803), vol. 5: "Voluta Edentula," pl. CLXV, and "Arca Noae," pl. CLVIII (n. p.).

74 P. J. P. Whitehead, "Zoological Specimens from Captain Cook's Voyages," *Society for the Bibliography of Natural History* 5/3 (1969): 161–201 (185).

75 Banks to the duchess of Portland, 10 June 1782; Natural History Museum, Botany Library, Dawson Turner Copy, 2: 138–39.

76 Duchess of Portland to Banks, Whitehall, 12 June 1782; Mitchell Library, DOC 1120 a and b; State Library of New South Wales, Papers of Sir Joseph Banks, series 72 [accessed online at http://www2.sl.nsw.gov.au/banks/series_72/72_134.cfm].

77 Duchess of Portland to Richard Pulteney, Margate, 21 August 1784; Pulteney Correspondence, Linnean Society.

78 Banks to Johan Alströmer, 16 November 1784; San Francisco, California State Library, Sutro Library, Banks Collection, LO 1:24; quoted in Duyker and Tingbrand, *Daniel Solander*, 413.

79 In 1784 Banks claimed that he would soon be publishing Solander's botanical work: "Solander's name will appear on the title page beside mine, since everything was written through our combined labour. While he was alive, hardly a single sentence was written while we were not together. Considering that all the descriptions were made while the plants were fresh, nothing remains to do than to complete the drawings which have still not been altogether finished and to add synonyms from books that we did not have with us or have been published since. All that remains is so little that it can be completed in just a few months if only the engravers can find the time to put the finishing touches to it" (quoted in Duyker and Tingbrand, *Daniel Solander*, 413). As Solander's biographers suggest, perhaps Banks was much more dependent on Solander than he could admit, since Banks never completed what he claimed was the work of only a few months (Duyker and Tingbrand, *Daniel Solander*, 268). Historians continue to puzzle over Banks's inability to publish the *Florilegium*, the illustrated botanical book describing the plants Banks and Solander encountered on their Cook voyage, which hovered on the cusp of readiness for 200 years until the plates were finally published as *Banks' Florilegium* by the British Museum (Natural History) and Alecto Historical Editions in 1988 as a limited edition of 100 sets.

6 THE DISPERSAL OF THE COLLECTION

1 *London Chronicle*, no. 4472 (Saturday, 16 July to Tuesday, 19 July 1785).

2 *Morning Chronicle and London Advertiser*, no. 5053 (25 July 1785). Concerning Bulstrode, this paper announced nine months later (no. 5285, 24 April 1786): "The following summer there are to be many alterations made at Bullstrode; during that time the Duke means to go to Welbeck."

3 John Lightfoot to Thomas Pennant, 18 July 1785; quoted in Jean K. Bowden, *John Lightfoot: His Work and Travels* (Kew and Pittsburgh: Royal Botanic Gardens / Hunt Institute for Botanical Documentation, 1989), 43.

4 Quoted in Bowden, *John Lightfoot*, 43–47. The duchess's will reads: "I desire all my China Japan Shells and Prints may be sold" (Last Will and Testament; Kew, The National Archives, Probate 11/1133).

5 Elizabeth Montagu to Elizabeth Carter, no date, in *Mrs. Montagu, "Queen of the Blues": Her Letters and Friendships from 1762 to 1800*, ed. Reginald Blunt from material left to him by Montagu's great-great-niece,

6 Simon Dewes, *Mrs. Delany* (London: Rich & Cowan [1940]), 299. Dewes is a pseudonym for John St. Clair Muriel, a prolific writer of novels, memoirs, and biographies.

7 See A. S. Turberville, *A History of Welbeck Abbey and its Owners*, 2 vols. (London: Faber, 1938–39), 2: 32: "She died in debt, and an early sale of her very miscellaneous collection was expected."

8 Since their finances were intertwined, especially those involving Welbeck Abbey, Bulstrode, and properties that the duchess had inherited from her mother with a life interest, there is much discussion in her correspondence with her son about debts, rents in arrears, and taxes. Typical is the following from the duchess to her son: "I am very desirous that the agreement & lease should be settled with the utmost accuracy & precision for both our sakes that there may be no deffect or doubt. I think for both our sakes you should purchase the arrears at the [illegible] of the agreement, the debts of which I am to pay the interest shou'd be particularly stated, Welbeck with the appurtenances and Bullstrode with the appurtenances should be ascertain'd. I beg you will indulge me that a provision be made for the speedy payment of the balance due to you. It maybe necessary for you to have some provision for leases which are to be executed under my power" (29 June 1771; University of Nottingham Library, PWF 842). See also the rest of the correspondence of the early 1770s (PWF 834, 843, 846).

9 Sylvia Harcstark Myers (*The Bluestocking Circle: Women, Friendship, and the Life of the Mind in Eighteenth-century England*, Oxford: Oxford University Press, 1990, 90), in describing the marriage contract between the Cavendish-Harley and the Bentinck families, which was satisfactory for both parties, notes that "the financial disagreements between mother and son were the story of the next generation." With her daughters she had a very close relationship: they visited her often, filling Bulstrode with grandchildren.

10 Horace Walpole, "July 1782," in *The Last Journals of Horace Walpole*, ed. John Doran, 2 vols. (London: J. Lane, 1910), 2: 448

11 Walpole to the Rev. William Mason, 10 July 1782, in *The Yale Edition of Horace Walpole's Correspondence*, ed. W. S. Lewis, 48 vols. (New Haven: Yale University Press, 1937–83), 29: 264–65; Walpole, "July 1782," in *The Last Journals of Horace Walpole*, 2: 448.

12 Lady Mary Coke, 23 December 1777; in "MS Journals," quoted in *The Yale Edition of Horace Walpole's Correspondence*, 33: 486, n. 7; Walpole to Lady Ossory, 10 August 1785, ibid., 33: 486.

13 Lady Mary Coke, 26 July 1785, quoted in *The Yale Edition of Horace Walpole's Correspondence*, 33: 484 n. 28. Emphasis mine.

14 "To her friend Mrs Delany She had bequeathed an exquisite portrait of Petitot . . . & also Raphael's Mice from the Royal Collection, two Mice (different) in watercolours, also called by Raphael, but certainly not, and an enamelled Snuffbox" (Horace Walpole, *The Duchess of Portland's Museum*, New York: Grolier Club, 1936, 9). See also Mary Delany, *The Autobiography and Correspondence of Mary Granville, Mrs. Delany: With Interesting Reminiscences of King George the Third and Queen Charlotte*, ed. Lady Llanover, 6 vols. (London: Bentley, 1861–62), 6: 272.

15 M. Preston to Mrs. Frances Hamilton, October 1787; Letter XX in *Letters from Mrs. Delany . . . to Mrs. Frances Hamilton*, third edn. (London: Longman, Hurst, Rees, Orme, and Brown, 1821), 99–100.

16 Mrs. Scott to Elizabeth Montagu, 23 July 1785, and Elizabeth Montagu to Elizabeth Carter, no date, in *Mrs. Montagu, "Queen of the Blues"*, 2: 192.

17 Bowden, *John Lightfoot*, 47.

18 Myers, *The Bluestocking Circle*, 269.

19 John Lightfoot to Thomas Pennant, 17 August 1785; quoted in Bowden, *John Lightfoot*, 46.

20 Montagu Correspondence, 9 September 1785; *Mrs. Montagu, "Queen of the Blues"*, 2: 193.

21 Walpole took a Swiss visitor to see the duchess's "'museum' of virtu, gems, and shells in her house in the Privy Garden" (Walpole to Horace-Bénédict de Saussure, November 1768, in *The Yale Edition of Horace Walpole's Correspondence*, 41: 165, n. 5).

22 "Preface," in *A Catalogue of the Portland Museum, lately the property of the Duchess Dowager of Portland* (London, 1786), iii–iv.

23 John Lightfoot to James Bolton, 13 October 1785; quoted in Bowden, *John Lightfoot*, 115.

24 John Lightfoot to Thomas Pennant, 14 March 1786; quoted in Bowden, *John Lightfoot*, 114.

25 Peter Dance, "The Authorship of the Portland Catalogue (1786)," *Journal of the Society for the Bibliography of Natural History* 4 (1962): 30–34.

26 Bowden, *John Lightfoot*, 114.

27 *A Catalogue of the Portland Museum*, vi.

28 Dance, "The Authorship of the Portland Catalogue," 32.

29 Bowden, *John Lightfoot*, 114.

30 *Gazetteer and New Daily Advertiser*, Thursday, 27 April 1786.

31 *Morning Herald*, Tuesday, 25 April 1786.

32 *London Chronicle*, Tuesday, 18 April 1786.

33 Lady Dorothy Seymour and her husband were portrayed by James Gillray in satiric prints, two of which were entitled "Sir Richard Worse-than-sly exposing his wife's bottom; o fye!," and "A peep into Lady W!!!!!y's seraglio," both published by William Humphrey, George Humphrey's brother. For more about the Worsleys, see Nigel Aston, "Worsley, Sir Richard, Seventh Baronet (1751–1805)," in *Oxford Dictionary of National Biography*, Oxford University Press, 2004; online edn.: www.oxforddnb.com.ezproxy1.lib.asu.edu/view/article/29986 [accessed 8 September 2009].

34 *Morning Herald*, Thursday, 25 April 1786, and Friday, 26 April 1786.

35 Caroline Powys, 9 May 1786; London, British Library, Add. MS 42,169, ff. 93–94.

36 Frances Burney, *Cecilia; or, The Memoirs of an Heiress* [1782], ed. Margaret Anne Doody (New York: Oxford University Press, 1999).

37 Burney, *Cecilia*, 31–32.

38 *The Yale Edition of Horace Walpole's Correspondence*, 12: 273, and 33: 518–19.

39 On 5 July 1786; *The Yale Edition of Horace Walpole's Correspondence*, 33: 484.

40 Walpole, *The Duchess of Portland's Museum*, 5–10.

41 Horace Walpole to Richard Bentley, August 1756; *The Yale Edition of Horace Walpole's Correspondence*, 35: 270–71.

42 *The Yale Edition of Horace Walpole's Correspondence*, 33: 394.

43 For a reassessment of Walpole, see George E. Haggerty, "Queering Horace Walpole," *Studies in English Literature, 1500–1900* 46 (2006): 543–61.

44 In *The Universal Conchologist*, second edn. (London, 1789), Martyn describes a few of the major collectors, who also happened to be dealers: "Mr. Thomas Sheldon's small museum has also a just claim to our particular notice, in which is to be found such a numerous and highly preserved assortment of Shells correctly and elegantly arranged, and intermixed with such a variety of fine specimens of other sea productions, as altogether present an appearance truly respectable." Of Forster, he wrote: "For exquisite taste and judgment in the various subjects of Conchology, Mineralogy, and every other species of fossil bodies, perhaps no collector has yet more distinguished himself than Mr. Jacob Forster, to whose constant application in the pursuit of every thing rare and beautiful in these branches, it is chiefly owing, that such matchless specimens now adorn his own, as well as other principal cabinets of natural history in this kingdom"; and of Humphrey: "Genius, abilities, and experience in the knowledge of Shells, are happily united in Mr. George Humphries; consequently his private collection possesses every requisite, attendant on such superior advantages" (19).

45 See the priced copy of *A Catalogue of the Portland Museum*, which belonged to Joseph Banks and is now in the British Library, microfilmed and available online through a digital database: Eighteenth Century Collections Online (Gale Learning). See pp. 179–90 for Dillon's and Humphrey's expenditures during the last days of the auction, when the more important specimens were put up for sale. See Peter Dance, *A History of Shell Collecting* (Leiden: Brill, 1986), 75.

46 George Humphrey, *Museum Calonnianum . . . Consisting of an Assemblage of the most Beautiful and Rare Subjects in Entomology, Conchology, Ornithology, Mineralogy, &c.* (London, 1797), iii–iv.

47 See William Swainson's claims (*A Treatise on Malacology; or, the Natural Classification of Shells and Shell-Fish*, London: Longman, 1840, 15) that Humphrey's catalogue was a crucial (even foundational) text in the transition from Linnaean to Lamarckian systematics, being "the most extensive improvement upon everything of the kind which had hitherto been done." "There can be no doubt, in fact, that this little unpretending pamphlet, published in this country merely as an exhibition catalogue, found its way to France, and served as the main foundation, although unacknowledged, for the subsequent system of Bru-

guière, if not of Lamarck and Cuvier." Tom Iredale dismisses these claims as ill-founded remarks in "The Truth about the Museum Calonnianum," in *Festschrift zum 60. Geburtstage von Professor Dr Embrik Strand*, 5 vols. (Riga, 1937), 3: 414.

48 Laskey made a booklet consisting of his article, "Account of North British Testacea," presented to the Wernerian Natural History Society on 23 January 1809; published in 1811, it is interleaved with hand-drawn illustrations and annotated with comments, one dated 1826. These comments were written into the margins of p. 32.

49 Edward Donovan, *The Natural History of British Shells*, 5 vols. (London, 1799–1803), 5: Plate CLXIV, Plate CLXV, Plate CLVI, Plate CLIV, n. p.

50 Guy L. Wilkins, "The Cracherode Shell Collection," *Bulletin of the British Museum (Natural History) Historical Series* 1/4 (1957): 121–84 (134).

51 Dance, *A History of Shell Collecting*, 104. See also Tom Iredale, "The Truth about the Museum Calonnianum," 3: 408–19 (413).

52 *Museum Calonnianum* (London, 1797), 17. When this volute was in Broderip's possession, William Swainson, who, in addition to his skills as a naturalist was an expert natural history illustrator, drew several volutes for his multi-volumed publication entitled *Exotic Conchology; or, Figures and Descriptions of Rare, Beautiful, or undescribed Shells* (1821–22, 1834–45). Though the *Voluta aulica* does not seem to have been illustrated, a specimen of the *Voluta Gambaroonica* is figured and labeled as belonging to the Broderip collection. Such a specimen was sold at the Portland auction to Dillon for £23, and again at the Calonne auction, lot 278, which is described as "this is the only perfect specimen of this beautiful species known . . . at the sale of her Grace's museum was bought for M. de Calonne, M. P. 4024" (*Museum Calonnianum*, 17). If the whole Broderip collection was purchased by the British Museum, this specimen may prove to be another shell from the duchess's collection. See *Exotic Conchology*, and Dance, *A History of Shell Collecting*, 92.

53 Captain J. Laskey, *A General Account of the Hunterian Museum, Glasgow* (Glasgow, 1813), 100, 104.

54 Unlike the fine art works, the natural history specimens, because they were not unique, are hard to trace. The painstaking process of identifying some of the decorative pieces, which are also not necessarily identifiable as unique, that were sold at the Portland auction has begun. Recovered are some of the jewelry, snuffboxes, cameos, candelabra, and china, which were exhibited together with these items and those that the duchess left to her heir, the third duke, such as the ebony cabinet – Walpole's "thousand pound" cabinet, for instance – in an exhibition at the Harley Gallery on the grounds of Welbeck Abbey in 2007.

55 *A Catalogue of the Portland Museum*, iii.

7 LOST LEGACIES

1 Rebecca Stott, *Duchess of Curiosities: The Life of Margaret, Duchess of Portland* (Welbeck: Pineapple Press / Harley Gallery, 2006), 6, 39.

2 This list of shell catalogue authors is merely representative, not exhaustive, and focuses on those writers who produced catalogues of British shells, since the goal is to see how the duchess's specimen hunting is depicted in them. For a thorough examination of conchological authors who cite Solander's manuscripts, see Guy L. Wilkins, "A Catalogue and Historical Account of the Banks Shell Collection," *Bulletin of the British Museum (Natural History) Historical Series* 16 (1955): 71–119, one of three remarkable articles on the history of important shell collections in the British Museum, now the Natural History Museum.

3 "Manuscript Catalogue of British Shells in the Collection of J. T. Swainson," Natural History Museum, 72 AOS.

4 She may have spent some of the summer of 1783 in Margate, even though I have not found evidence in her correspondence that she did so.

5 Duchess of Portland to Mary Delany, 31 July 1784; in Mary Delany, *The Autobiography and Correspondence of Mary Granville, Mrs. Delany: With Interesting Reminiscences of King George the Third and Queen Charlotte*, ed. Lady Llanover, 6 vols. (London: Bentley, 1861–62), 6: 223.

6 Thomas Martyn, *The Universal Conchologist*, 4 vols. (London, 1784–89), 1: 20.

7 Solander was long dead at this point, but this did not stop the duchess from classifying shell specimens and from weighing in on discussions about proper names.

8   Duchess of Portland to J. T. Swainson, 29 November 1784, transcript by Nora F. McMillan, Swainsonia, Zoology Library, Natural History Museum, London. Nora F. McMillan, "John Timothy Swainson the Second (1756–1824)," *Archives of Natural History* 14/3 (1987): 289–96. McMillan explains that these letters "were pasted inside the da Costa volume" (292).

9   "Manuscript Catalogue of British Shells in the Collection of J. T. Swainson."

10   "Manuscript Catalogue of British Shells in the Collection of J. T. Swainson."

11   With the onset of the Napoleonic Wars and the exhaustion of the novelty interest in the South Pacific, interest in British shells moved away from a bizarre obsession (see da Costa and Pulteney's correspondence about this) to a valued activity. Quite like the inward turn of tourism from the Grand Tour to Picturesque tours of the West Country, Wales, the Lake District, and Scotland, the turn from exotic to domestic shells carried with it a nationalist subtext – to learn to appreciate and value things British.

12   Swainson may have toyed with the idea of publishing a catalogue of his shell collection because several of his shells were drawn by Agnew and Lewin. Several of these drawings were then engraved by Montagu's "friend" Eliza Dorville and included in George Montagu, *Testacea Britannica; or, Natural History of British shells, marine, land, and fresh-water, including the most minute: systematically arranged and embellished with figures*, 2 vols. (London, 1803), and Montagu, *Supplement to Testacea Britannica* (London, 1808).

13   Richard Pulteney, *Catalogues of the Birds, Shells, and Some of the more Rare Plants of Dorsetshire. From the new and enlarged edition of Mr. Hutchins's History of that County* (London, 1799), 39, 25, 27.

14   Pulteney, *Catalogues of . . . Dorsetshire*, 25, 28. Edward Donovan (*The Natural History of British Shells*, 5 vols., London, 1799–1803, vol. 5) wrote similarly and actually said that he consulted Solander's manuscript: "Dr. Solander, who, it is well known to the scientific conchologist, intended to have published a catalogue of that Museum [Portland], it appears, on a reference to his posthumous papers, called this species *edentula*; a name which, without detracting from the merit of that able naturalist, it must be allowed is by no means applicable" (vol. 5, pl. CLXV, Voluta Lacuis). Also, "from a ms. note in one of the copies of Lister's work, in the library of Sir Joseph Banks, we find the late Dr. Solander, intended to have named it specifically *fusca*, had he lived to publish his new arrangement of Conchology" (vol. 5, pl. CLVIII, Arca Noae).

15   William George Maton, MD, and the Reverend Thomas Rackett, "A Descriptive Catalogue of British Testacea," *Transactions of the Linnean Society* 8 (1807): 17–250 (51).

16   Montagu, *Supplement*, ii, 146; Montagu, *Testacea Britannica*, 1: 29, 51, 52.

17   Montagu, *Testacea Britannica*, 1: 40–41.

18   Montagu, *Testacea Britannica*, 1: xiii, 58, 51, xi.

19   Montagu, *Testacea Britannica*, 1: 59 and 195; Montagu, *Supplement*, 89, preface iii, and 153.

20   J. Laskey, "Account of North British Testacea" (Edinburgh, 1811); reprinted in *Memoirs of the Wernerian Natural History Society* (1839): 370–417 (370–71). This paper was originally presented to the Wernerian Natural History Society, Edinburgh, on 23 January 1809.

21   The Natural History Museum's annotated copy of Laskey's Wernerian Society article, interleaved with handwritten notes (presumably Laskey's notations), interleave f. 38v.

22   Laskey, "Account of North British Testacea," 376.

23   Montagu, *Testacea Britannica*, 1: x; *Supplement*, i–ii.

24   Richard Pulteney to da Costa, 10 June 1776; Natural History Museum, Pulteney Correspondence, ff. 9–9v.

25   Montagu, *Supplement*, 125.

26   See Linda Colley, *Britons: Forging the Nation, 1707–1837* (New Haven and London: Yale University Press, 1992).

27   Pulteney, *Catalogues of . . . Dorsetshire*, 27.

28   William Turton, "Assisted by his Daughter", *A Conchological Dictionary of the British Islands* (London, 1819), viii.

29   For Darwin as the "exemplary scientific naturalist," see Robert E. Kohler, *Landscapes and Labscapes: Exploring the Lab–Field Border in Biology* (Chicago and London: University of Chicago Press, 2002), 33.

30 "Preface," in *A Catalogue of the Portland Museum, lately the property of the Duchess Dowager of Portland* (London, 1786), iii.

31 William Swainson, *Taxidermy: Bibliography and Biography* (London: Longman, Orme, Brown, Green, & Longmans, 1840), 77–79.

32 W. Swainson, *A Treatise on Malacology; or, the Natural Classification of Shells and Shell-Fish* (London: Longman, 1840), 18.

33 Turton, *A Conchological Dictionary*, vii. The connection between Banks's death and Turton's access to Solander's slips may be more than coincidental.

34 William Turton, *Conchylia Insularum Britannicarum: The Shells of the British Islands, Systematically Arranged* (London, 1822), 3.

35 Gordon McOuat, "Cataloguing Power: Delineating 'Competent Naturalists' and the Meaning of Species in the British Museum," *British Journal for the History of Science* 34/1 (2001): 1–28. McOuat's narrative about the internal politics of the British Museum and the battle over Linnaean and Lamarckian systems differs considerably from Guy Wilkins's, especially in their description of John G. Children's position on Lamarck.

36 Guy L. Wilkins, *The Cracherode Shell Collection* (London, 1957), 128–29.

37 For the tensions between observational, field-based scientific practice and experimental, laboratory-based scientific activities, see Kohler, *Landscapes and Labscapes*. See also his sustained analysis of natural history expeditions and collecting in the United States: *All Creatures: Naturalists, Collectors, and Biodiversity, 1850–1950* (Princeton and Oxford: Princeton University Press, 2006).

38 As Quentin Wheeler, entomologist and advocate for taxonomy, has written: "Taxonomy has always been and shall remain essential for credible biology." In defense of taxonomy based on observation, he wrote: "In the face of this biodiversity crisis, the need to rebuild expertise and infrastructure for taxonomy is paramount. This taxonomic imperative has theoretical, practical and even ethical implications. . . . What we accomplish in taxonomic work in this century will be a priceless legacy to all the generations of scientists, natural historians, and educated humans that follow." Quentin D. Wheeler ("Taxonomic Triage and the Poverty of Phylogeny," *Philosophical Transactions of the Royal Society*, 359 [2004]: 571–83 (571).

39 For fantasies of rescue, see Walter Benjamin on collecting: "one of the finest memories of a collector is the moment when he rescued a book to which he might never have given a thought, much less a wishful look, because he found it lonely and abandoned on the market place and bought it to give it its freedom – the way the prince bought a beautiful slave girl in *The Arabian Nights*. To a book collector, you see, the true freedom of all books is somewhere on his shelves" (Walter Benjamin, "Unpacking my Library: A Talk about Book Collecting," in *Illuminations*, ed. Hannah Arendt (New York: Schocken Books, 1968), 59–67 (64)).

INDEX

NOTE: Page numbers in italics refer to illustrations; those followed by *n* and a number refer to information in a note.

abalone shells 14
accumulation and collecting 52, 53, 59, 60
Adanson, Michel 97
aesthetics
    and the natural world 36–40
    and science in shell collecting 75, 88, 95–99
    shell grottoes 100
    textual representation of shells 202
affect and value of objects 61
African expeditions and collecting 64–65, 73–74
agency and objects 9
Agnew, J.
    British Conchology illustrations 6, 17, 19, 46, 47, 86
    as gardener and naturalist at Bulstrode 44–46, 64, 76, 92, 117, 241
    shell drawings and watercolours 45, *45*, 76, 264
Allen, David Elliston 62, 74–75, 272–73n.17, 293n.1
Altick, Robert 137
amateur activity *see* natural history as amateur interest
anatomical classification of shells 97–98
Anning, Mary 277n.108
anthropology
    Cook's view of Pacific islanders 154–56
    and objects in social contexts 20
    and significance of shells 9–10, 116
    see also exchange cultures and shells
Appadurai, Arjun 61, 178, 270n.22–23
aristocracy *see* elite cultures of collecting
art and natural history
    amateur artists and illustrators 33, 38, 44, 65
    at Bulstrode 42, 44–46, 101
    decorative art and craft work 33–36
        shellwork 32, 33, 34, *34*, 36, 75, 87–88
    and observational skills 33
    see also textual representation of shells and collections
artists and engravers for publications
    Conchology serial 185–86, 192, 195, 197
    Drury's *Exotic Insects* 38
    Martyn's "seminary" for illustrators 204–5
    see also Brown, Pierre; Dorville, Elizabeth
Ashmolean Museum, Oxford 150, 151, 208–9
auction *see Catalogue of the Portland Museum*; shell collection of duchess of Portland: auction
Aurelian Society 38, 73
Austen, Jane: *Emma* 51, 57

Banks, Sir Joseph vii, *42*
    on accuracy of Mrs. Delany's *hortus siccus* 36
    Bulstrode visit viii, 41, 134–35
    and Cook's voyages viii–ix, 132, 133, 147, 148, *148*, 149, 150–51, 159–60, 175–76

criticism of Brown's drawings 186
and duchess and acquisition of shells 134–35, 147
and innuendo in Pennant's book 187, 188
and networks of exchange 117, 118, 128, 133, 135, 147
shell collection 146, 147, 176–77, 224, 267
and Smeathman's expedition 65, 73, 74
and Solander's unfinished work 211–12, 254
and Turton's access to Solander's manuscripts 263–64
Barclay, Mrs. John 63
Barrington, Hon. Jane 48
Bath, marquess and marchioness *see* Thynne
Baudrillard, Jean 53, 59, 60
Benjamin, Walter 58, 59, 303n.39
Bennett, Charles, 4th earl of Tankerville 136, 241, 263
Bentinck, Lord Edward Charles Cavendish 219, 220, 221
Bentinck family *see* Portland
Beretta, Marco 10, 12
Biagioli, Mario 285n.5, 286n.33
birds at Bulstrode Park 48–50
Bligh, Elizabeth 63, 138
books *see* textual representation of shells and collections
Borlase, Rev. William 190, 200
Boscawen, Admiral Edward 74, 134
botanical interests
  aesthetic and religious responses 40
  and artistic activity 36–37
  "botanizing" 32, 62
  Linnaean botanical classification
    duchess's use of 100–01
    salacious terminology 103–4
Botanical Magazine 36
Boucher, François 244
Bougainville, Louis Antoine de 151
Boulter, Daniel: Yarmouth "museum" 172–73
Bowden, Jean 226
Bowes-Smyth, Arthur 158–59
Boys, Henry 259, 260
Boys, William 118, 184–85, 253, 257, 259, 260
Brazil: Mawe's shell collecting 153
British Library, London 27, 94
British Museum, London 30, 58
  access to specimens for *Conchology* publication 189, 192–94
  Broderip collection 242
  Enlightenment exhibition and Pacific shells 267
  Harley papers 27, 223
  Pacific voyages and acquisition of specimens 135, 142, 147, 150, 151
  Portland Vase 27

Solander's manuscript classifications 210, *210*
see also British Library; Natural History Museum
British shells
  attitudes towards collectors 50–51, 128, 130–31, 207, 259–60
  collecting practices 75, 76–87
  identitification and provenance 260
  Napoleonic Wars and renewal of interest in 178, 253, 259
  see also da Costa: *The British Conchology*; Donovan: *The Natural History of British Shells*; Montagu: *Testacea Britannica*; Pennant: *British Zoology*
Broderip, William John 242
Brown, Pierre (Peter) 82, 185–86, 199
Buchan, Alexander 156
Budgen, John Smith 202
Bullock, William: museum 173–74
Bulstrode, Buckinghamshire *41*
  china closet 27
  Duke of Portland as heir 46, 218, 221
  fire and demise 249
  and study of natural history viii, 28, 41–46, 118
    animals in grounds 48–50
    botanical classification 100–01
    cabinets for shell display 92
    collection of snails and shells 76, 77
    and duchess's patronage 42–46, 62
Bunn, James H. 60
Buonanni, Filippo 102, 103
Burgoyne, Mrs. Montagu 36
Burney, Edward Francis: illustration to *Catalogue of the Portland Museum* 54, *55*, 244
Burney, Frances: *Cecilia* 4, 233–35
Bute, Countess of 48, 63
Bute, Lord 48
buying and selling of shells 15, 16, 30–32
  dealers and sources for purchase 31–32, 131–32, 135–38, 140, 142
  and Pacific voyages 20–21, 151, 242
    Cook's death and value of specimens 166, 168, 172–73, 174–75, 242
    dealers' instructions for collecting and preserving shells 152–54
    Forsters' scientific specimens 165–66, 176–77
    oversupply and decline in value 175–78, 259
    Seymer's problems in purchasing 138–39, 140, 142
    ships' officers and crew and dealers in shells 74, 131, 132–38, 139–40, 149–51, 153–54, 163–65
    transactions with Pacific islanders 161–63, 178

Seymer's preference for purchase over exchange 125–27
valuation and prices 131, 139, 140–43
see also commodities

cabin boys as shell collectors 153–54
cabinets of curiosities 46, 48, 57, 60
cabinets for shell storage and display 92, 94
Calonne, Charles Alexandre de 5
  auction purchases 115, 236–37, 241, 242
  Portland shells in collection 241–42
  sale of collection and *Museum Calonnianum* catalogue 171, 237, 241
capitalism and collecting 60–61
captains and sale of shells 74, 132–38, 149–51
card trays for storing shells 75, 90–92, 93
Catalogue of the Portland Museum, A (auction catalogue) 2, 4–5
  authorship vii, 223–27
  classification of shells 109, 111, 209–10
  and "disorderly" view of collection 54, 55, 244–45
  healthy sales 4, 232
  as inadequate record of collection 10, 12
  incorporation of unfinished catalogue 209–10, 212–13, 255, 264
  individual shells in 12, 13, 14, 109, 111, 115
  limitations 12, 244
  Lister's drawings 207
  playing card containers with shells 75, 91
  Pulteney's reordering along Linnaean lines 226–27, 228–29
Cavendish family *see* Portland
Charlotte, queen of Great Britain 222
civilization: Cook's view of Pacific islanders 154–56
classification of collections 99–111, 201
  dealers' lack of knowledge 140
  debates on anatomical classification 97–98
  duchess's collection 57, 64, 69, 75, 99–101, 106–11
    advice from duchess and Solander 184–85
    dismantling of Linnaean classification for auction 224–25
    friends' deferral to expertise of 99–100, 108–9
    and neglect of taxonomy in history of science 265–66
    Solander and cataloguing of shell collection 69, 106, 107, 109, 111, 209–13, 225, 226, 254, 255
    systematic ordering of collections 56, 59, 60
  introduction of "French" systems 263, 264
  need for selective approach 262–63

shell collectors 13–14, 38–39, 96, 99–111
  and shortcomings of Linnaean classification 101–2, 104–6, 107–8, 243
  Solander as authority on 210–11, 260
  undermining in print 255–56
  and systematic collecting 59–60, 69, 98
  transformation from aesthetic to scientific objects 75
  see also Linnaeus; taxonomy
cleaning shells 75, 87–90, 152
Clerke, Lieutenant Charles
  death 132, 177
  on islanders' diet of fish 154
  and procurement of shells 5, 74, 132, 133–34, 147, 149–50
  on red feathers and trade with islanders 162
Coke, Lady Mary 221
collecting 27–60
  attitudes towards collectors 50–53, 56
  and duchess 53–57, 244–45, 254–59, 260, 265
  consumption and modes of collecting 57–61
  curiosities 46, 48, 57, 60
  elite cultures of 46–50, 52
  ephemeral nature and fate of collections 266, 267
  need for selective approach 262–63
  and neglect in academic studies 29, 52–53
  as popular activity 30–32
  provenance and integrity of collections 243–44
  sailors and crew and collection of specimens 147–79, 163–65
  as social practice 51, 52–53, 64, 83–84, 92–93
  see also natural history collections; objects; shell collection of duchess of Portland; shell collectors; "virtuoso" and connoisseur collectors
Collinson, Peter 64
colonialism and "indigenous things" 21
commodities
  accumulation and collecting 52, 53, 59–60
  Marxist theory and collecting 53, 60–61
  natural history specimens 30–32, 126–27, 131–32, 152
  Pacific shells 132–43
  valuation 131, 139, 140–43, 178, 242
  prices for cleaning shells 90
  sale of shells 15, 16, 131–32
  study of 20
  see also buying and selling of shells; exchange value and commodity consumption; objects
Conchology; or, The Natural History of Shells (serial publication) 107, 186, 188–99, *193*, *196*, *198*, 202

authorship issue 189–92, 200
difficulty of obtaining specimens 192–94
images of duchess's shells 16–17
subscriptions and promotion of 194–95
suspension and reasons for failure 197–99
connoisseurs *see* "virtuoso" and connoisseur collectors
consumption and collecting 53
modes of collecting 57–61
see also buying and selling of shells; commodities; exchange value and commodity consumption
Cook, Alexandra vii
Cook, Captain James 133, *148*, 169
death and celebrity status and memorabilia 143, 166–75
on Pacific islanders and standard of civilization 154–56
voyages and procurement of shells vii, viii–ix, 5, 15, 16, 20, 65–66, 132–43, 147–79
circulation in collecting circles 137–38
on crew's indiscriminate trade in artifacts 166
for duchess's collection 74, 115–16, 135, 147
impact of death on value of specimens 166, 168, 172–73, 174–75
see also Pacific voyages and natural history collecting
cooks as shell collectors 153–54
Cordiner, Dr. Charles 87
Costa, Emanuel Mendes da *see* da Costa
Cowley, Hannah: *The Belle's Stratagem* 52
Cowper, William 168
Cracherode, Rev. Clayton Mordaunt 91, 241–42, 267
crafts *see* art and natural history
Cullum, Sir John 218–19
Cuming, Hugh 259, 261
curiosity collecting 46, 48, 57, 60
Curtis, William 36, 41
Flora Londinensis 30
Cypraea Aurora (orange cowry) 15–16, *15*

da Costa, Emanuel Mendes
on aesthetic and scientific treatment of shells 96–97
on anatomical classification of shells 97–98
on attitudes towards British shell collectors 50–51
auction purchases 236
on authorship of auction catalogue 225–26
The British Conchology 38–39, 51, 117, 128, 130, 191–92, 199, 200, 201, *248*, 261
career as conchologist 130–31
on catalogues of shell sales 142
on cleaning shells 90
and *Conchology* serial publication 7, 107, 189–99, 200

correspondence viii
on "dead shells" 84, 85
on dredging for shells 85
Elements of Conchology 76, 104–5, *105*, 106, 197, 199–200
imprisonment for debt 130, 190–91, 199–200
Jewish heritage 130
on Linnaeus's inappropriate terminology 104–6, 201
and Lister's works and new editions 207, 208
Montagu's loans to 122
and Pulteney 85, 106, 128–30, 131, 142, 199, 207, 294n.7
secret shared passion for British shells 259–60
shell collection's reputation 200
social hierarchy and networks of exchange 128–32
and sourcing rare books 207
Dance, Peter
on aesthetic appeal of shells 88
on Cook's voyages and collecting market 137, 138, 298n.3
on Martyn's work 201, 204
on Portland Museum catalogue and auction vii, 226, 227, 236
d'Argenville, Antoine-Joseph Dezallier 102, 103
Darwin, Charles 261
Darwin, Erasmus 103
de Clifford, Hon. Dowager Lady 48
"dead shells" 84, 85
dealers *see* buying and selling of shells
Dear, Peter 278n.109
Deard, Mr. (shell dealer) 131–32, 135
decorative arts
neglect of collections in scholarship 29
and popular interest in nature 33–36
see also shellwork
Delany, Rev. Patrick 44
Delany, Mary vii, *43*, 244, 268n.2
aesthetic response to nature and specimens 40
on Banks's visit to Bulstrode viii, 134–35
botanical collages and mosaics 36–37, *37*, 44
on cabinets for shell storage 94
on cleaning shells 89–90
on dealings with ships' captains 134
on Deard's shop 135
and duchess's art collection 137
duchess's death and lack of provision for 217, 219, 221–23
and duchess's "disorderly" collection 55–56
on duchess's visit to Banks 287–88n.59

307

on Ehret 101
and natural history at Bulstrode 41, 48, 49, 62
on playing cards for shell storage 90–91, 93
residence at Bulstrode 43–44, 63, 92, 118
shell collection 87, 93, 99
on sociability of shell hunting 83–84
D'Entrecasteaux, Antoine de Bruni 151, 155
descriptive writing in nineteenth century 279n.12
Dewes, Simon 220
Dezallier d'Argenville, Antoine-Joseph 102, 103
Dillon, Mr. (shell dealer) 115, 236–37
Dillwyn, Lewis Weston 226, 264
display practices 87–94
diving for shells in Pacific islands 153, 161
Donovan, Edward 85, 200–01, 210–11
    duchess's shells in collection 17, 74, 237, 238–40, 239, 241
    Instructions for collecting and preserving 88, 89
    The Natural History of British Shells 17, 18, 63, 72, 74, 78–79, 200, 237, 238–40, 239, 241
Dorville, Elizabeth 63, 258–59, 297n.63
Douglas, Rev. John 167
dredging for shells 85, 87, 153, 241, 253
Drury, Dru
    auction purchases 236
    correspondence with duchess 64
    on da Costa's imprisonment 199–200
    Illustrations of . . . Exotic Insects 33, 38
    Illustrations of Natural History 35
    and Smeathman's expedition 73, 74
"Duchess of Curiosities, The" (exhibition, Welbeck, 2006) vii
duplicate shells
    and exchange practices 12, 109, 118, 119, 122, 123–27
    and museums and societies 243

East India Company 126
Edgar, Thomas 154, 162, 163, 177–78
Ehret, Georg Dionysis 42, 64, 101, 118
elite cultures of collecting 46–50, 52
    perceptions of duchess 53–57, 96, 99, 258, 265
    see also "virtuoso" and connoisseur collectors
Ellis, John 64, 74
Elmsley, Peter 189
embodied practices and shell collecting 75
English shells *see* British shells
engravers *see* artists and engravers
entomology 33, 38, 139–40

epidermis of shells 97
ethnography and shell collecting 10
    see also anthropology
exchange cultures and shells
    and duchess's friendships and networks 13, 65, 68–69, 81, 116–32, 252
        Walker's miniscule shells 184–85
    European societies and shell exchange 10, 20–21, 116
    Pacific islanders 9–10, 14–15, 116
        Europeans' trade with 161–63, 178
    protocols of exchange 123–27
    and social hierarchies 116–17, 120–21, 123, 127–32
    see also loans of shells
exchange value and commodity consumption 61, 131–32
    and auction of duchess's collection 109, 111
    Pacific islanders and desirable commodities 162–63
    and protocols of exchange 123–25, 126–27, 128–29
    valuation of shells from Pacific voyages 131, 139, 140–43, 168, 242
extra shells *see* duplicate shells

Fabricius, Johann Christian 284n.113
Fane, Lady: shell grotto 99, 100
featherwork 33, 34–35
feminine and shells 269n.8
fetishism and collecting 53, 58, 59, 60
field collectors 261–62
Findlen, Paula 117, 285n.5
fine art collections 29
First Fleet (1787–88) 151, 158
flowers
    drying and pressing 34
    see also botanical interests
folding doors and storage cabinets 94
food *see* shellfish as food
Fordyce, Dr. George 136, 236, 237, 239, 241
Fordyce, Margaret 63, 202
Forster, Elizabeth Humphrey 63
Forster, Georg viii–ix, 131
    and scientific collection of specimens 148, 149–50, 159, 160–61, 163, 164–66
    and shells in duchess's collection 132, 136
Forster, Jacob 128, 236
Forster, Johann Reinhold viii–ix, 128, 131
    on Pacific islanders' diet 154
    pamphlet on collecting 76, 152
    scientific collection of specimens 148, 149–51, 159, 160–61, 163–64, 165–66, 176–77
    and shells in duchess's collection 132, 136

Foster, Ingham 122, 128, 137
   auction purchases 236
   friendship with da Costa 190, 200
   and Lister's books 207–8
   purchase of shells for Seymer 138–39, 142
Fothergill, Dr. John 64, 73–74, 200, 202, 242
Freud, Sigmund 53, 58, 59
Furneaux, Captain Tobias 156, 177

Gardeners Kalendar, The 30
Gell, Alfred 9
George III, king of Great Britain 222, 223
gifts
   duchess's gifts to friends 65, 68–69, 80–81, 107, 109, 110, 117–18, 119–20, 125–27, 132–34
   and protocols of exchange 123, 124–27
Gilpin, William 36–37, 272n.9
Gore, Second Lieutenant John 156
Gould, Stephen Jay 104, 106
grottoes 32, 99, 100
Gualtieri, Niccolò 102, 195

Hamilton, Mary 91, 92, 93
Hanson, Craig 52
Harley, Edward, 2nd duke of Oxford (duchess's father): manuscript collection 27, 223
Harris, Moses 35, 38
Hawkesworth, Sir John 148, 157, 158, 159, 166
"heaps of shells" 154–59
Heron, Mrs. Thomas 63
Hoare, Michael 151
Hollar, Wenceslaus 28
Huddesford, William 184, 207, 208–9
Hudson, William 122, 186–87
Hume, Lady Amelia 48
Humphrey, George (junior) viii, 31, 97, 131
   and auction of duchess's collection 224, 230
   auction purchases 81, 115, 236, 264
   as author of auction catalogue 225–26
   bankruptcy and sale of collection 171
   classification irregularities 102, 107, 140, *141*
   on cleaning shells 90
   "Collecting and Preserving all Kinds of Natural Curiosities" 152
   and *Conchology* serial publication 189–99
   duchess's loans to 178
   duchess's purchases from 74, 135, 136, 140, 147
   friendship with da Costa 200
   friendship with J. T. Swainson 135, 253, 280n.17
   good nature and reputation 135–36
   Museum Calonnianum and sale of Calonne's collection 237, 241
   Museum Humfredianum 190
   and networks of exchange 128
   on oversupply of Pacific shells 177, 259
   on practicalities of shell hunting 76, 78
   transcription of Solander's notes 211
   and valuation of shells 140, *141*, 142
Humphrey, George (senior) 31–32
Hunter, John 236, 242
Hunter, William 242
Hunterian Museum, Glasgow 242, 243
   and duchess's shells 10, *11*, 174

illustrations *see* textual representation of shells and collections
insect collections 33, 38, 139–40
International Code of Zoological Nomenclature: "Principle of Priority" 265
Iredale, Tom 190
"it" novels 17

Jackson, J. Wilfrid 190
Jackson, Mr. (shell dealer) 140
Jacob, Margaret 63

Kearsley, Mr. (publisher) 233
knowledge-making and interest in natural history 32, 33
Koerner, Lisbet 283n.95
Kohler, Robert E. 272n.14
Kopytoff, Igor 270n.23
Kula exchange 9–10, 116

Labillardière, Jacques 155
Laird, Mark 62
Lamarckian classification 263, 264–65
Laskey, Captain James 174, 237, 242–43, 257–58, 259, 260
Latour, Bruno 271n.28
Law, John 273n.25
Le Cocq, Mrs. (shell hunter) 13, 82, 83, 97
Lever, Sir Ashton vii, 135
   museum 30, *31*, 39, 151, 172, 223
Lewin, John William: *British Conchology* 17, 46
Lewis, W. S. 56
Lightfoot, Rev. John viii, 64, 76, 137
   advises Pennant on auction lots 225, 230
   assists duchess with classification 100, 106, 107, 109

309

compiles auction catalogue 109, 224–26
continues Solander's unfinished work on catalogue
  212–13
and death of duchess 217, 218–19, 225
and duchess's shell hunting trips at Weymouth 12–13,
  62, 82–83
Flora Scotica dedication to duchess 293–94n.2
on Mrs. Delany 223
residence at Bulstrode as chaplain 42–43, 92, 118
Linnaeus, Carl
  duchess in competition for specimens 134
  Fundamenta Testaceologiae 104
  shell collection 14
  Systema Naturae 100–01, 102–3, 106
    shell classification and shortcomings 101–2, 107–8,
      243
  taxonomy
    ambition of project 261, 262
    and *Catalogue of the Portland Museum* 224–25,
      226–27, 228–29
    and debate on classification of shells 97, 98
    and duchess's classification of collection 56, 57, 59,
      64, 69, 100–01, 106–9, 224
    female anatomy and shell terminology 103–4,
      269n.8
    and introduction of "French" systems 263, 264–65
    Laskey's fictional account of shell classification
      242–43
    and popular interest in natural history 30, 32, 57,
      96, 97, 98
    Pulteney's works on 65
    shortcomings of shell taxonomy 101–2, 104–6,
      107–8, 243
    Solander's revision 109, 209, 257, 261, 262
Linnean Society viii, 118, 244
Lister, Martin
  Historiae Animalium Angliae 207
  Historiae conchyliorum 51, 102, 184, 207–9
Llanover, Lady (Augusta Hall) 222
loans of shells 117
  difficulties of obtaining specimens for illustration
    192–94, 202–3
  duchess's loans for illustration 82, 178–79, 183,
    184–88, 189, 202, 205
  and misunderstandings 122, 123, 186–87
  and protocols of exchange 119–20, 121, 123–24
  Pulteney's loans at duchess's death 13–14, 122, 212
  see also exchange cultures and shells
Longleat archives viii

Louisa Ulrika, queen of Sweden 5, 63
Loutherbourg, Philippe Jacques: *Omai* 168
Lynch, Michael 273n.25

McMillan, Nora F. 85, 252, 280n.17
Macnamara, Captain 134
Malinowski, Bronislaw 9, 116
Margate, Kent 83, 117, 134, 251
Marra, John 131, 290n.10
Martyn, Thomas
  admiration for duchess's collection 5, 7
  auction purchases 236
  disappointment at shells from Cook's third voyage 177
  duchess's loan of shells 178, 185, 202, 205
  duchess's purchases from 135
  "seminary" to train illustrators 204–5
  on J. T. Swainson 251–52
  The Universal Conchologist 15, 16, 17, 63, 66–67, 102,
    115, 201–5, *203*, 206
    loans of shells and scarcity of specimens 202–4, 205
Marxist theory and collecting 53, 60–61
material culture
  collected objects as 9, 17, 29, 57, 75
  and history of natural history 10, 12
  natural history and knowledge-making 32, 33
  see also commodities; objects
Maton, William George
  on *Conchology* serial publication 198–99
  on da Costa 106, 131, 201, 295n.20
  on Martyn's *Universal Conchologist* 205
  and Montagu 259
  and Pulteney's notebooks and papers 226
  references to duchess in survey article 250–51, 254–55
Mauss, Marcel 9
Mawe, John 63, 85, 152–54
Mawe, Sarah 63
Mendes da Costa, Emanuel *see* da Costa
microscopic shells 184–85
Miller, Philip: *The Gardener's Dictionary* 30
Montagu, Elizabeth vii, 5, 30, 74, 244, 276n.75
  and Bulstrode 62
  defers to duchess's knowledge of shells 99–100
  on duchess's birds 49–50
  featherwork 34–35
  on Mrs. Delany's fate 222, 223
  on press response to duchess's death 219–20
Montagu, George 201, *250*, 253
  network of shell collectors 259
  and Pulteney 122, 191

Testacea Britannica 63, 76, 82, 184–85, 297n.63
    attitude towards duchess and collection 250–51, 254–59, 260
Montagu, John *see* Sandwich
More, Hannah 168
Mulgrave, Constantine John Phipps, 2nd baron of 232
museum studies: theories on collecting 29, 57–61
museums and societies and collections 243, 266, 267
Myers, Sarah Harcstark 299n.9

Napoleonic Wars and shell collecting 178, 253, 259
native shells *see* British shells
natural history
    as amateur interest 22, 62–69
    and art and craft activities 33–36
        see also shellwork
    and artistic and aesthetic expression 36–40, 88, 100
    attitudes towards collectors 50–53, 98–99, 256
    buying and selling of specimens 30–32
    and duchess's studies at Bulstrode 28, 41–46, 62–65
        animals in grounds 48–50
    identification and classification 56, 57, 59, 69, 75, 96, 99–111
        debate on classification of shells 97–98, 255–56, 259
        need for selective approach 262–63
        reference texts 38–39, 64, 69, 75, 102–3, 106–7, 128
    publications on 30
        acknowledgment of amateur contribution 63, 66–67, 257–58
        and duchess's posthumous reputation 251–54, 256
    and "scientific" collecting 57, 95–99
    women's involvement and contribution 63–64
    see also botanical interests; Bulstrode: and study of natural history; natural history collections
natural history collections 10, 12
    and aesthetic response 39–40, 95–96, 100
    auction sales of collections 137, 223
    commodification of objects 32, 131–32
    ephemeral nature and fate of collections 266, 267
    exchange and patronage networks 22, 116–32, 135
    and exploration and colonialism 21
    negative attitudes towards collectors 50–53, 56–57
        perceptions of duchess 53–57, 96, 244–45, 254–59, 260, 265
    as neglected subject in scholarship 29, 52–53, 266
    Pacific and other voyages and supply of specimens 147–51, 152–54, 158–59, 159–61
    provenance and integrity of collections 243–44

    sponsorship of collecting expeditions 5, 64–65, 73–74
    see also shell collection of duchess of Portland
Natural History Museum, London
    Banks's shell collection *146*, *147*, 267
    Botany Library and Agnew's work 44–46, *45*
    catalogue of Cracherode's collection 241–42
    and duchess's collection 10, *11*, 179
    naturalists' archives viii
    Zoology Library and Lewin and Agnew's *British Conchology* 6, 17, 46
    see also British Museum, London
needlework and popular interest in nature 33, 36
newspapers *see* press

objects
    and agency 9
    auctions and dead people's belongings 234, 235–36
    consumption and modes of collecting 57–61
    movement through social contexts 12, 20–21, 175, 266–67
    transformation in collections 9, 17, 29, 57, 75
    see also social lives of shells
    use in place of duchess in print 256
    see also commodities; material culture
Omai 170–71, *170*
orange cowry (*Cypraea Aurora*) 15–16, *15*

Pacific islanders
    assistance in shell collecting 161
    barter and exchange with voyagers 161–63, 178
    and Cook and Omai memorabilia 168–71
    European collection of native costume and artifacts 173–74, 175–76, *176*
    exchange cultures 9–10, 14–15, 116
    shellfish diet and Cook's views on 154–56
    shellwork 15–16, *149*, *150*, 154, 156
    views on Europeans 157
Pacific Studies and exchange cultures 9–10, 21, 116
Pacific voyages and natural history collecting
    advice on collecting and preserving specimens 152–53
    captains and officers and buying and selling of specimens 74, 132–38, 149–51
    exploitation of vulnerable buyers 139–40
    lack of interest in natural history in third voyage 177–78
    natural history specimens 147–51, 152–54, 158–59, 159–61
    oversupply and decline in value 175–78, 259
    and procurement of shells 132–43, 147–79, 242

crew and collection and sale of specimens 131, 149,
    158–59, 163–65
and duchess's collection 74, 115–16, 119, 132–36,
    147, 171, 260–61
and exchange and value 20–21
Forsters' scientific collecting 148, 149–51, 159,
    160–61, 163–66, 176–77
Seymer's collection 65–66
Seymer's problems in purchasing 138–39, 140, 142
social lives of shells 14–16
surviving shells from 267
use of local knowledge 161
textual representation of shells from 201–5
textual sources on viii–ix
Parkinson, Sydney 135, 159
patronage and shell networks 22, 116–32, 135
    exclusion of women 135
    Montagu's circle 259
    see also Portland, dowager duchess of: patronage
Pearce, Susan 29, 57–59, 60
pellicle of shells 97
Pennant, Thomas viii, 41, 91
    acknowledgment of duchess's expertise on dredging
        87
    auction purchases 225, 230, 236
    British Zoology 17, 30, 82, 106–7, 183, 184, 185–88,
        237, 261
        sexual innuendo in 187–88
    duchess's loan of shells to 82, 178, 183, 185–87
    duchess's refusal to loan rare books 187–88
    loans and gifts to duchess 117, 122, 185, 186–87
    on *Vermes* classification 101
Perkin, Harold 116
Pickersgill, Lieutenant Richard 149–50, 164
playing card storage trays 75, 90–92, 93
Pocock, Miss (shell hunter) 63, 277n.108
polishing shells 88, 89–90, 96
"polite science" and popular interest in natural history 22,
    30–40
Polynesia and social life of orange cowry 15–16
Pomare (Tahitian chief) 168–70
Pomet, Pierre 10
Pomian, Krysztof 46
Portland, Margaret Cavendish Bentinck, dowager duchess
    of 8, 26
    bankruptcy speculation 220
    botanical classification 100–01
    collection of fine and decorative arts 27–28, 218, 219,
        244

auction purchases 137
death and dispersal of 4, 54, 55, 221, 235
commemorative images at auction 230, *231*
and *Conchology* serial publication 189, 191, 199, 200
correspondence and archives vii–viii
    see also under Pulteney; Seymer
critique of Linnaean classification 107–8
death and immediate after effects 217, 218–23
"disorderly" view of collection 53, *54*, 55–56,
    244–45
expenditure on collections 28, 218
inherited names 271n.2
legacy and diminished reputation 244–45, 249–67
    reasons for 265–67
natural history collection 27, 28, 29
    involvement with collection 62–63
    sale and dispersal on death 5, 219, 223
patronage
    Bulstrode and natural history studies 42–46, 62
    dedications to 184, 185, 207, 293–94n.2
    funding of expeditions 5, 64–65, 73–74
    loan of shells and expertise for illustrations and
        books 82, 178–79, 183–88, 189, 202, 205, 213
    and Seymer's son 126
    and shell networks 116–32, 135
    relations with eldest son 220
    see also shell collection of duchess of Portland
Portland, William Bentinck, 2nd duke of 27
Portland, William Henry Cavendish-Bentinck, 3rd duke of
    alterations to Bulstrode 46
    auction and preservation of heirlooms 235
    duchess's death and inheritance 218, 221
    financial problems 220–21
    gesture to Mary Delany 222
    influence on behalf of duchess 126, 136
Portland Vase 27, 219, 235
Powys, Caroline 233
Powys, Caroline (Mrs. Philip Lybbe) 48–49, 50, 233
Pratt, Mary Louise 21
press
    on auction of collection 3, 4, 217, 230–33
    speculation on duchess's death 218, 219–20
"primitive society": Cook on Pacific islanders 154–55
"Principle of Priority" of International Code of Zoological
    Nomenclature 265
provenance
    importance for British shells 260
    and integrity of collections 243–44
    in Martyn's *Universal Conchologist* 202, 203

prestige of duchess's collection 174–75, 237, 239, 241
    problems of proving provenance 5, 241–45
psychoanalysis and collecting 53, 58
publications
    and popular interest in natural history 30
    see also textual representation of shells and collections
Pulteney, Richard 66, 99, 191
    access to reference books 103
    catalogue of natural history of Dorsetshire 65, 191,
        253–54, 255, 256, 259, 261, 262
    and da Costa 85, 128–30, 131, 142, 199, 200, 201,
        207, 294n.7
    secret passion for British shells 259–60
    friendship and correspondence with duchess vi–viii,
        68–69, 106–8, *114*, 136
    acknowledgment of duchess in published catalogue
        250, 253–54, 255
    gifts and exchange of shells 65, 69, 107, 109, *110*,
        117, 118–22, 132–34
    seeks advice on classification 108–9
    shells on loan to duchess at her death 13–14, 122,
        212
    and Linnaean classification 108
    on da Costa's criticisms 106
    in Montagu's work 255–56
    and Montagu 255–56, 259
    Pacific shells and revival of interest in collecting 132
    The Progress of Botany in England 103
    reordering of auction catalogue along Linnaean lines
        226–27, *228–29*
    reputation as naturalist 255–56
    and Seymer 90, 188
        as intermediary between duchess and Seymer 117,
            121–22, 125–26
        and protocols of exchange 123–25
    on Solander's labour on duchess's shell catalogue 209

Rackett, Rev. Thomas
    on *Conchology* serial publication 198–99
    on da Costa 106, 131, 201
    on Martyn's *Universal Conchologist* 205
    references to duchess in survey article 250–51,
        254–55
religious response to nature 40
Repton, Humphrey 46
Reynolds, Sir Joshua 235
Robertson, Hannah 88–89
"romantic collecting" 58
Rousseau, Jean-Jacques 62, 244

Royal Society, London 135, 151
Rumph, Georg Eberhard 102

sailors *see* Pacific voyages and natural history collecting
sale of shells *see* buying and selling of shells; commodities;
    shell collection of duchess of Portland: auction (1786)
Salmond, Anne 157
Samwell, David (surgeon's mate) 173
Sandwich, John Montagu, earl of 148, *148*
science
    acknowledgment of amateur contribution 63, 98–99,
        257–58
    and aesthetics in shell collecting 75, 88, 95–99, 100
    exclusion of natural history collecting 29, 57
    Pacific voyages and Forsters' scientific specimens 148,
        149–51, 159, 160–61, 163–66, 176–77
    proximity to arts in eighteenth century 33, 36
    "scientific" naturalists and duchess's reputation 251,
        254–65
    see also classification of collections
Scott, Sarah 222
self and modes of collecting 57–58, 60
selling shells *see* buying and selling of shells; commodities
Sepp, J. C. 195
seriality and systematic collecting 59
Seward, Anna 168
sexuality
    Freudian theory and collecting 53
    and shells 187–88, 269n.8
        Linnaean terminology 104, 106
        Pennant's *British Zoology* 187–88
Seymer, Edward 126
Seymer, Henry viii, 65–69, *67*, 178, 188, 203, 259
    advice on *Conchology* project 194, 199
    aesthetic response to shells 95–96, 99
    artistic skills 33, 35, 65
    and classification 96, 98, 140
        criticisms of Linnaean system 108
        on duchess's second-hand knowledge 284n.113
    on cleaning shells 90
    friendship and correspondence with duchess 65,
        68–69, 121–22
    and gifts and protocols of exchange 123–27
        with duchess 117, 120, 125–27
        with Pulteney 123–25
    pitfalls of acquiring Pacific specimens 138–40, 142
Shadwell, Thomas: *The Virtuoso* 51, 52
Shaw, Dr. Thomas 74, 99, 100
Shelburne, Lady (Sophia Carteret Petty) 48

Sheldon, Thomas 236
shell art *see* shellwork
shell collection of duchess of Portland 6–7, 10–17, *11, 14, 16, 18–19*
   acquisition of shells 115–43
      collecting British shells at Bulstrode 76
      dredging practices 85, 87, 241
      exotic shells from expeditions and dealers 64–65, 74, 115–16, 134, 135–36, 140, 260–61
      from Cook and Pacific voyages 74, 115–16, 119, 132–36, 147, 171, 260–61
      and networks of exchange 13, 22, 81, 116–32, 135
      purchases through dealers and seamen 132–38
      shell-hunting trips in Britain 12–13, 62–69, 74, 80–85
      social influence and unofficial channels 136
   admiration and praise for 5, 7
   auction (1786) and dispersal 2, 3–5, 7, 10, 14, 213, 217–45
      breaking up of collection 217
      commemorative images of duchess 230, *231*
      descriptions of auction day 230–36
      dismantling of species classification 224–25
      interest in duchess's death 218–23
      press and public interest in 3–4, 230–32
      prestige of provenance from 174–75, 237, 239, 241
      problems of proving provenance 5, 241–45
      purchasers and later owners 17, 115, 225, 232–33, 236–45, 260
      see also Catalogue of the Portland Museum, A (auction catalogue)
   cataloguing of collection 7, 209–13
      acknowledgment in published works 254
      ambition of project 109, 261, 262, 263
      and dissemination of knowledge 209
      incorporation into auction catalogue 109, 111, 209–10, 212–13
      as unfinished project 5, 7, 109, 209, 211–13, 254
   and funding of expeditions 5, 64–65, 73–74
   gifts and exchanges with friends 65, 68–69, 80–81, 107, 109, *110*, 117–18, 119–20, 125–27, 132–34
   illustrations and images of 16–17, 178–79, 206–7
   legacy and diminished reputation 244–45, 249–67
   loans of rare shells for illustration 82, 178–79, 183, 184–88, 189, 202, 205
   references in published texts 249–67
   remainder of collection 10, *11*, 174, 243–44, 267
   scientific approach 74–75
      anatomy of mollusks 97
      classification of shells 56, 57, 59, 64, 69, 75, 99–100, 106–11, 224
      preparation for display 75, 87–94
      see also cataloguing of collection *above*
      sorting and storage of shells 75, 90–94
   textual sources and history of collection viii–ix, 12
shell collectors
   attitudes towards collectors and British shells 50–51, 128, 130–31, 178, 207, 253, 259–60, 261
   collecting processes 75
      instructions and equipment for 76, 78–80
      shell hunting 76–87
   expeditions and procurement of shells 5, 64–65, 73–74, 261
      see also Pacific voyages and natural history collecting
   George Humphrey (sr) 31–32
   networks and exchange of shells 22, 116–32, 135
      care in packaging and handling 185
      duchess and networks of exchange 13, 65, 68–69, 81, 116–22, 252
      duchess and Walker's miniscule shells 184–85
      duchess's exclusion as woman 135
      duplicate shells 12, 109, 118, 119, 122, 123–27
      loans for published illustrated 82, 178–79, 183, 184–88, 189, 202, 205
      Montagu's circle 259
      protocols of exchange 119–20, 121, 123–27
      and regimes of value 20–21, 116, 123–25
      social hierarchies and networks of exchange 116–17, 120–21, 123, 127–32
   preparation for display 87–94
      identification and classification 13–14, 38–39, 64, 69, 75, 96, 99–111, 140, 259
      removal of creatures and cleaning and polishing 75, 87–90, 96, 152
      sorting and storing shells 75, 90–94
   preservation of integrity of collections 243–44
   scientific and aesthetic approaches 75, 88, 95–99, 100
   scientific versus amateur authorities 254–65
   see also loans of shells; shell collection of duchess of Portland
shell grottoes 32, 99, 100
shellfish as food
   for European voyagers 156–57, 158–59
   Pacific islanders' diet and standard of civilization 154–55
shellwork 32, 33, *34*, 34, 36, 75, 87–88
   from Pacific islands 15–16, 149, *150*, 154, 156

"Short Instructions for Collecting Shells" (pamphlet) 79–80
Shteir, Ann 50
Sibthorpe, Dr. (collector and traveller) 74
Skinner, Mr. (auctioneer) 4, 224
Sloane, Sir Hans 28, 58, 249
Sloboda, Stacey vii, 276nn.74&75
Smeathman, Henry 5, 64–65, 73–74
Smith, Charles 175
Smith, Sir James Edward vi, 219
snails and shells 36, 42, 62, 63, 185
    collection of exotic specimens 153
    specimens at Bulstrode 76, 77
Soane, Sir John 59
social class
    and natural history collecting 63–64
    social hierarchy and networks of exchange 116–17, 120–21, 123, 127–32
social lives of shells 9, 12, 14–16, 17
    as commodities and exchange 15–16, 20–21, 81, 116, 131–32, 137–38
    importance of provenance 175, 202
social practice: collecting as 51, 52–53, 64, 83–84, 92–93
Society of Aurelians 38, 73
Society Islands (Tahiti) 160, 161
Solander, Daniel viii, 41, 68, 118, 147
    and acquisitions for British Museum 94, 134, 142
    as authority on taxonomy 210–11, 260
        reputation and references in publications 253–54, 255, 257, 263–64
    Banks's plans to publish his work 211–12
    classification and cataloguing of duchess's collection 69, 106, 107, 109, 111, 209–13, 225, 226, 254, 255
    and Cook's voyages 132, 133, 134–35, 148–49, *148*, 150–51
    critique of Linnaean molluscan taxonomy 101–2
    da Costa's correspondence with 130
    and networks of exchange 128, 134
        assistance in classification 184, 185
    "Ostrea peregrina" manuscript *210*
    revision of Linnaeus's molluscan taxonomy 109, 209, 257, 261, 262
    shell collection 267
"South Seas" shells *see* Pacific voyages and natural history collecting
souvenirs and self 57–58, 59–60
Spary, Emma 10, 12, 29, 271n.28, 285n.5, 290n.9
Stamford, Henrietta Cavendish-Bentinck, countess of 219, 221

Stewart, Susan 58, 59–61
storage of shells 75, 90–94
Stott, Rebecca vii, 41, 249
Strathern, Marilyn 9
Sturkenboom, Dorothée 63
Swainson, John Timothy viii, 251–53
    Agnew's illustrations for 17, 46
    at Bulstrode 41–42, 76
    duchess's gifts and exchanges 13, 80–81, 117–18, 199, 252, 264
    friendship with Humphrey 135, 253, 280n.17
    meets duchess in Margate 83, 251
    Montagu's recognition of 257
    notebook of British shells 13, 16, 80–81, *80*, 85
    references to duchess 250, 251, 252–53
Swainson, William 261, 280n.17
    on cabinets for shell storage 94
    on George Humphrey 135
    on making card trays 91–92
    on selective approach to collecting 262–63
    on shell cleaning 90
    on shell hunting and equipment 76, 78–79, 84, 85, 87, 279n.12
    on types of collections 95, 98–99
    Voluta illustrations 301n.52
systematic collecting 59, 60, 69, 98

Tahiti *see* Society Islands (Tahiti)
Tankerville, Charles Bennett, 4th earl 136, 241, 263
taxidermy 33, 36
Te Taniwha, Horeta 157
textual representation of shells and collections 183–213
    as aid to identification and classification 38–39, 64, 69, 75, 102–3, 106–7, 128
        Conchology's serial approach and new specimens 188–89
    as collaborative enterprise 189–99, 207–8, 213
    dedications to duchess 184, 185, 207, 293–94n.2
    illustrations and images of duchess's collection 16–17, *16*, 206–7
        patronage and loan of rare shells 82, 178–79, 183, 184–88, 189, 202, 205, 213
    as inadequate guide to collections 10, 12
    limited access to rare books 103, 188, 207
    references to duchess and posthumous reputation 249–67
    shortage of qualified illustrators and Martyn's seminary 204
    South Pacific shells 201–5

Thomas, Nicholas 9, 21, 155–56, 164, 166
Thornton, Robert 40, 103
Thurlow, Edward 232–33
Thynne, Elizabeth, Viscountess Weymouth and marchioness of Bath viii, 81–82, 219, 221
Thynne, Thomas, 3rd Viscount Weymouth and 1st marquess of Bath 81–82
Tobin, George 168–70
Tonga and social life of orange cowry 15–16
Torrens, Hugh 63
Tradescant, John 58
Transactions of the Linnean Society: Maton's and Rackett's survey article 254–55
Trobriand Islanders: exchange culture 9–10, 116
Tunstall, Marmaduke 64–65, 73–74, 135, 151
Turton, William 250–51, 261, 263–65

University of Nottingham library viii

Valenti, Michael Bernhard 196
value regimes
  and collecting 61, 266–67
    and auction of duchess's collection 109, 111
  and exchange of shells 20–21, 116, 123–25
  provenance and social context 174–75
  and shells from Pacific voyages 14–16, 131, 139, 140–43, 168, 172–73, 242
  trade with Pacific islanders 161–63, 178
  see also buying and selling of shells; exchange value and commodity consumption
"vanished collection" problem 10
Vermes class and divisions 101–2
Vertue, George: Portland shells 16, *16*
Vickery, Amanda 33, 41
"virtuoso" and connoisseur collectors 46–50, 60, 98, 261
  auctions of collections on death 137–38
  Martyn's acknowledgment of loans 202
  negative and satirical views of 51–52, 56
  perceptions of duchess 53–57, 96, 99, 258, 265
  publications aimed at 194, 198–99, 202, 241

Wales, William 165
Walker, George 118, 236, 253, 259
  Minute and Rare Shells 184–85, 261
Walker, Mrs. Isaac 63
Wall, Cynthia 55, 279n.12

Wallingford, Lady 221
Walpole, Horace
  auction purchases 4, 235, 236
  on duchess's collecting career and practices 27–28, 56, 134
  on third duke of Portland 220–21
  on Welbeck Abbey auction 235–36
Way, Kathie 243
Webber, John: *James Cook* 168–70, *169*
Weiner, Annette 9
Welbeck Abbey, Nottinghamshire vii, viii, 28, 220, 221
  Walpole's auction visit 235–36
West, James and Mrs. 136, 137, 142
Weymouth, Viscount and Viscountess *see* Thynne
Weymouth, Dorset 81
  duchess's shell-hunting trips 12–13, 62, 65, 80–83
    dredging for shells 85, 241, 253
Wheeler, Quentin 266
Whitaker, Katie 46, 48
White, Benjamin 189
White, Gilbert: *The Natural History and Antiques of Selbourne* 30
White, John 157
Whitehall house, London viii, 119, *216*
  auction at 3, 55, 223–24, 230–32
  cabinets for shell display 92
  "disorderly" view of collection 54, 55–56, 244–45
Whitehead, P. J. P. 190
Wicksteed, J. 296n.32
Wilkes, Mr. (shell collector) 259
Wilkins, Guy 91, 241–42, 264, 294n.2
women as collectors
  barriers and discrimination 135
  and Linnaean terminology 103–4
  negative views of duchess 53–57, 244–45, 254–59, 260, 265
  and social class 63–64
  views on gender and collecting 278–79n.6
Woodward, John 51–52, 149
Worsley, Lady (Seymour Dorothy Fleming) 232, 233
Worsley, Sir Richard 233
Wunderkammern see cabinets of curiosities

Yeats, Mr. (friend of Seymer) 140
Yeats, Mr. (shell collector) 14, 15
Young, Mary Julia: *The East Indian; or, Clifford Priory* 39–40